D1752320

66
Advances in Biochemical Engineering/Biotechnology

Managing Editor: T. Scheper

Editorial Board:
W. Babel · H. W. Blanch · C. L. Cooney
S.-O. Enfors · K.-E. L. Eriksson · A. Fiechter
A. M. Klibanov · B. Mattiasson · S. B. Primrose
H. J. Rehm · P. L. Rogers · H. Sahm · K. Schügerl
G. T. Tsao · K. Venkat · J. Villadsen
U. von Stockar · C. Wandrey

Springer

Berlin
Heidelberg
New York
Barcelona
Hong Kong
London
Milan
Paris
Singapore
Tokyo

Bioanalysis and Biosensors for Bioprocess Monitoring

With contributions by
B. K. Alsberg, M. Beyer, D. Broadhurst, B. Christensen,
H. M. Davey, R. J. Gilbert, R. Goodacre, N. Kaderbhai,
D. B. Kell, C.-F. Mandenius, A. C. McGovern, J. Nielsen,
M.-N. Pons, J. J. Rowland, K. C. Schuster, K. Schügerl,
G. Seidel, A. D. Shaw, B. Sonnleitner, J. Taylor,
E. M. Timmins, C. Tollnick, H. Vivier, M. K. Winson,
A. M. Woodward

Springer

Advances in Biochemical Engineering/Biotechnology reviews actual trends in modern biotechnology. Its aim is to cover all aspects of this interdisciplinary technology where knowledge, methods and expertise are required for chemistry, biochemistry, microbiology, genetics, chemical engineering and computer science. Special volumes are dedicated to selected topics which focus on new biotechnological products and new processes for their synthesis and purification. They give the state-of-the-art of a topic in a comprehensive way thus being a valuable source for the next 3–5 years. It also discusses new discoveries and applications.

In general, special volumes are edited by well known guest editors. The managing editor and publisher will however always be pleased to receive suggestions and supplementary information. Manuscripts are accepted in English.

In references Advances in Biochemical Engineering/Biotechnology is abbreviated as Adv. Biochem. Engin./Biotechnol. as a journal.

ISSN 0724-6145
ISBN 3-540-66052-6
Springer-Verlag Berlin Heidelberg New York

Library of Congress Catalog Card Number 72-152360

This work is subject to copyright. All rights are reserved, whether the whole or part of the material is concerned, specifically the rights of translation, reprinting, reuse of illustrations, recitation, broadcasting, reproduction on microfilm or in any other way, and storage in data banks. Duplication of this publication or parts thereof is permitted only under the provisions of the German Copyright Law of September 9, 1965, in its current version, and permission for use must always be obtained from Springer-Verlag. Violations are liable for prosecution under the German Copyright Law.

© Springer-Verlag Berlin Heidelberg 2000
Printed in Germany

The use of general descriptive names, registered names, trademarks, etc. in this publication does not imply, even in the absence of a specific statement, that such names are exempt from the relevant protective laws and regulations and therefore free for general use.

Typesetting: Fotosatz-Service Köhler GmbH, Würzburg
Cover: Design & Production, Heidelberg
SPIN: 10691489   02/3020 – 5 4 3 2 1 0 – Printed on acid-free paper

Managing Editor

Professor Dr. T. Scheper
Institute of Technical Chemistry
University of Hannover
Callinstraße 3
D-30167 Hannover/FRG
E-mail: scheper@mbox.iftc.uni-hannover.de

Volume Editor

PD Dr. B. Sonnleitner
Department of Chemistry
Zürich University of Applied Sciences
P. O. Box 805
CH-8401 Winterthur/Switzerland
E-mail: bernhard.sonnleitner@zhwin.ch

Editorial Board

Prof. Dr. W. Babel
Section of Environmental Microbiology
Leipzig-Halle GmbH
Permoserstraße 15
D-04318 Leipzig/FRG
E-mail: babel@umb.ufz.de

Prof. Dr. H. W. Blanch
Department of Chemical Engineering
University of California
Berkely, CA 94720-9989/USA
E-mail: blanch@socrates.berkeley.edu

Prof. Dr. C. L. Cooney
Department of Chemical Engineering
Massachusetts Institute of Technology
25 Ames Street, Room 66-350
Cambridge, MA 02139-4307 /USA
E-mail: ccooney@mit.edu

Prof. Dr. S.-O. Enfors
Department of Biochemistry and
Biotechnology
Royal Institute of Technology
Teknikringen 34, S-100 44 Stockholm/Sweden
E-mail: olle@biochem.kth.se

Prof. Dr. K.-E. L. Eriksson
Center for Biological Resource Recovery
The University of Georgia
A214 Life Science Building
Athens, GA 30602-7229/USA
E-mail: eriksson@uga.cc.uga.edu

Prof. Dr. A. Fiechter
Institute of Biotechnology
Eidgenössische Technische Hochschule
ETH-Hönggerberg
CH-8093 Zürich/Switzerland
E-mail: ae.fiechter@bluewin.ch

Prof. Dr. A. M. Klibanov
Department of Chemistry
Massachusetts Institute of Technology
Cambridge, MA 02139/USA
E-mail: klibanov@mit.edu

Prof. Dr. B. Mattiasson
Department of Biotechnology
Chemical Center, Lund University
P.O. Box 124, S-221 00 Lund/Sweden
E-mail: bo.mattiasson@biotek.lu.se

Prof. Dr. S. B. Primrose
21 Amersham Road
High Wycombe
Bucks HP13 6QS/UK

Prof. Dr. P. L. Rogers
Department of Biotechnology
Faculty of Life Sciences
The University of New South Wales
Sydney 2052/Australia
E-mail: p.rogers@unsw.edu.au

Prof. Dr. K. Schügerl
Institute of Technical Chemistry
University of Hannover
Callinstraße 3,
D-30167 Hannover/FRG
E-mail: schuegerl@mbox.iftc.uni-hannover.de

Dr. K. Venkat
Phyton Incorporation
125 Langmuir Lab.
95 Brown Road
Ithaca, NY 14850-1257/USA
E-mail: venkat@clarityconnect.com

Prof. Dr. U. von Stockar
Laboratoire de Génie Chimique et
Biologique (LGCB)
Départment de Chimie
Swiss Federal Institute
of Technology Lausanne
CH-1015 Lausanne/Switzerland
E-mail: stockar@igc.dc.epfl.ch

Prof. Dr. H. J. Rehm
Institute of Microbiology
Westfälische Wilhelms-Universität Münster
Correnstr. 3, D-48149 Münster/FRG

Prof. Dr. H. Sahm
Institute of Biotechnolgy
Forschungszentrum Jülich GmbH
D-52425 Jülich/FRG
E-mail: h.sahm@kfa-juelich.de

Prof. Dr. G. T. Tsao
Director
Lab. of Renewable Resources Eng.
A. A. Potter Eng. Center
Purdue University
West Lafayette, IN 47907/USA
E-mail: tsaogt@ecn.purdue.edu

Prof. Dr. J. Villadsen
Department of Biotechnology
Technical University of Denmark
Bygning 223
DK-2800 Lyngby/Denmark

Prof. Dr. C. Wandrey
Institute of Biotechnology
Forschungszentrum Jülich GmbH
D-52425 Jülich/FRG
E-mail: c.wandrey@fz-juelich.de

Attention all "Enzyme Handbook" Users:

Information on this handbook can be found via the internet at

http://www.springer.de/chem/samsup/enzym-hb/ehb_home.html

At no charge you can download the complete volume indexes Volumes 1 through 17 from the Springer www server at the above mentioned URL. Just click on the volume you are interested in and receive the list of enzymes according to their EC-numbers.

Professor Dr. A. Fiechter

Laudatio

This volume is dedicated to Dr. Armin Fiechter, Professor Emeritus of Biotechnology at the ETH Zürich and former managing editor of Advances in Biochemical Engineering/Biotechnology and Journal of Biotechnology and editor and member of Advisory Boards of several international periodicals on the occasion of his 75th birthday.

Armin Fiechter is one of the pioneers in biotechnology – recognized worldwide for his important contributions to various fields of biotechnology. Professor Fiechter's research covers a broad area. He carried out pioneering work in several fields. From the beginning, he stressed the necessity of interdisciplinary and international cooperation. He especially promoted cooperation between engineering and biological research groups and helped to overcome the hurdles and borders between these groups. His active role as a teacher of young scientists led to the well known "Fiechter School". Some well-known researchers in industry and science come from his laboratory. His more than 500 publications document his research activities in different areas of biotechnology.

The quantitative evaluation of biological regulation was especially difficult, because reproducibility of the measurement of the dynamical processes was unsatisfactory in the 1960s. One of the first long-term continuous cultivation of baker's yeast in a chemostat system in combination with aseptic operation and use of pH- redox- and oxygen-electrodes was realized by his group. The sterility was obtained by O-ring sealing. The sterilizable pH-, redox- and oxygen electrodes were developed in the industry with his co-operation. The sealing of the stirrer shaft with a sliding sleeve and the use a marine propeller in combination with a draft tube (compact loop reactor, COLOR) for maintaining ideal mixing and for better mechanical foam control was also developed in cooperation with his group. One of the key issue was the better process control by means of in situ monitored pH- and redox-values and dissolved oxygen concentration in the cultivation medium under aseptic operation. Various instruments (FIA, HPLC, GC, MS) were adapted for on-line monitoring of the concentrations of key components and computer programs were developed for automatic data evaluation and control. In this compact loop reactor and by means of advanced measuring and control systems highly reproducible measurements became possible.

Professor Fiechter succeeded to show using the improved chemostat technique that glucose and oxygen influence various yeast stains differently. Beside the catabolite repression (glucose effect) a second regulation type exists which is controlled by the dynamic substrate flux (glucose). This causes different types of physiological phenomena such as diauxie, secondary monoauxie or atypical changes in growth and ethanol production continuous cultures. Sonnleitner and Kaeppeli in his group developed an overflow model to explain these phenomena. Overflow reaction is common not only in yeast, but in bacteria as well. In addition, they investigated the cell cycle by means of the analysis of stable synchronous growth, which was maintained in the high performance chemostat system. It was possible to recognize the trigger-function of trehalose for the onset of budding and the testing of the secretion and reuse of metabolites during the budding.

Investigations of the processes with different strains and reactor types under close control are necessary for the transfer of biological processes from a laboratory to an industrial scale (scale up). Most of the early biochemical engineering research was restricted to the investigation of oxygen transfer and carried out with model media without micro-organisms. Systematic pilot plant investigations were performed with various micro-organisms and different types of reactors up to 3000 l volume in Hönggerberg by the Fiechter research group. The reactor performances were compared and optimal process operations were evaluated. The high process performance of the compact loop reactor was proved.

In addition to this technical oriented development, a broad field of applied biological research was at the center of interest in Fiechter's laboratory. The development of bioreactors, bioprocess monitoring and control served as a means of obtaining more information on the biology of microorganisms and improving the process performance.

The investigation of the physiology of baker's yeast was a central issue in this laboratory. Evaluation of the details of the cell cycle and the importance of the overflow phenomenon are discussed above. However, other microorganisms, such as the strictly respiratory yeast, *Trichosporon cutaneum*, and bacteria, such as *Escherichia coli*, were investigated and applied for reactor characterization as well. *Zymomonas mobilis* surpasses baker's yeast with regard to alcohol production by a factor of five. In the high performance reactor under aseptic conditions extremely high ethanol productivities (250 ml l^{-1} h^{-1}) were obtained in Fiechter's laboratory.

As early as 1983, a cell culture group was established and in the following 10 years serum- and protein-free cultivation media were developed by means of a systematic analysis of key C-sources, intermediate and final metabolites and their influence on the growth and product formation. Lactate formation was identified as an overflow phenomenon caused by a respiratory bottleneck, incomplete medium composition, glucose excess, and stress factors. In continuous cultivation of CHO cells with cell recycling generation times of 12 h were obtained. By means of a Process Identification and Management System (PIMS), which was developed by his group, automatic on-line analysis and control of animal tissue cultivation became possible. In cooperation with Weissmann, recombinant Interferon was produced by *Escherichia coli* in a 3000 l reactor for clinical investigations in 1980.

Of his many research activities only few have been mentioned: In the frame of the SCP project, Cytochrome P-450 studies were carried out in connection with the investigation of hydrocarbon metabolisms of yeasts. Enzymes from thermophilic bacteria (*Bac. stearothermophilus*) were identified and isolated. In connection with biodegradation of lignin, new enzymes were identified and isolated. In the framework of the microbial-enhanced oil recovery project Rhamnolipid biotensides were produced by genetically modified *Pseudomonas aeruginosa*. A process for the production of Lipoteichonacid (LTA) was developed and the anticarcinogenic compound was produced in a 3000 l reactor. Outside of industry, no other academic research group gained so many important results on the pilot plant scale. These and many other results help us in transferring biotechnological processes from the laboratory to the industrial scale.

Because of his broad spectrum of activities and successful research he was invited into several countries and where he acted as visiting professor. He became a member of the Supervisory Board of GBF (Central Biotechnology Research Laboratory of Germany), Braunschweig, a member of the Board and Interim Director of the Institute of Surface- and Biotechnology of the Fraunhofer-Society, Stuttgart, a member of the Swiss Academy of Engineering Sciences, a founding member of the European Federation of Biotechnology, a member of the IUPAC Commission on Microbiology, an honorary member of DECHEMA, president of the Swiss Microbial Society, etc.

We, his colleagues and former students thank him for his enthusiasm and continuous support in biotechnology also after his retirement. By dedicating this volume of *Advances in Biochemical Engineering/Biotechnology* to Professor Fiechter, the authors of this volume and many other colleagues around the world want to honor his outstanding achievements in the broad field of biotechnology and wish him good health.

Hannover, July 1999 Karl Schügerl

Preface

This special volume on "bioanalysis and biosensors for bioprocess monitoring" has a twofold target.

Firstly, it is dedicated to the 75th birthday of Armin Fiechter, who was a major driving force among the pioneers to the progress of biochemical engineering. Not only the aseptic connection technique with septa and needles still used until today was established by him, but also the development of the first sterilizable pH-electrodes with W Ingold is also credited to him. He made in-vivo bioanalysis a topic of general interest, for instance by setting up the first chemostat in Switzerland. It was again Armin Fiechter who pushed the use of non-invasive exhaust gas analysis in the late 1960s and promoted development and exploitation of in-situ sensors and on-line analytical instruments in bioprocessing, among other means, by founding a spin-off company. In his laudatio, Karl Schügerl extends the list of his merits and achievements.

On the other hand, this volume is the first product of a core group working in the first Task Group "synopsis of conventional and non-conventional bioprocess monitoring" of the first Section of the EFB, namely the Section on Biochemical Engineering Science. All the various monitoring techniques are so determinant and central that the EFB decided to found the Working Party on Measurement and Control, as one of the last Working Parties, as late as 1988. The Section, however, was founded in 1996 in order to facilitate communication and cooperation among biochemical engineers and scientists so far organized, or should I say split up, into various different Working Parties. It was strongly felt that the business of measurement (modeling) and control could not be confined to the respective Working Party, it was and is so important for all the colleagues associated with bioreactor performance or down stream processing that a broadening of the horizon was actively sought.

Within the Section, several Task Groups are playing the role of workhorses.

A synopsis of monitoring methods and devices was missing from the beginning. The interest in obtaining up-to-date information and exchanging mutual experience with older and up-to-date bioprocess monitoring tools became obvious before, during and after several advanced courses organized and run by the predecessors of the present Section. The conclusion soon became clear, but the realization came later, and here is the first report from the Task Group!

Certainly, these few contributions cover a great variety of achievements, bring some success stories, discuss some potential pitfalls and discuss several practical experiences. It is clear that this synopsis is non-exhaustive; it is also obvious that we have failed to include contributions specifically focused on downstream processing and product qualification problems or targeted to bioreactor performance characterization. However, it was important to show, with a first report, that there are people active in these fields and, hopefully, continuing to be so and attracting more people to join them in this work.

The contributions to this special volume were selected in order to show the present dynamics in the field of bioprocess monitoring. Some quite conventional methods are addressed, other contributions focus on more fuzzy things such as electronic noses or chemometric techniques. One contribution illustrates the potential with a precise example of cephalosporin production. Three of them have dared to "look" inside cells using different methods, one by the analysis of (microscopic) images, one by trying to estimate the physiological state, and the third by analyzing the metabolic network. This gives a rough but good idea of how sophisticated analytical tools – (bio)chemical ones hand in hand with mathematical ones, – give rise to a better understanding of living systems and bioprocesses.

Along with monitoring and estimation we also focus on modeling and control of bioprocesses in the future. Perhaps, other Task Groups will evolve to accomplish this. In the field of monitoring and estimation, we face the great challenge of realizing an appropriate technology transfer of many scientific highlights described in this volume into everyday industrial applications. A big gap in knowledge and experience still makes the decision between "must" and "nice to have" not easy. I hope that this special volume initiates many successful steps towards this goal.

Winterthur, June 1999   B. Sonnleitner

Contents

Instrumentation of Biotechnological Processes
B. Sonnleitner 1

Electronic Noses for Bioreactor Monitoring
C.-F. Mandenius 65

Rapid Analysis of High-Dimensional Bioprocesses Using Multivariate
Spectroscopies and Advanced Chemometrics
A. D. Shaw, M. K. Winson, A. M. Woodward, A. C. McGovern,
H. M. Davey, N. Kaderbhai, D. Broadhurst, R. J. Gilbert, J. Taylor,
É. M. Timmins, B. K. Alsberg, J. J. Rowland, R. Goodacre, D. B. Kell 83

On-Line and Off-Line Monitoring of the Production
of Cephalosporin C by *Acremonium chrysogenum*
G. Seidel, C. Tollnick, M. Beyer, K. Schügerl 115

Biomass Quantification by Image Analysis
M.-N. Pons, H. Vivier 133

Monitoring the Physiological Status in Bioprocesses on the Cellular Level
K. C. Schuster 185

Metabolic Network Analysis – A Powerful Tool in Metabolic Engineering
B. Christensen, J. Nielsen 209

Author Index Volumes 51–66 233

Subject Index 241

Instrumentation of Biotechnological Processes

Bernhard Sonnleitner

University of Applied Sciences, Winterthur, Switzerland
E-mail: bernhard.sonnleitner@zhwin.ch

Modern bioprocesses are monitored by on-line sensing devices mounted either in situ or externally. In addition to sensor probes, more and more analytical subsystems are being exploited to monitor the state of a bioprocess on-line and in real time. Some of these subsystems deliver signals that are useful for documentation only, other, less delayed systems generate signals useful for closed loop process control. Various conventional and non-conventional monitoring instruments are evaluated; their usefulness, benefits and associated pitfalls are discussed.

Keywords. Conventional and non-conventional sensors and analytical instruments, On-line bioprocess monitoring, Software sensors, Dynamics of measurements, Real time estimation, Interfacing aseptic sampling

1	Process Monitoring Requirements	3
1.1	Standard Techniques (State of Routine)	3
1.2	Biomass	4
1.3	Substrates	5
1.4	Products, Intermediates and Effectors	5
2	On-Line Sensing Devices	6
2.1	In Situ Instruments	6
2.1.1	Temperature	6
2.1.2	pH	7
2.1.3	Pressure	8
2.1.4	Oxygen	10
2.1.4.1	Oxygen Partial Pressure (pO_2)	10
2.1.4.2	Oxygen in the Gas Phase	11
2.1.5	Carbon Dioxide	12
2.1.5.1	Carbon Dioxide Partial Pressure (pCO_2)	12
2.1.5.2	Carbon Dioxide in the Gas Phase	13
2.1.6	Culture Fluorescence	14
2.1.7	Redox Potential	15
2.1.8	Biomass	16
2.1.8.1	Comparability of Sensors	17
2.1.8.2	Optical Density	17
2.1.8.3	Interferences	18
2.1.8.4	Electrical Properties	21
2.1.8.5	Thermodynamics	21

2.2	Ex Situ, i.e. in a Bypass or at the Exit Line	23
2.2.1	Sampling	23
2.2.1.1	Sampling of Culture Fluid Containing Cells	24
2.2.1.2	Sampling of Culture Supernatant Without Cells	25
2.2.2	Interfaces	25
2.2.3	Flow Injection Analysis (FIA)	25
2.2.4	Chromatography such as GC, HPLC	28
2.2.5	Mass Spectrometry (MS)	29
2.2.6	Biosensors	31
2.2.6.1	Electrochemical Biosensors	32
2.2.6.2	Fiber Optic Sensors	33
2.2.6.3	Calorimetric Sensors	33
2.2.6.4	Acoustic/Mechanical Sensors	34
2.2.7	Biomass	34
2.2.7.1	Dynamic Range – Dilution	34
2.2.7.2	Electrical Properties	35
2.2.7.3	Filtration Properties	35
2.3	Software Sensors	35
2.4	Validation	36
3	**Off-Line Analyses**	**38**
3.1	Flow Cytometry	38
3.2	Nuclear Magnetic Resonance (NMR) Spectroscopy	39
3.3	Field Flow Fractionation (FFF)	41
3.4	Biomass	41
3.4.1	Cell Mass Concentration	43
3.4.2	Cell Number Concentration	43
3.4.3	Viability	45
3.4.4	Cellular Components or Activities	45
3.5	Substrates, Products, Intermediates and Effectors	45
4	**Real Time Considerations**	**46**
4.1	Dynamics of Biosystems	47
4.2	Continuous Signals and Frequency of Discrete Analyses	49
5	**Relevant Pitfalls**	**49**
5.1	α,β-D-Glucose Analyzed with Glucose Oxidase	50
5.2	CO_2 Equilibrium with HCO_3^-	50
5.3	Some Remarks on Error Propagation	51
5.4	The Importance of Selecting Data To Keep	52
6	**Conclusions**	**53**
	References	**54**

1
Process Monitoring Requirements

Cellular activities such as those of enzymes, DNA, RNA and other components are the primary variables which determine the performance of microbial or cellular cultures. The development of specific analytical tools for measurement of these activities in vivo is therefore of essential importance in order to achieve direct analytical access to these primary variables. The focus needs to be the minimization of relevant disturbances of cultures by measurements, i.e. rapid, non-invasive concepts should be promoted in bioprocess engineering science [110, 402]. What we can measure routinely today are the operating and secondary variables such as the concentrations of metabolites which fully depend on primary and operating variables.

In comparison to other disciplines such as physics or engineering, sensors useful for in situ monitoring of biotechnological processes are comparatively few; they measure physical and chemical variables rather than biological ones [248]. The reasons are manifold but, generally, biologically relevant variables are much more difficult and complex than others (e.g. temperature, pressure). Another important reason derives from restricting requirements, namely

- sterilization procedures,
- stability and reliability over extended periods,
- application over an extended dynamic range,
- no interference with the sterile barrier,
- insensitivity to protein adsorption and surface growth, and
- resistance to degradation or enzymatic break down.

Finally, material problems arise from the constraints dictated by aseptic culture conditions, biocompatibility and the necessity to measure over extended dynamic ranges which often make the construction of sensors rather difficult.

Historically, the technical term "fermenters" is used for any reactor design used for microbial or cellular or enzymatic bioconversions and is basically synonymous with a vessel equipped with a stirring and aeration device. (High performance) bioreactors, however, are equipped with as large as possible a number of sensors and connected hard- or software controllers. It is a necessary prerequisite to know the macro- and microenvironmental conditions exactly and to keep them in desired permissive (or even optimal) ranges for the biocatalysts; in other words, the bioreaction in a bioreactor is under control [307, 401].

1.1
Standard Techniques (State of Routine)

There are undoubtedly a few variables that are generally regarded as a must in bioprocess engineering. Among these are several physical, less chemical and even less biological variables. Figure 1 gives a summary of what is nowadays believed to be a minimum set of required measurements in a bioprocess. Such a piece of equipment is typical for standard production of material, see, e.g.

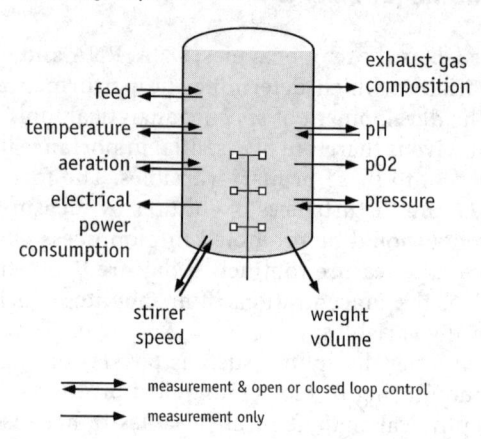

Fig. 1. Common measurement and control of bioreactors as generally accepted as routine equipment

[347]. However, the conclusion that these variables are sufficient to characterize the microenvironment and activity of cells is, of course, questionable.

Besides some environmental and operational variables, the state variables of systems must be known, namely the amounts of active biocatalyst, of starting materials, of products, byproducts and metabolites.

1.2
Biomass

Biomass concentration is of paramount importance to scientists as well as engineers. It is a simple measure of the available quantity of a biocatalyst and is definitely an important key variable because it determines – simplifying – the rates of growth and/or product formation. Almost all mathematical models used to describe growth or product formation contain biomass as a most important state variable. Many control strategies involve the objective of maximizing biomass concentration; it remains to be decided whether this is always wise.

The measure of mass is important with respect to calculating mass balance. However, the elemental composition of biomass is normally ill defined. Another reason for determining biomass is the need for a reference when calculating specific rates (q_i): $q_i = r_i/x$. An ideal measure for the biocatalysts in a bioreaction system of interest would be their activity, physiological state, morphology or other classification rather than just their mass. Unfortunately, these are even more difficult to quantify objectively and this is obviously why the biomass concentration is still of the greatest interest.

1.3
Substrates

Cells can only grow or form products when sufficient starting material, i.e. substrates, is available. The presence of substrates is the cause and growth or product formation is the effect. One can solve the inverse problem, namely conclude that biological activities cease whenever an essential substrate is exhausted, and so omit the measurement of the substrate, provided the progress of growth (i.e. biomass) and/or product formation is known [215]. This is not a proper solution because there are many more plausible, and also probable, reasons for a decrease in bioactivities than just their limitation by depletion of a substrate. It is, for instance, also possible that too much of a substrate (or a product) inhibits or even intoxicates cellular activities. In such a situation, the above conclusion that a substrate must be depleted when growth or product formation ceases, no longer holds. One must, then, solve the direct problem, namely analyze the concentrations of relevant substrates, in order to pin-point the reasons for missing bioactivities. From an engineering point of view, this measure should be available instantaneously in order to be able to control the process (via the concentration of the inhibitory substrate). The technical term for such an operating mode is nutristat: a well-controlled level of a relevant nutrient causes a steady state.

In environmental biotechnology, in particular, the objective of a bioprocess can be to remove a "substrate", e.g. a pollutant, as completely as possible rather than making a valuable product. In this case, the analytical verification of the intention is, of course, mandatory for validation.

The classical methods to determine substrate concentrations are off-line laboratory methods. This implies that samples are taken aseptically, pre-treated and transported to a suitable laboratory, where storage of these samples might be necessary before processing. The problems associated with these procedures are discussed below. There is only one general exception to this, namely, the gaseous substrate oxygen, for which in situ electrodes are generally used.

1.4
Products, Intermediates and Effectors

The product is almost the only reason why a bioprocess is run. The main concern is in maximizing the profit which depends directly on the concentration and/or volumetric productivity and/or of the purity of the product. It is therefore interesting to know the values which require measurement. The classical methods to determine product concentrations are typically off-line laboratory methods and the above statements for substrate determinations are valid here, too.

One may need to account for labile intermediates as found, for instance, in penicillin production [196, 304]. Then, on-line analyses will best avoid artifacts due to storage of materials even though the samples are cooled to 4 or 5 °C.

In summary, bioprocess science needs more quantitative measurements. It is insufficient to know that something happens, we need to know why and how [260].

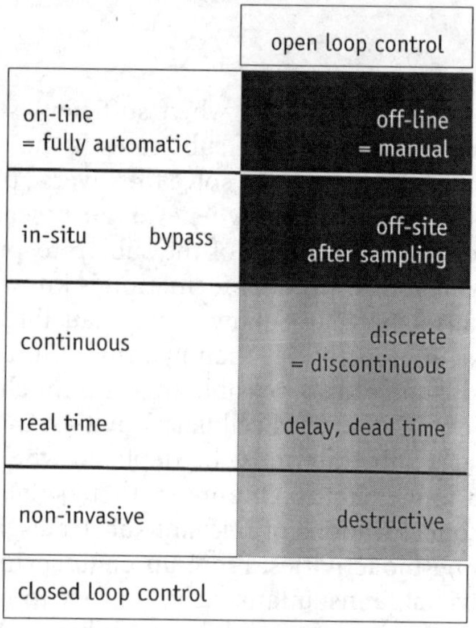

Fig. 2. Terminology of types of signals and signal generation

2
On-Line Sensing Devices

On-line is synonymous for fully automatic. No manual interaction is necessary to obtain the desired results. However, this statement is not intended to promote a blind reliance on on-line measuring equipment. Depending on the site of installation, one discriminates further between in situ, which means built-in, and ex situ, which can mean in a bypass or in an exit line; in the latter case, the withdrawn volumes are lost for the process (Fig. 2). Depending on the mode of operation of the sensing device, one can discriminate between continuous and discontinuous (or discrete) signal generation; in the latter case, a signal is repeatedly generated periodically but, in between, there is no signal available.

2.1
In Situ Instruments

2.1.1
Temperature

Generally, the relationship between growth and temperature (approximated by the Arrhenius equation at suboptimal temperatures) is strain-dependent and shows a distinct optimum. Hence, temperature should be maintained at this level by closed loop control. Industry seems to be satisfied with a control precision of ±0.4 K.

Temperature can be the variable most often determined in bioprocesses. In the range between 0 and 130 °C, this can be performed using thermoelements

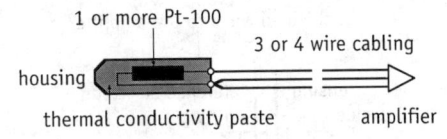

Fig. 3. Schematic design of temperature sensors. One or more thermoresistors are packed into a stainless steel housing. 3- or 4-strand cabling is recommended

or by thermometers based on resistance changes, e.g. of a platinum wire (then this sensor is called a Pt-100 or Pt-1000 sensor; the resistance is either 100 or 1000 Ω at 0 °C; Fig. 3). This is, although not linear per se, one of the most reliable but not necessarily most accurate measures in bioprocesses. The necessary calibration references (standards) are usually not available. Temperature is most often controlled. With a sound control system it is possible to obtain a precision of 1–10 mK in laboratory scale bioreactors [398].

2.1.2
pH

pH is one of the variables often controlled in bioprocesses operated in bioreactors because enzymatic activities and, therefore, metabolism is very sensitive to pH changes. The acidification derives in most cases predominantly from the ammonia uptake when ammonium ions are provided as the nitrogen source: NH_3 is consumed and the proton left over from the NH_4^+ causes a drop in pH.

In shake flask cultures, there is only one reasonable possibility to keep pH within a narrow range, namely the use of a very strong buffer, usually phosphate buffer. This is the major reason why culture media often contain a tremendous excess of phosphate. Insertion of multiple pH probes and titrant-addition tubes into shakers has, however, been proposed and marketed [66].

The pH of process suspensions is measured potentiometrically using electrodes filled with liquid or gel electrolytes. A brief comparison of properties is given in the literature [123]. Glass electrodes develop a gel layer with mobile hydrogen ions when dipped into an aqueous solution. pH changes cause ion diffusion processes generating an electrode potential. Lithium-rich glasses are well suited for this purpose. The potential is measured in comparison to a reference electrode which is usually a Ag/AgCl system since calomel would decompose during sterilization (strictly speaking above 80 °C). The electric circuit is closed via a diaphragm separating the reference electrolyte from the solution (Fig. 4).

Spoilage of the reference electrolyte is one of the major problems during long-term cultivations. Monzambe et al. [292] and Bühler (personal communication) have reported discrepancies of one pH unit between in situ on-line and off-line measurements which were caused by black clogging of the porous diaphragm. Either acidification or pressurization of the electrolyte was suitable to restrain this.

Alternatives to the glass electrode are optical measurements of pH [4, 79] or exploitation of pH-sensitive field effect transistors, a so-called pH-FET [378]

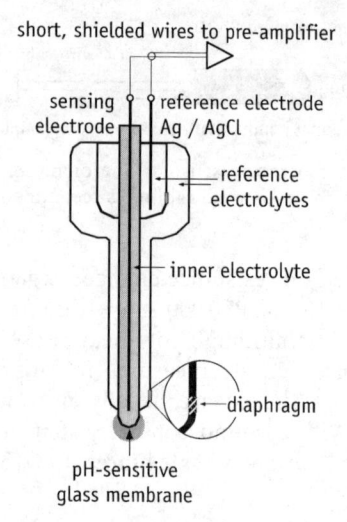

Fig. 4. Schematic design of a sterilizable pH electrode made of glass. The pH-sensitive glass which develops a gel layer with highest mobility for protons is actually only the tip (calotte) of the electrode. Electrolytes can contain gelling substances. Double (or so-called bridged) electrolyte electrodes are less sensitive to poisoning of the reference electrode (e.g. formation of Ag_2S precipitates)

(Fig. 5); however, these alternatives are not yet mature enough to be routinely used. pH can be maintained within a few hundredths of a pH unit, provided mixing time is sufficiently small. Interestingly, many scientists "control" the pH by exclusively adding alkali. Addition of acid is often not foreseen. But if pH is well controlled it is rewarding to monitor the pH controller output signal as well because it reveals the activities of the culture with respect to production and consumption of pH-active substances, i.e. (de)protonized molecules such as organic acids or ammonium ion. This can be very valuable information which usually remains unused.

In pH-controlled cultivations, the amount of titrant added to the culture can be used to calculate the (specific) growth rate provided a useful model is available (a typical inverse problem). Bicarbonate affects the stoichiometry between titrant and biomass but does not prevent determination of growth rates [187]. This approach works even though non-linear relationships hold between biomass and, for instance, lactic acid concentrations [3].

2.1.3
Pressure

The direct dependence of microorganisms on pressure changes is negligible provided they do not exceed many bars [18, 186, 211, 474]. However, the partial pressure of dissolved gases and their solubility is indirectly affected and must, therefore, be at least considered if not controlled. A data sampling frequency in the range of a few 100 ms is appropriate for direct digital pressure control (DDC) in laboratory scale bioreactors.

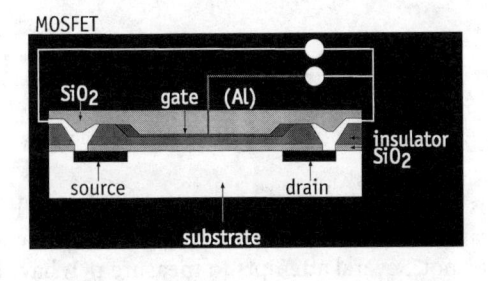

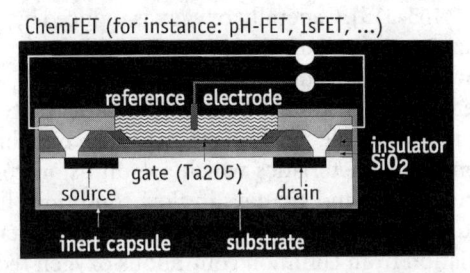

Fig. 5. Schematic design of a usual metal oxide field effect transistor (MOSFET; *top*) and of an ion-sensitive field effect transistor (IsFET, *bottom*). The voltage applied to the gate – which is the controlling electrode – determines the current that flows between source and drain. The substrate is p-Si, source and drain are n-Si, the metal contacts are made from Al, and the insulators are Si_3N_4. Instead of a metallic gate, a pH-FET has a gate from nitrides or oxides, for instance Ta_2O_5. Depending on the pH of the measuring solution, the voltage at the interface solution/gate-oxide changes and controls the source-drain current. Generally, in bio-FETs (which are also biosensors, of course) an additional layer of immobilized enzymes, whole cells, antibodies, antigens or receptor is mounted on top of the gate; the reaction must, of course, affect the pH by producing or consuming protons to be detectable with this transducer. Note that the reference electrode is still necessary; this means that all problems associated with the reference pertain also to such a semiconductor-based electrode

In addition, the reduction of infection risks by a controlled overpressure is advantageous. During sterilization, pressure is of paramount interest for safety reasons. A variety of sterilizable sensors exists, e.g. piezo-resistive, capacitive or resistance strain gauge sensors (Fig. 6), but not all of them are sufficiently temperature compensated.

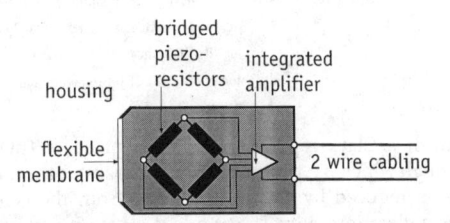

Fig. 6. Schematic design of a pressure sensor. A flexible stainless steel membrane interfaces the pressure-sensitive elements (bridged piezo-resistors) from the measuring liquid. Some products contain the amplifier electronics in the housing and are (somehow) temperature compensated. The shown 2-strand cabling mode resulting in a current signal is very convenient

2.1.4
Oxygen

2.1.4.1
Oxygen Partial Pressure (pO_2)

Oxygen solubility is low in aqueous solutions, namely 36 mg l^{-1} bar^{-1} at 30 °C in pure water. Mass transfer is, therefore, determinant whether a culture suffers from oxygen limitation or not. Several attempts to measure pO_2 have been made in the past, see, e.g. [46, 106, 163, 315]. Generally, oxygen is reduced by means of a cathode operated at a polarizing potential of 600–750 mV which is generated either externally (polarographic method) or internally (galvanic method). A membrane separates the electrolyte from the medium to create some selectivity for diffusible substances rather than nondiffusible materials (Fig. 7). The membrane is responsible for the dynamic sensor characteristics which are diffusion controlled. Less sensitivity to membrane fouling and changes in flow conditions have been reported for transient measuring techniques, where the reducing voltage is applied in a pulsed mode, a deviation from common continuous oxygen reduction [451].

A control loop for low pO_2 (< 100 ppb) based on a fast but non-sterilizable sensor (Marubishi DY-2) was devised by Heinzle et al. [160].

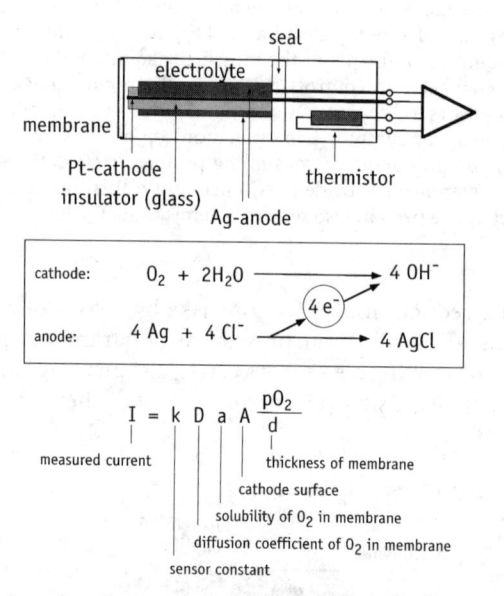

cathode: $O_2 + 2H_2O \xrightarrow{4e^-} 4\,OH^-$

anode: $4\,Ag + 4\,Cl^- \longrightarrow 4\,AgCl$

$$I = k\ D\ a\ A\ \frac{pO_2}{d}$$

- I: measured current
- k: sensor constant
- D: diffusion coefficient of O_2 in membrane
- a: solubility of O_2 in membrane
- A: cathode surface
- d: thickness of membrane

Fig. 7. Schematic design of a Clark-type oxygen partial pressure (pO_2) electrode. A (sandwiched) membrane through which oxygen must diffuse separates the measuring solution from the electrolyte. Oxygen is reduced by electrons coming from the central platinum cathode which is surrounded by a glass insulator. The anode is a massive silver ring usually mounted around the insulator. This design, a so-called polarographic electrode, needs an external power supply. For oxygen, the polarization voltage is in the order of 700 mV and the typical current for atmospheric pO_2 is in the order of 10^{-7} A. A built-in thermistor allows automatic correction of the temperature-dependent drift of approximately 3% K^{-1} at around 30 °C

Merchuk et al. [276] investigated the dynamics of oxygen electrodes when analyzing mass transfer, and they reported whether and when an instantaneous response occurs. A semiempirical description of diffusion coefficients was provided by Ju and Ho [198]. *Bacillus subtilis* cultures change the product concentration ratio between acetoin and butanediol rapidly in the range of $pO_2 \approx 80-90$ ppb [286]. This fact could be used for the characterization of the oxygen transport capabilities of bioreactors.

2.1.4.2
Oxygen in the Gas Phase

Measurements of oxygen in the gas phase are based on its paramagnetic properties. Any change in the mass concentration of O_2 affects the density of a magnetic field and thus the forces on any (dia- or para)magnetic material in this field. These forces on, for example, an electrobalance can be compensated electrically and the current can be converted into mass concentrations: further conversion into a molar ratio, e.g. % O_2, requires the knowledge of total pressure (Fig. 8).

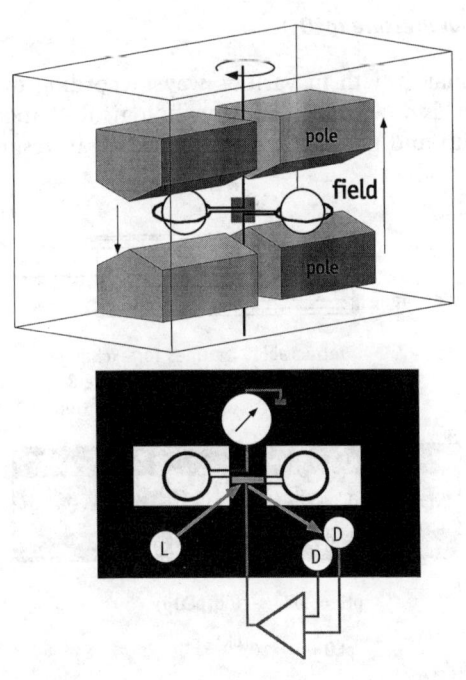

Fig. 8. Schematic design of a paramagnetic oxygen analyzer. A diamagnetic electrobalance is placed in a permanent magnetic field. Whenever the paramagnetic oxygen enters this space, the field lines intensify and exert a force on the diamagnetic balance trying to move it out of the field. This force is compensated by powering the electric coils around the balance so much that it does not change its position in the field. The current is proportional to the mass of paramagnetic matter (i.e. oxygen) in the measuring cell, i.e. a concentration and not a (relative) fraction or content

The effect of oxygen on metabolism is better known than the effects of other nutrients. For instance, Furukawa et al. [119] reported on a long-term adaptation of *Saccharomyces cerevisiae* to low oxygen levels and Pih et al. [325] observed a clear relationship between pO_2 and catabolic repression, catabolic inhibition, and inducer repression for β-galactosidase during growth of *Escherichia coli*. Wilson [459] based on-line biomass estimation on dynamic oxygen balancing.

Analysis of O_2 as well as CO_2 in exhaust gas is becoming generally accepted and is likely to be applied as a standard measuring technique in bioprocessing. It is possible to multiplex the exhaust gas lines from several reactors in order to reduce costs. However, it should be taken into account that the time delay of measurements with classical instruments is in the order of several minutes, depending on the efforts for gas transport (active, passive) and sample pretreatment (drying, filtering of the gas aliquot).

2.1.5
Carbon Dioxide

2.1.5.1
Carbon Dioxide Partial Pressure (pCO_2)

CO_2 affects microbial growth in various ways according to its appearance in catabolism as well as in anabolism. Morphological changes (e.g. [97]) and variations in growth and metabolic rates [195, 310] in response to pCO_2 have

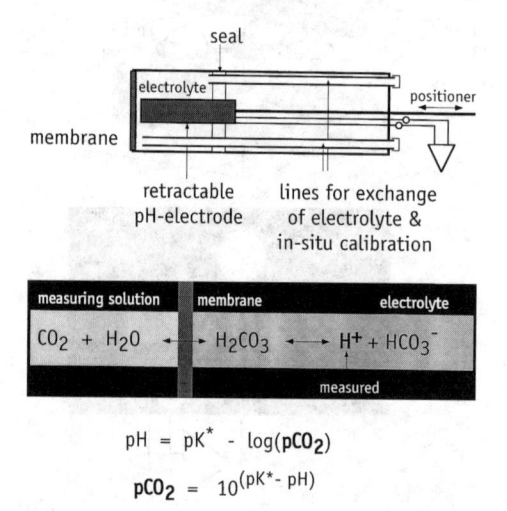

$$pH = pK^* - \log(pCO_2)$$

$$pCO_2 = 10^{(pK^* - pH)}$$

Fig. 9. Schematic design of a carbon dioxide partial pressure (pCO_2) electrode. CO_2 diffuses through the membrane into or out of the electrolyte where it equilibrates with HCO_3^- thus generating or consuming protons. The respective pH change of the electrolyte is sensed with a pH electrode and is logarithmically proportional to the pCO_2 in the measuring solution. Since the electrolyte may become exhausted, one can replace it through in/out lines. These can also be used to re-calibrate the pH electrode. Therefore, the electrode is retractable by means of a mechanical positioner

been demonstrated. pCO_2 can be measured indirectly: the pH value of a bicarbonate buffer, separated from the medium by a gas-permeable membrane, drops whenever CO_2 diffuses into this compartment and vice versa (Fig. 9); pH depends on the logarithm of pCO_2 [334]. Either a glass electrode or optical principles [439] can be used for pH determination.

The response of the pCO_2 sensor is not exclusively CO_2 dependent [91]. Yegneswaran et al. [477] modeled the effect of changes in physical conditions on the pCO_2 signal. A step up in external pH resulted in a pCO_2 downward spike and vice versa. Pressure shifts in the range of 1–2 bar caused pCO_2 fluctuations to an extent of >10%. Mass transfer is assumed to control the dynamics of CO_2 equilibration. The bicarbonate buffer solution must be replaced regularly due to its limited capacity. Otherwise, the equilibration will be prolonged and base line drifts occur. This was one of the reasons why Mettler Toledo (formerly Ingold) took this electrode off the market.

2.1.5.2
Carbon Dioxide in the Gas Phase

CO_2 in the gas phase can be determined by means of its significant infrared absorbance (Fig. 10) at wave lengths (λ) <15 µm, particularly at 4.3 µm [289], or by acoustic means. Integrated photoacoustic spectroscopy and magnetoacoustic (PAS/MA) technology for combined CO_2 and O_2 analysis has rapid response time and a small sample volume is sufficient. The acoustic methods are accurate, stable over long periods and very simple to use.

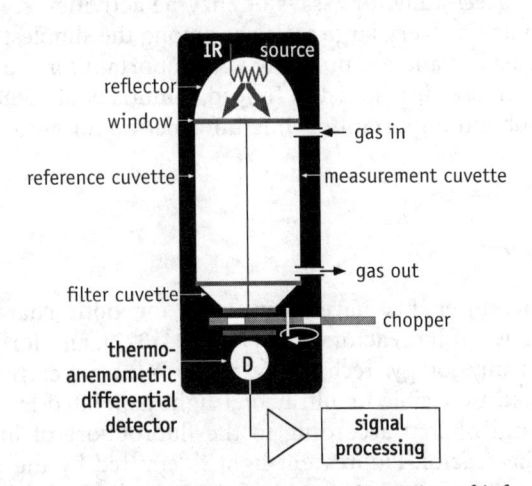

Fig. 10. Schematic design of a CO_2 analyzer based on absorption of infrared (IR) radiation. An IR generator illuminates both the measuring and the reference cuvette. The latter is used to adapt the measuring range and is often filled with just a noble gas (zero). The remaining radiation then passes a filter cuvette which can be filled with interfering gas that absorbs all radiation energy at the respective wavelength in both light paths equally. A light chopper (electrically driven with a few 100 Hz) lets the light alternatively pass from the measuring and from the reference path. A thermoanemometric detector quantifies the arriving IR radiation which is inversely proportional to the CO_2 present in the cuvettes

Molin [287] grouped certain types of food-related bacteria according to their CO_2 resistance and Jones and Greenfield [195] reviewed the inhibition of yeasts, distinguished by metabolic and membrane effects. Supercritical CO_2 – an interesting extraction fluid – was found to be moderately tolerated by yeasts [186, 239]. It is most likely that an optimum CO_2 level exists which is generally accepted for mammalian cells but also reported for bacteria, e.g. the growth rate of *Escherichia coli* [228, 346] or for biomass yield, glucose uptake and ethanol production of *Zymomonas mobilis* [310]. Hirose [170] considered the biochemical effects of O_2 supply and CO_2 removal and concluded that further physiological studies are needed to promote better understanding of the mechanisms involved. Xylose metabolism of *Candida* and *Pichia* yeasts is also affected by CO_2 [235] as well as growth of other yeasts [226].

Park et al. [320] and Rothen et al. [355] assumed a linear correlation between biomass growth rate and carbon dioxide evolution rate (CER) and exploited this model for the estimation of cell concentration, an elegant tool for processes using technical media such as highly colored molasses-mineral salts medium with large amounts of particles. Note that this is a typical solution to an inverse problem: substrate consumption is a cause and CO_2 evolution is an effect; one measures the effect and estimates the cause. Similarly, the cell concentrations of *Streptococcus thermophilus* in co-culture with *Lactobacillus thermophilus* were determined due to its ability to metabolize urea in milk to CO_2 and ammonia [407]. CO_2 was reported to serve as a control variable in cultures of *Candida brassicae* and allowed O_2 and, thus, ethanol to be maintained automatically at a constant level [424]. Furthermore, CO_2 measurements have been used successfully for assays of enzyme activities, e.g. [41, 362]. CO_2 flux measurements on a very large scale are among the simplest measurements that can be carried out and are probably also important for economic reasons, for instance, in the brewing industry. Indeed, Simutis et al. [388–390] selected this method to obtain important on-line information for automatic control of such processes.

2.1.6
Culture Fluorescence

Fluorescence measurements have been used for both characterization of technical properties of bioreactors, e.g. [140, 234, 372], and for basic scientific investigations of physiology. Technically, either intra- or extracellular fluorophores are excited by visible or ultraviolet light generated by a low-pressure mercury lamp and filtered according to the fluorophore of interest prior to emission into the reactor. Fluorescent light is emitted by the excited fluorophores at a longer characteristic wavelength. Only the backward fluorescence can be collected with appropriate (fiber) optics, is most likely filtered, and the residual light is detected by a sensitive photodetector (Fig. 11). Descriptions of typical sensors are given by Beyeler et al. [29] and Scheper [368, 371]. Intensity measurement is prone to many interferences and disturbances from the background. These drawbacks can be avoided by measuring the fluorescence lifetime but this is more demanding [16, 269, 406].

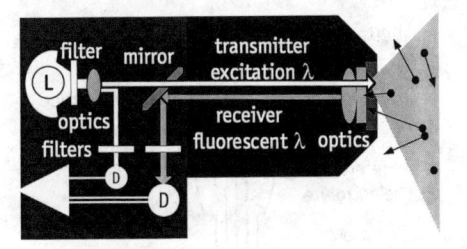

L   mercury lamp
D   photomultipliers or photodiodes
filters might be replaced by tunable monochromators
dichroitic mirror not needed when fiber optics are used

Fig. 11. Schematic design of a fluorescence sensor. A strong light source creates radiation with low wavelengths. Optics like lenses and filters extract and focus the desired excitation light which is sent through the window into the measuring solution. Only a small fraction of the fluorescent light arrives at the window, passes this, and is collected by appropriate optics and fed to a sensitive detector (usually a photomultiplier). Variations in the light source intensity can be compensated by a comparative measurement. When optical fibers are used inside the instrument, the dichroitic mirror shown is obsolete

Most investigators have measured NAD(P)H-dependent culture fluorescence but other fluorophores are also interesting. Humphrey [182] gave a (non-exhaustive) survey of the historical evolution of fluorescence measurements for bioprocess monitoring. All these data have to be interpreted carefully. Quantification appears difficult even though attempts at a theoretical analysis of involved effects have been made [411, 453]. Calibrations are tricky since the quenching behavior of cell material and the chemical composition of the medium contribute substantially and time-variably to the measured signal [363]. Further, the production of interfering fluorophores must be considered [179, 281]. Turbidity of the culture suspension should be low and the bubble distribution should remain constant [29].

NAD(P)H-dependent culture fluorescence has mainly been exploited for metabolic investigations, e.g. [199, 227, 339–341, 410]. The signal is sensitive to variables such as substrate concentration or oxygen supply. Thus, all attempts to exploit this signal as a biomass sensor [478] have been limited to conditions where no metabolic alterations occur [257, 395, 396]. It is well known that a mechanistic or causal-analytical interpretation of the signal trajectory in secondary metabolite cultivations can be very difficult [303].

The outstandingly rapid principle of fluorescence measurements served excellently for the controlled suppression of ethanol formation during continuous baker's yeast production [280].

2.1.7
Redox Potential

Bioprocess media and culture liquids contain many different components which can exist in a reduced and an oxidized form as redox couples. The resulting redox potential, as measured by a redox electrode, is related to an "overall

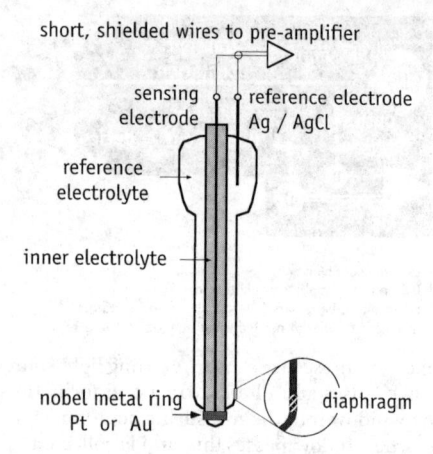

Fig. 12. Schematic design of a redox electrode. It strongly resembles the pH glass electrode. The active measuring element is a noble metal, usually constructed as a ring around the tip of the electrode

availability of electrons" rather than to a specific compound. The extracellular redox measurement is very instructive, specifically under microaerobic conditions where the pO_2 sensor signal becomes inaccurate [460]. The signal generation is faster than that of pO_2 because the diffusion step is omitted [111].

Redox potential is measured potentiometrically with electrodes made of noble metals (Pt, Au) (Fig. 12). The mechanical construction is similar to that of pH electrodes. Accordingly, the reference electrode must meet the same requirements. The use and control of redox potential has been reviewed by Kjaergaard [218]. Considerations of redox couples, e.g. in yeast metabolism [47], are often restricted to theoretical investigations because the measurement is too unspecific and experimental evidence for cause–effect chains cannot be given. Reports on the successful application of redox sensors, e.g. [26, 191], are confined to a detailed description of observed phenomena rather than their interpretation.

The application of a redox sensor in a control loop has been reported by Memmert and Wandrey [274] who controlled xylanase production of *Bacillus amyloliquefaciens* by defined oxygen limitation: redox electrodes refer essentially to dissolved oxygen concentration below 10 mmol l^{-1} O_2. This property was also promoted to determine the quality of anaerobic processes [403].

2.1.8
Biomass

Since an on-line generated signal for biomass concentration is decisive for control purposes a series of sensors and methods that can be automated have appeared in recent decades. Many of them rely on optical measuring principles, others exploit filtration characteristics, density changes of the suspension as a consequence of cells, or (di)electrical properties of suspended cells. Some of the proposed methods have been used off-line, not on-line, as a standard. However, most of the approaches that are discussed below can be adapted for on-line application, either in situ or, more generally, ex situ by using a small sample stream of culture which is (then named bypass [69]) or is not returned to the reactor (wasted), see Sect. 2.2.

2.1.8.1
Comparability of Sensors

A direct comparison of some representative sensors to estimate biomass in bacterial and yeast cultures was made by Nipkow et al. [309], by Fehrenbach et al. [107], by Konstantinov et al. [219] and, more recently, by Wu et al. [465]. These studies are of importance because the sensors were mounted in situ and used in parallel. Most of the sensors measured the optical density (OD), one the autofluorescence of the cultures (fluorosensor) and another was a capacitance sensor (ßugmeter).

2.1.8.2
Optical Density

Current commercially available optical density (OD) sensors are based on the determination of either transmission, reflection or scatter of light, or a combination thereof. The theoretical background as to why these OD measurements reflect the biomass concentration are rather manifold and complicated, and would constrain the application tremendously if not many simplifications could be reasonably applied [345, 468]. A direct a priori calculation of dry weight concentration from any OD measurement cannot be expected to be realistic, but the systems can be calibrated from case to case. Ries [345] derived some technically relevant proposals: the primary beam of the light source should be narrowly focused and be of high power (laser source) because of the low ratio of intensities of scatter to primary light and a high fraction of the scatter should be in a forward direction. Theoretically, for bacteria not exceeding a typical length of 3 µm, the visible wavelength should be chosen, for larger organisms the infrared. Large plant cells can also be estimated with turbidimetric methods [428] or insect cell cultures [21]. Tunable sensors are currently not yet routinely available and the wavelength choice of the vendors seems to be a compromise which also takes into account the fact that many media absorb increasingly with decreasing wavelength: green filters, IR diodes, laser diodes or lasers between 780 and 900 nm in others (Fig. 13).

Fiber sensors with high quality spectrophotometers outside the reactor in a protected room are a valuable but probably expensive alternative [74]. Inexpensive variants can be made by using stabilized light emitting diodes (LEDs emitting at around 850 nm) or arrays thereof [154]; modulation with a few 100 Hz ("light chopping") should be used in order to minimize influences from ambient light [479].

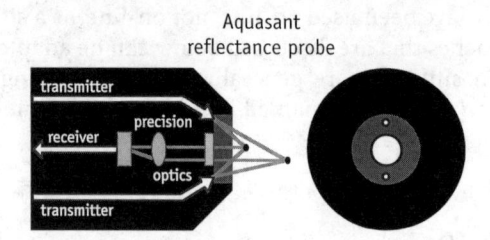

Fig. 13. Schematic design of the Aquasant probe. This is a sensor for optical density measuring the reflected light. Precision optics focus and collect the incident and the reflected light. Internally, light is guided through optical fibers. *Left*: cross section; *right*: front view

2.1.8.3
Interferences

Interferences from gas bubbles or particulate matter other than cells (Hong et al. [175] and Desgranges et al. [86] even report on a spectrophotometric cell mass determination in semi-solid cultivations) are common to almost all sensors but different methods are available to circumvent and minimize such problems.

The FundaLux system, for instance, aspirates a liquid aliquot with a Teflon piston into an external glass cell, allows a (selectable) time (typically 2 min) to degas, measures transmission in comparison to an air blank, and releases the aliquot back to the reactor; an interesting feature – specific to this instrument – is the repetitive cleaning of the optical window by the moving Teflon piston. Some problems with infections have been communicated with this device since the measuring cell is external to the bioreactor and the sensor is probably insufficiently sterilized in situ.

Geppert and Thielemann [125] and Geppert et al. [126] have used a similar method but a different instrument to measure a suspension aliquot outside the bioreactor and reported a fairly good linear correlation between OD and biomass concentration for some bacteria and yeasts.

A sensor based on the same principle of sample degassing in a void volume but mounted completely inside the reactor (Foxboro/Cerex; Fig. 14) has been described by Hopkins and Hatch [178]. The minimal time interval between individual measurements is 30 s. Both 90° scatter and transmission measurements can be made simultaneously. A linear correlation between OD and *Saccharomyces cerevisiae* density from 0.1 to 150 g l^{-1} has been claimed for this instrument [152].

Simple transmission measurements with inexpensive components were made to estimate the local specific interfacial area of a suspended phase (i.e. of gas bubbles) in a bioreactor [473].

The LT 201 (ASR/Komatsugawa/Biolafitte) instrument (Fig. 15) attempts to keep gas bubbles out of the optical path by mounting a cylindrical stainless steel screen around this region and positioning the sensor at a certain angle into a less turbulent zone in the reactor. Hibino et al. [168] and Yamane et al. [471, 472]

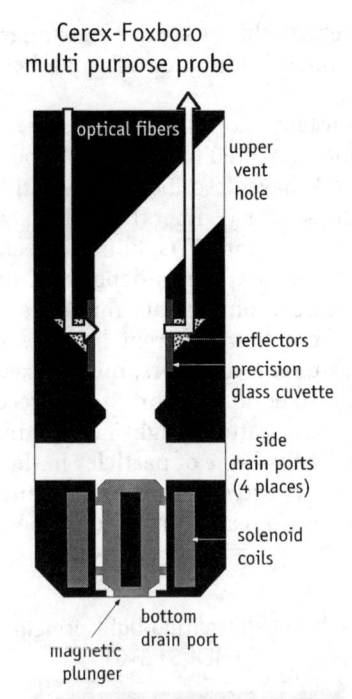

Fig. 14. Schematic design of the Cerex probe. This is a sensor for optical density and mounted vertically in situ. Suspension enters the side drain ports deliberately and can be trapped inside the sensor by powering the solenoid coils: the magnetic plunger closes the side ports. In the meantime, the trapped dispersion degasses and bubbles disappear through the upper vent hole. After some time, the optical density reading is "declared representative". The next cycle starts with opening the side drain ports

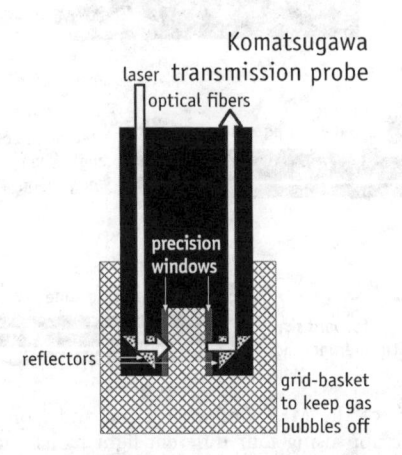

Fig. 15. Schematic design of the Komatsugawa probe. This is a sensor for measuring light transmission. It is powered with a laser that has enough energy to also measure highly dense cultures. Optical fibers send and collect light. Around the measuring zone, a stainless steel grid basket is mounted. Its function is to let cells pass and, at the same time, exclude gas bubbles. This is why the mesh size of the grid must be selected according to the type of cells being measured

reported good experience with this or similar instruments and found the signal so reliable that they exploited it for automation and control of fed-batch processes [202].

Other OD sensors are totally subject to the interference of bubbles; however, filters allow the signal noise (created by bubbles) to be dampened more or less effectively. Iijima et al. [183] have described a sensor that measures both transmission (1 fiber) and 90° scatter (2 fibers) which may allow a compensation mathematically. The MEX-3 sensor (BTG, Bonnier Technology Group; Fig. 16) compensates internally for errors due to deposition on the optical windows, temperature or aging of optical components; this is made possible by evaluating quotients of intensities from four different light beams (straight and cross beams from two emitters to two detectors; multiplexed). The Monitek sensor has a special optical construction (prior to the receiver; so-called spatial filtering system) to eliminate scattered light not originating from particles or bubbles in the light path. The volume of particles in the medium is determined by calculating the ratio of forward scattered to transmitted light. Other sensors – used in different industrial areas – are equipped with mechanical wipers.

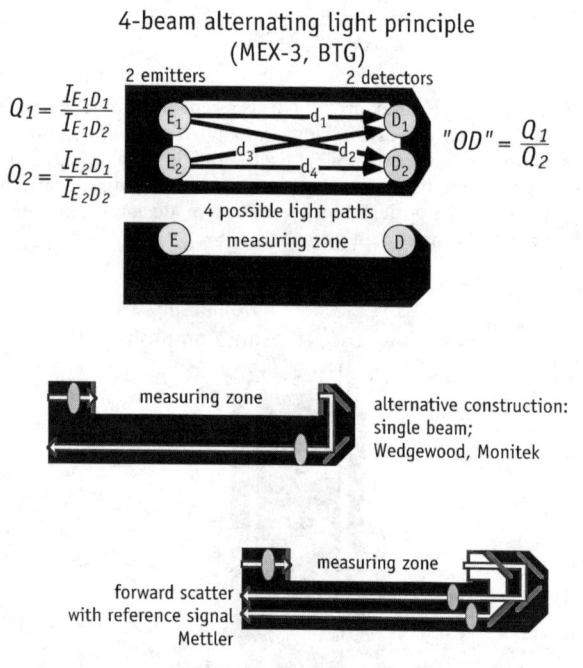

Fig. 16. Schematic design of the MEX probe (*top*: top and front view). This is a sensor measuring light transmission using four different light paths with two emitters and two detectors. The emitters are alternately switched on and off. The electronics determine the ratios of received intensities, Q_1 and Q_2. The created signal is again a ratio of these values which is virtually independent of fouling of the window surfaces. Alternative constructions are shown below: a single-beam sensor and a variant allowing the comparison of the transmitted light with forward scattered light (Mettler sensor)

Aquasant claims to have minimized interferences with depositions on the optical window by a special design of the precision receiver optics of the AF 44 S sensor (which can be confirmed by our own practical experiences).

In the various studies, the different sensors were found to be significantly different with respect to sensitivity, linearity and signal to noise ratio. None of the sensors lost sensitivity completely due to surface growth on the optical window. The signals never correlated perfectly linearly with biomass concentration. Only little disturbance was observed from changing environmental illumination (artificial or sun light when used in glass reactors).

2.1.8.4
Electrical Properties

The measurement principle of the biomonitor (formerly called ßugmeter) relies on the fact that the capacitance of a suspension at low radio frequencies is correlated with the concentration of the suspended phase of fluid elements that are enclosed by a polarizable membrane, i.e. intact cells [11, 150, 213, 265, 266]. The capacitance range covered is from 0.1 to 200 pF, the radio frequency some 200 kHz to 10 MHz. A severe limit to this principle is the maximally acceptable conductivity (of the liquid phase) of approximately 8 mS cm^{-1} in earlier versions which has been significantly improved (to the order of 24 mS cm^{-1}) recently. This conductivity is, however, easily reached in the more concentrated media necessary for high cell density cultures. Noise is also created by gas bubbles. This could theoretically be reduced by using very rapid spike elimination algorithms. The method of applying periodic cleaning pulses to the electrodes in order to remove (potential) surface fouling (by attached organisms) in situ superimposed significant spikes to the signals of high impedance sensors mounted in the same reactor (such as pH or redox electrodes) thus corrupting the respective controllers. Not cleaning the electrodes results in fouling within some days even in defined media.

2.1.8.5
Thermodynamics

An elegant, completely non-invasive method is to exploit the heat generated during growth and other metabolic activities of organisms which is also proportional to the amount of active cells in a reaction system [36]. Under well-defined conditions, calorimetry can be an excellent tool for the estimation of total (active) biomass [31, 32, 258, 448], even for such slow growing organisms as hybridoma cells or for anaerobic bacteria growing with an extremely low biomass yield [373, 416, 447]. The method is so inherently sensitive that cell cycle dependent events can also be analyzed [13].

Bioreactions are exothermic. The net heat released during growth represents the sum of the many enzymatic reactions involved. Reasonably, this measure depends on both the biomass concentration and the metabolic state of the cells. Its general use in biotechnology has been reviewed by von Stockar and Marison [415]. A theoretical thermodynamic derivation for aerobic growth gives a prediction for the heat yield coefficient $Y_{Q/O}$ of 460 kJ (mol O_2)$^{-1}$ and it was experimentally confirmed to be an excellent estimate because the average value found in many different experiments was 440 ± 33 kJ (mol O_2)$^{-1}$ [31].

Three different approaches are chiefly applied: micro-, flow and heat flux calorimetry. Heat flux calorimetry is certainly the best choice for bioprocess monitoring (Fig. 17) [264]. In a dynamic calorimeter, the timely change of temperature is measured and various heat fluxes (e.g. heat dissipated by stirrer, or lost due to vaporization of water) need to be known in order to calculate the heat flux from the bioreaction:

$$K_P \frac{dT}{dt} = q_R + q_A - q_L - q_G \tag{1}$$

where K_P(J K^{-1}) is the total heat capacity of the system, q_R is the reaction-derived heat, q_A derives from agitation and q_L and q_G are losses via surfaces and vaporization of water due to aeration. The fractions q_A, q_L and q_G must be electrically calibrated prior to inoculation. Unfortunately, the losses are unlikely to be constant. In a heat flux calorimeter, one needs to determine the timely change of the temperature difference between reactor and jacket $T_R - T_J$ (K); random heat losses can be neglected and the systematic contributions to heat generation q_A or removal q_G can be determined prior to inoculation. The global heat transfer coefficient k_w(W K^{-1} m^{-2}) can be simply determined by electrical

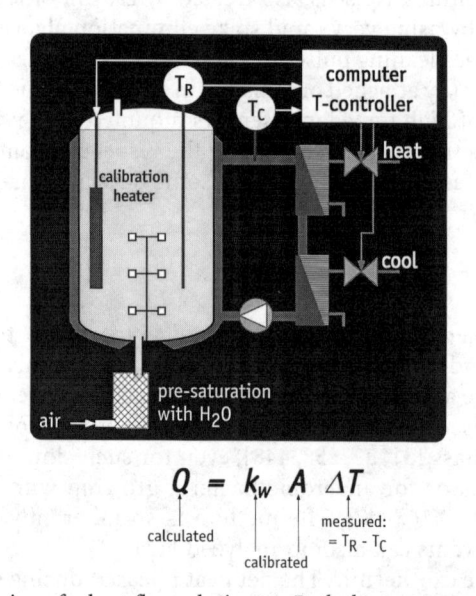

Fig. 17. Schematic design of a heat flux calorimeter. Both the temperature in the reactor and in the circuit (or jacket) are measured as sensitively and reproducibly as possible. A well-tuned temperature controller keeps the reactor temperature constant by feeding the circuit with warmer or colder water or oil. The circulating water or oil can be taken from either a chilled and a heated reservoir or, as shown, be heated or cooled via external heat exchangers. Calibration is made possible via an electric heater of known power

calibration and the heat exchange area A (m^2) is usually constant; both parameters can be lumped together ($k_w A$):

$$q = k_w A (T_R - T_J) \tag{2}$$

where q(W) is the heat exchanged. The temperature controller needs appropriate tuning; van Kleeff [447] has reported practical tips for tuning a very simple system.

In flow calorimeters, samples of a culture grown in a bioreactor are continuously pumped through the measuring cell of a microcalorimeter. The sensitivity of the differential signal between the reaction vessel and the reference vessel is comparable to that obtained from microcalorimetry, e.g. [193]. From a practical point of view, they are quite flexible because they can be connected to any reactor but, due to transfer times in the minute(s) range, gas and substrate limitations must be considered.

Heat flux calorimeters are bioreactors equipped with special temperature control tools. They provide a sensitivity which is approximately two orders of magnitude better than that of microcalorimeters, e.g. [33, 258]. The evaluation and description of microbial heat release is based on a heat balance; heat yields and the heat of combustion of biological components are central parameters for quantification [70]. Measurements obtained so far have been used to investigate growth, biomass yield, maintenance energy, the role of the reduction degree of substrates, oxygen uptake [414] and product formation [272].

Approaches for practical exploitation have been made. Fardeau et al. [105] proposed integrated thermograms as a measure for biodegradability and Lovrien et al. [255] used microcalorimetric analysis of *Klebsiella* sp. growth for indirect sugar determination. The exploitation of calorimetry for biomass estimation has been compared with other methods [36]. Although often proposed, there are only a few reports on the control of processes based on calorimetric data, e.g. by Silman [387]. Metabolic uncoupling in *Saccharomyces cerevisiae* under various C/N ratios have also been investigated [231].

Entropy is closely related to heat (enthalpy) and energy. If all the ATP available from catabolic processes were used for anabolism (chemical synthesis), up to ten times more cellular material could be produced. First investigations of this large outflow of entropy from growing cells have been made by Bormann [40]; however, classical thermodynamics are hardly applicable to complex, non-equilibrium metabolic systems and must be extended [458].

2.2
Ex Situ, i.e. in a Bypass or at the Exit Line

2.2.1
Sampling

Samples removed from the reactor in some way can be analyzed with devices that are not (yet) suitable or available to be mounted in situ, but this is a reasonable way around the problem. Depending on the analyte of interest, i.e. whether it is soluble or (in) the dispersed phase, one needs to sample either the entire culture liquid or just the supernatant. The latter can be acquired using, for example, a filter. In this case, the filter is usually also the sterile barrier.

2.2.1.1
Sampling of Culture Fluid Containing Cells

When the dispersed phase, usually but not necessarily the cells, is of interest, no separation of phases need take place during sampling. The system must be opened in such a way that no infections can enter the reaction space either during sampling or between the sampling events. This requires the use of non-return valves and probably some repetitive sterilization procedure of the valve and exit line(s), as depicted in Fig. 18.

However, having the technical solution to avoid infections is not enough: the reaction mixture is not separated yet and certainly continues to react until at least one component becomes limiting. The sample so taken would not be representative of the interior of the reactor, and appropriate measures need to be taken to assure representativity. Time is critical; for further details, see Sect. 4.1 and Fig. 29. Cooling, heating, poisoning or separating phases may be a solution, but this cannot be generalized; it depends specifically on the actual intentions. If enzyme ac-

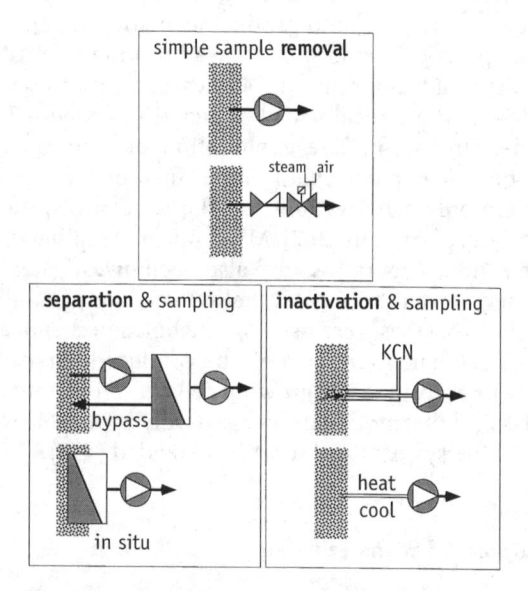

Fig. 18. Schematic design of various interface types used to acquire samples from a monoseptic bioreactor. *Top*: whole-culture aliquots are withdrawn either just using a pump or from a pressurized vessel through valves; these may or may not be re-sterilized with steam and dried with air repetitively after each sampling event. *Left*: cell-free supernatant is created using filters, either mounted in situ or in a bypass. *Right*: whole-culture aliquots are somehow removed and inactivated either using temperature changes or by adding inhibitory (toxic) components at a known rate (this is very important because this component dilutes the sample, yet is not shown here in this sketch)

tivities are to be determined, heating may be the worst choice, if just biomass concentration is to be determined, immediate filtering may be the best choice.

2.2.1.2
Sampling of Culture Supernatant Without Cells

Whenever the analyte of interest is soluble in the liquid phase or part of the gas phase, sample removal via a filtering device is the most reasonable solution. Filters mounted in situ are ideal provided they do not foul within an unreasonably short period. If this is the case, a filter operating in bypass must be used because this can be repeatedly exchanged with a freshly prepared one. In our experience, bypass filters should be operated with a high tangential flow, say ≥ 2 m s^{-1} of superficial liquid velocity. Then, a useful lifetime of a few weeks can be achieved even in cultures of filamentous organisms.

Separation of phases by using flotation or gravitational (at 1 g) sedimentation takes too much time to be useful and, furthermore, does not permit complete phase separation.

2.2.2
Interfaces

The interface between the reaction site, in the case of mono-cultures in bioprocessing a monoseptic space, and the site of analysis is of decisive importance for two reasons: (1) the monoseptic space must be protected from contamination, and (2) the sample specimen must be "transported" to the analytical device without significant change in composition; since transport always takes some finite time, one must – for the sake of representativity – assure that the reactions continuing during transport are negligible or, at least, well known. This goal can be achieved by various strategies with better or lesser effort/effect ratio. Rapid sampling is, in any case, advantageous. A couple of methods separate catalysts from reactants and another approach is poisoning or inactivating the catalyst by either addition of a strongly inhibitory material (e.g. heparin or KCN, which both may well interfere with the analytical method) or temperature variations such as heating or cooling [268]. These aspects are not restricted to samples from monoseptic bioprocesses, they are equally important for environmental analyses [127].

Membranes do not only serve as the analytical interface. Schneider et al. [377] have demonstrated that hydrophobic membranes, for example, made from PTFE, mounted either in situ or in bypass, can also be used for preparative removal of ammonium from an animal cell culture.

2.2.3
Flow Injection Analysis (FIA)

Ruzicka and Hansen [359] characterized flow injection analysis (FIA) as: "... information gathering from a concentration gradient formed from an injected, well-defined zone of a fluid, dispersed into a continuous unsegmented stream of a carrier ...". Accordingly, basic components of FIA equipment are a transport system consisting of tubing, pumps, valves and a carrier stream into which a technical system injects a sample [147, 358]. A (bio)chemical reaction, which is typical for the substance to be measured, usually occurs during the flow and products or residual (co)substrates are measured by the sensing system. In addition, physical sample treatment such as extraction [293, 449, 481], separation [449] or diffusion [242] can also be easily implemented. The detector is most often an optical or electrical device (Fig. 19), but it can be based, for instance, on enzymatic or immunological reactions [1, 101, 102, 225, 308, 374, 417, 421], thermistor, e.g. [81], redox measurement [143], or even based on other analytical devices such as a mass spectrometer, e.g. [155, 210], or a biosensor, e.g. [144, 256], or microbial electrodes [319, 437]. Fiber optics are also used [423, 470].

FIA does not generate continuous signals but there are several important advantages: a high sampling frequency (up to $> 100\ h^{-1}$), small sample volumes, low reagent consumption, high reproducibility and total versatility of sensing methods. Separation of compounds by high performance liquid chromatography (HPLC) prior to FIA analysis has even been reported [475]. Kroner [222] reports on the good automation properties of FIA used for enzyme analysis. The stability of enzymes in the sensing systems often limits the use of bio-

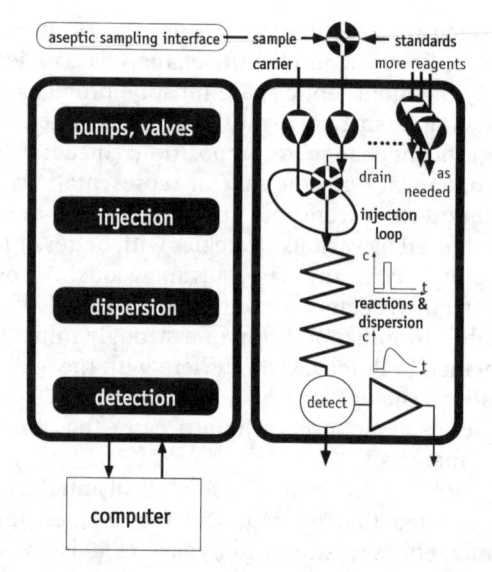

Fig. 19. Schematic design of a flow injection analysis (FIA) system. A selection valve (*top*) allows a selection between sample stream and standard(s). The selected specimen is pumped through an injection loop. Repeatedly, the injection valve is switched for a short while so that the contents of the loop are transported by the carrier stream into the dispersion/reaction manifold. In this manifold, any type of chemical or physical reaction can be implemented (e.g. by addition of other chemicals, passing through an enzyme column, dilution by another injection, diffusion through a membrane, liquid-liquid extraction, etc.; not shown). On its way through the manifold, the original plug undergoes axial dispersion which results in the typical shape of the finally detected signal peak

sensors substantially even though improvements have been reported [94]. No interference with the sterile barrier is likely since the entire apparatus works outside the sterile space. Special emphasis must be given to the sampling device interfacing the sterile barrier (see above).

FIA easily permits validation requirements to be met because alternative measuring principles can be run in parallel. This helps to exclude systematic errors which might originate from the complex matrix. For instance, Carlsen et al. [54] reported an example in which two different FIA methods for penicillin V monitoring have been compared.

A comprehensive survey of various applications with bacteria, yeasts and fungi on a laboratory and pilot scale can be found in reports by Nielsen [302] or Decastro and Valcarcel [80]. FIA has been used for on-line determination of glucose, e.g. [122], or to estimate biomass directly [19, 305] or indirectly by means of an extended Kalman filter [441]. Schmitz et al. [376] even determined chemical oxygen demand (COD) from waste water stream using FIA in the range of 30 to 23000 mg l^{-1} within only 3 to 7 min. Filippini et al. [113] compared FIA with an in situ enzyme electrode during continuous cultivation of *Saccharomyces cerevisiae*. FIA is also useful in environmental sciences such as water monitoring [8, 464] and has become increasingly important in downstream processing [59, 278]. FIA has been applied to detect microorganisms indirectly by measuring the concentration of a mediator which is reduced by the organisms [89]. Amino acids, such as L-lysine, have been measured [53, 326] and even intracellular enzymes can be determined on-line [5, 380]. A separation of peptides can be afforded by miniaturized capillary electrophoresis [99, 263]. Rapid analyses of antibiotics have been realized by a combination of supercritical fluid extraction and FIA [44]. DNA and RNA have been quantified in extracts [49]. Metabolic studies of a lactic acid production based on glucose, lactose, galactose, lactate, and protein determinations after nutrient pulses have been reported by Nielsen et al. [305] and Shu et al. [385]. In addition, acetate has recently been determined on-line using an FIA technique [436]. An important development is its combination with cytometry, see e.g. [237, 238, 360], or the estimation of nucleic acids [452].

Biosensors are being increasingly used as detectors in FIA systems [284, 285, 322, 379, 476]. The drawbacks of biosensors as direct in situ sensors, namely their low dynamic range, their lack of ability to survive sterilization, their limited lifetime, etc. are no longer valid ex situ because the analyzer interfaces the biosensor which can be changed at any time and FIA can provide samples in optimal dilution. The need for chemicals and reagents can be drastically reduced when employing biosensors, specifically when the entire system is miniaturized [48].

An outstanding property of FIA is its range of application. It can be viewed as a general solution-handling technique rather than a distinct sensor; this causes high flexibility with respect to analytical methods. A high degree of automation is, however, necessary and desirable [171, 172]. FIA can be expected to become one of the most powerful tools for quantitative bioprocess monitoring in the near future provided that non-linear calibration models are also used and that data evaluation techniques improve [43, 114, 166, 185, 259]. Wu and Bellgardt [466, 467] were able to detect faults in the analytical system automatically. The present tendency is towards using multi-channel FIA systems that work either in parallel or with sequential injection [20, 283, 446, 469], miniaturization of FIA devices [48, 144, 420, 421], and automation [112, 171, 172, 321].

Only recently, flow injection has been used as an interface to the first on-line application of flow cytometry [480]. Gorlach et al. [137] used flow injection for high throughput mass spectrometric mapping.

Interestingly, FIA can also be operated without an injection and gives valuable results. To this end we stained the DNA within yeast cells removed at a minute flux from a reactor and were able to quantify the amount of DNA on-line thus giving evidence for the cell-cycle dependence of oscillations [397].

2.2.4
Chromatography such as GC, HPLC

A review of chromatographic methods is beyond the scope of this contribution. Both liquid chromatography (LC) and gas chromatography (GC) have been applied in numerous cases to off-line analyses of biotechnological samples but the on-line application has only recently been developed. The scope of chromatographic methods is the separation of the individual constituents of mixtures as they pass through columns filled with suitable stationary phases (Fig. 20). The

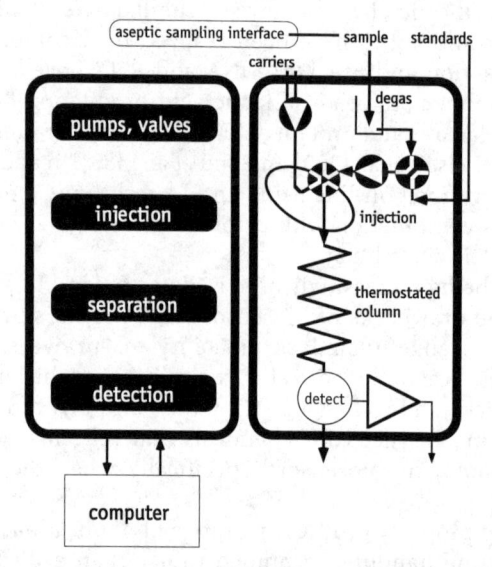

Fig. 20. Schematic design of linking a chromatograph on-line to bioprocesses. In principle, the design is almost identical to an FIA system. This is why FIA is often characterized as chromatography without a column. However, degassing of the sample is essential, in particular, when no internal standard is added (as in this sketch). In addition, the technical designs of injection valves differ and the injector to a gas chromatograph is heated to 200 or 250 °C which means it needs, therefore, a special construction

retention in the column is determined by the interaction between the individual constituents and the stationary phase. Miniaturized versions using micromachined instruments [90] or array detectors [425] have recently been reported.

In the following, a non-exhaustive list of some examples is given to illustrate the versatility of these methods. Comberbach and Bu'lock [64] measured ethanol in the bioreactor head space every 6 min using an electro-pneumatic sampling system connected to a gas chromatograph. McLaughlin et al. [270] provided cell-free samples of a butanol/acetone bioreaction by means of a tangential flow ultrafiltration membrane; they determined the components also in the head space using a gas chromatograph. Groboillot et al. [139] monitored the kinetics of acetaldehyde, ethanol, fuel alcohols and CO_2 after gas chromatographic separation. DaPra et al. [75] used GC data to control the influx rate of highly polluted waste water to an anaerobic filter; HPLC is reported to compare well with gas chromatographic results in anaerobic waste water treatment [482]. The first on-line capillary-GC monitoring of liquid cultivation samples was reported by Filippini et al. [111]; contrary to previous predictions, a capillary column survives many thousands of injections.

HPLC systems were helpful in monitoring cephalosporin production [174] and p-cresol degradation [392]. Other HPLC systems have been reported to serve for the control of penicillin production, namely via the precursor feed [62, 288], 3-chlorobenzoate conversion [375] or naphthalenesulfonic acid reduction [279], as well as amino acids [444, 445]. HPLC is also found to be useful when linked to biochemical assays [316]. Mailinger et al. [262] have discussed many important aspects of on-line HPLC systems.

It is possible after some adaptation and modification to link such apparatuses to bioreactors ("intelligent analytical subsystems"). This trend towards increased automation is not restricted to chromatographic methods (cf. FIA).

2.2.5
Mass Spectrometry (MS)

Mass spectrometry (MS) has been applied mainly for the on-line detection and quantification of gases such as pO_2, pCO_2, pN_2, pH_2, pCH_4 and even H_2S [282] or volatiles (alcohols, acetoin, butanediol). The detection principle allows simultaneous monitoring and, consequently, control of important metabolites.

The principles, sampling systems, control of the measuring device and application of MS for bioprocesses have been summarized by Heinzle [157, 158] and Heinzle and Reuss [162]. Samples are introduced into a vacuum ($< 10^{-5}$ bar) via a capillary (heated, stainless steel or fused silica, 0.3×1000 mm or longer) or a direct membrane inlet, for example, silicon or Teflon [72, 412]. Electron impact ionization with high energy (approx. 70 eV) causes (undesired) extensive fragmentation but is commonly applied. Mass separation can be obtained either by quadrupole or magnetic instruments and the detection should be performed by (fast and sensitive) secondary electron multipliers rather than (slower and less sensitive) Faraday cups (Fig. 21).

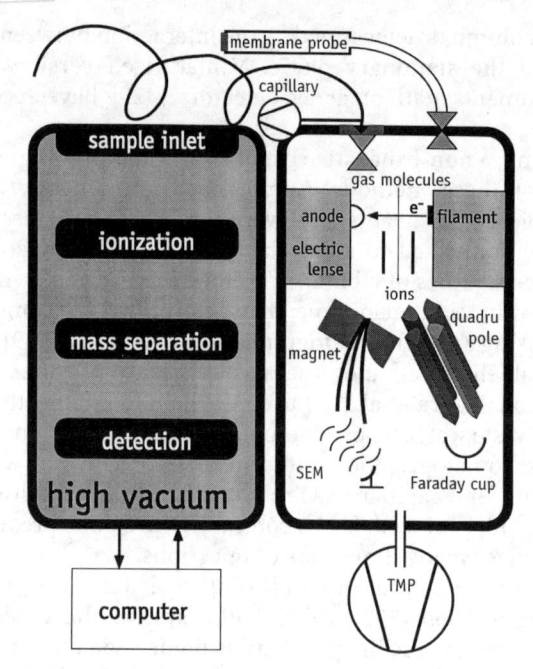

Fig. 21. Schematic design of a mass spectrometer connected on-line to bioprocesses. Two alternative uses are sketched and two alternative separation principles. *Top*: Pressure of a gas is converted down to approximately 1 mbar on its way through a capillary through which it is sucked using a mechanical pump. A fraction of this low-pressure gas can enter the high vacuum system of the mass spectrometer via a frit or tiny hole (several 10 to a few 100 µm in diameter). The alternative inlet is a direct membrane inlet. A thin (only a few µm thick) tight membrane, mechanically re-enforced to withstand the pressure gradient, is both a pressure and sterile barrier; the membrane is mounted in situ and interfaces cultivation liquid to high vacuum. Molecules entering the high-vacuum system are ionized (here shown as electron impact ionization) and electromagnetically focused into the mass separation space which can be either a magnetic sector field (*left*) or a quadrupole system (*right*). After mass separation, the ions of interest are quantified using either a highly sensitive secondary electron multiplier (SEM) or an inexpensive Faraday cup detector. The high vacuum is usually achieved using a turbo-molecular pump (TMP) cascaded to a mechanical pump (not shown)

Capillary inlet MS is the "scientific exception" [52] but it is routinely applied even in industry; most scientific publications report on membrane inlet applications. Almost all volatile substances can be analyzed from the gas phase using capillary inlet MS provided their partial pressure in the exhaust gas is ≥1 µbar [313]. Software is necessary but is available only for a few special cases [291].

An important problem with membrane inlet systems is quantification because the membrane behavior is more or less unpredictable [138, 352]; however, the membrane inlet can be even more rapid [56, 57].

Entire mass spectra of complex supernatants also containing unknown compounds have been evaluated as typical fingerprints to characterize the process state [159]. This comprehensive view of process data was also applied to pyrolyzed samples [68, 135, 364]. The analysis of such data makes, however, computation power indispensable, e.g. [336].

A control loop based on H_2 measurements has been set up by Lloyd and Whitmore [243] and Whitmore et al. [457] in order to prevent inhibition of methanogenesis: they controlled the addition of the carbon source to a thermophilic anaerobic digestion process. Even linked pO_2, pH and OUR control are reported based on direct mass spectrometric measurements [103, 312].

There are no reports currently available of an on-line application of the very powerful MALDI technique (matrix-assisted laser desorption/ionization time-of-flight mass spectrometry). However, this technique has become increasingly important for the analysis of complex molecules and quality control, for example, for glycosylated material [177] or peptides [104] or nucleotides [324].

Of course, mass spectrometry requires expensive equipment. But it should be taken into account that automatic multiplexing of different sample streams is possible and, in addition, a great variety of different substances can be determined simultaneously.

2.2.6
Biosensors

The rationale for using biosensors is to combine the high specificity of biological components with the capabilities of electronic tools (i.e. "usual" sensors). Biosensors consist of a sensing biological module of either catalytic (e.g. enzymes, organism) or affinity reaction type (e.g. antibodies, cell receptors) in intimate contact with a physical transducer. The latter finally converts the chemical into an electric signal [194]. Principles and typical examples are found in Karube et al. [205-207, 209], Delaguardia [85] and Jobst et al. [192]. Co-immobilization of enzymes can be advantageous when compared with sequential operation [24]. In general, the bio-part of the biosensor cannot be sterilized but, however, there is one report of an ethylene oxide sterilizable, implantable glucose biosensor [314]. Because of the small size and the array-type of construction of modern biosensors, it may well be that there is a carry-over of reaction product from one part of the array to the next thus falsifying the results. Urban et al. [438] have eliminated this problem by adding a second enzyme membrane layer containing catalase on top of the membrane containing oxidase enzymes (Fig. 22).

According to the diversity of possible applications many applications of biosensors have been published, e.g. [145, 317, 462, 463]. Often they concentrate especially on a certain group of biosensors but high redundancy is obvious. In the following sections, different sensor types, ordered by the transduction principle, are introduced.

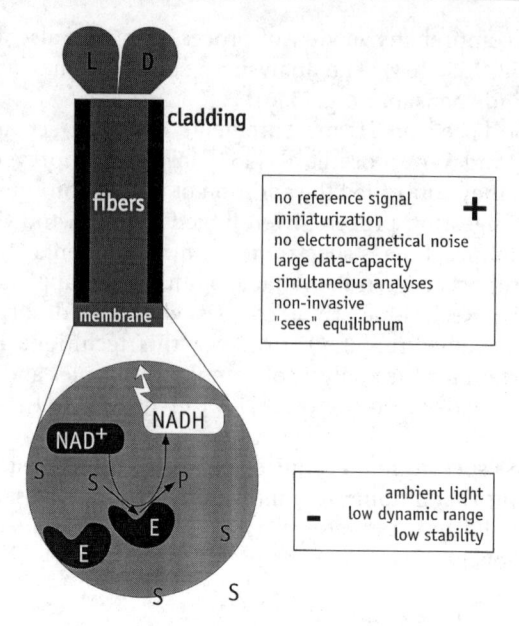

Fig. 22. Schematic design of one example of a biosensor, here an optical biosensor or "opt(r)ode". The general working principle is that a biological catalyst/recognition element is tightly coupled with a so-called transducer. In this special case, the transducer is an optical system, namely optical fibers connected to a light source and a detector (indicated with L and D). On top of the transducer a biological sensing element is mounted; in this case a membrane containing an (immobilized) enzyme which catalyzes a reaction that requires NAD as cofactor. If substrate and co-substrate enter the membrane, substrate will be converted to the product and, stoichiometrically, to NADH which can be quantified by fluorescence measurement (co-factor must not be limiting). Substrate and co-substrate as well as product can diffuse into and out of the membrane and are assumed to equilibrate with the measuring solution. This biosensor cannot be sterilized since the enzyme would not survive the procedure

2.2.6.1
Electrochemical Biosensors

Electrochemical transducers work based on either an amperometric, potentiometric, or conductometric principle. Further, chemically sensitive semiconductors are under development. Commercially available today are sensors for carbohydrates, such as glucose, sucrose, lactose, maltose, galactose, the artificial sweetener NutraSweet, for urea, creatinine, uric acid, lactate, ascorbate, aspirin, alcohol, amino acids and aspartate. The determinations are mainly based on the detection of simple co-substrates and products such as O_2, H_2O_2, NH_3, or CO_2 [142].

Amperometric transducers measure the current (flux of electrons) caused by oxidation or reduction of the species of interest when a voltage is applied between the working and the reference electrode. Often, oxygen serves as electron acceptor but interferences have encouraged development of new methods to avoid this, e.g. controlled oxygen supply and the application of mediators, such as ferrocene, e.g. [169]. Other determinations focus on the detection of H_2O_2 or NADH [9]. Sensors are currently being made smaller and smaller [297, 298].

Potentiometric transducers measure the potential between the sensing element and a reference element. Thus, in contrast to amperometric transducers, practically no mass transport occurs; the response depends on the development of the thermodynamic equilibrium. pH changes often correlate with the measured substance because many enzymatic reactions consume or produce protons.

Conductometric transducers consist of two pairs of identical electrodes, one of which contains an immobilized enzyme. As the enzyme-catalyzed reaction causes concentration changes in the electrolyte the conductivity alters and can be detected.

The conductivity of certain semiconductors such as field effect transistors (FETs) can be affected by specific chemicals. Ion-selective FETs (ISFETs) are metal oxide semiconductor FETs (MOSFETs) [25]. A great advantage is the small size (miniaturization) resulting in the possibility of combining several (identical or alternative) sensing units in multifunctional devices [206, 225] and having a reasonably short response time [386]. They have already been used for the detection of urea, ATP, alcohol, glucose and glutamate [142, 378].

2.2.6.2
Fiber Optic Sensors

Optical biosensors typically consist of an optical fiber which is coated with the indicator chemistry for the material of interest at the distal tip (Fig. 22). The quantity or concentration is derived from the intensity of absorbed, reflected, scattered, or re-emitted electromagnetic radiation (e.g. fluorescence, bio- and chemiluminescence). Usually, enzymatic reactions are exploited, e.g. [463].

These sensors are ideal for miniaturization, are of low cost and the fiber optics are sterilizable (even if the analyte is not!). The most limiting disadvantages are actually interferences from ambient light and the comparatively small dynamic ranges. Applications so far reported in the literature appeared for pH, e.g. [12], CO_2, e.g. [296], NH_3, e.g. [10], CH_4 and metal ions, see Guilbault and Luong [141], O_2, e.g. [323], glucose, H_2O_2 and lysine [333] and even for biomass, e.g. [201].

2.2.6.3
Calorimetric Sensors

This type of biosensor exploits the fact that enzymatic reactions are exothermic (5–100 kJ mol^{-1}). The biogenic heat can be detected by thermistors or temperature-sensitive semiconductor devices. A technical realization can be performed either by immobilizing enzymes on particles in a column around the heat-sensing device or by direct attachment of the immobilized enzyme to the temperature transducer. Applications to measure biotechnologically relevant substances have been: ATP, glucose, lactate, triglycerides, cellobiose, ethanol, galactose, lactose, sucrose, penicillin and others [30, 120, 142, 337]. Very sensitive thermopiles allow the limit of detection to be decreased significantly [17].

2.2.6.4
Acoustic/Mechanical Sensors

The piezo-electric effect of deformations of quartz under alternating current (at a frequency in the order of 10 MHz) is used by coating the crystal with a selectively binding substance, e.g. an antibody. When exposed to the antigen, an antibody–antigen complex will be formed on the surface and shift the resonance frequency of the crystal proportionally to the mass increment which is, in turn, proportional to the antigen concentration. A similar approach is used with surface acoustic wave detectors [142] or with the surface plasmon resonance technology (BIAcore, Pharmacia).

Generally, biosensors are tricky to handle. Due to the vulnerable biological element, e.g. an enzyme or a living microorganism [208, 209], they principally cannot be sterilized. They also suffer from changes in the environment, for example, changes in pH or formation of aggressive chemicals such as H_2O_2 [342]. They must therefore be used in a suitable environment, preferably after sample preparation as in an FIA system. This, eventually, allows simultaneous compensation of endogenous interferences [357]. The long-term stability under working conditions is often poor; however, suitably immobilized glucose oxidase is useful for more than one year. Only limited experience has been made under technical process monitoring conditions.

2.2.7
Biomass

2.2.7.1
Dynamic Range – Dilution

High density cultures present a problem to optical density measurements because of the inner filter effect (i.e. light intensity is lost due to absorbance and scatter by cells over the length of the light path). Lee and Lim [232] and Thatipamala et al. [431, 432] have described a way to circumvent and eliminate the dilution problem at higher cell densities. A larger dynamic range can be achieved by using concentrically compartmented flow-through cells with distilled water in the inner tube (which is called optical dilution) and the suspension to be measured in the outer tube (i.e. in the annular space). Variation of the diameter ratios of these tubes allows the OD readings to be kept between 0 and 0.5. However, the variation is not straightforward because it implies switching from one cuvette to another; this is only reasonable when the sample is run to waste and not returned to the reactor. A very elegant method to obtain a greater dynamic range for OD was used by Nielsen et al. [305]. They exploited the dilution equipment of flow injection analysis (FIA) and measured a steady state absorbance (not the transient as usual in FIA) which resulted in a reliable and reproducible signal. Cleaning of this analyte stream was found advantageous although deposits were not a problem when a small flow of highly diluted sample was used.

2.2.7.2
Electrical Properties

Determination of conductivity (or impedance) of a culture is another way to generate a signal which depends on biomass; of course, it also depends on other factors such as medium composition [67] more or less pronouncedly and is also subject to bubble interferences. Taya et al. [429] found little dependence on pH or sugar concentration and linear correlations with biomass concentrations of plant cells up to 10 g l^{-1}. Ebina et al. [96] reported on similar experiences with yeasts. Henschke and Thomas [164] determined even very low concentrations of cells in wine. Owens et al. [318] indirectly followed the growth of yeasts by absorbing the produced CO_2 and conductometric analysis of this product. So far direct electrochemical determination of cells does not seem to be very reliable [361]. Fuel cells [22] can measure the redox potential of a culture [149] but are comparably very slow and delayed. An electrochemical measurement of a mediator reduced by microbes in a flow injection system was reported by Ding and Schmid [89] to give good correlation with classical determinations, e.g. colony forming units (cfu), and to be rapid (a few minutes per analysis). This system was applied for on-line monitoring of an *Escherichia coli* culture. However, the dielectric properties of cells such as distinct deviations in the capacity values have been shown to depend on the growth stage of a culture [267].

2.2.7.3
Filtration Properties

Lenz et al. [233] and Reuss et al. [344] have further developed an automatic filtration device [300, 435] that allows the estimation of the biomass concentration in a relatively large sample (approx. 100 ml) according to its filtration properties. Each sample is filtered through a fresh filter and the flux of the filtrate is monitored as well as the build-up of the filter cake. This information – together with appropriate descriptive models – allows the calculation of the biomass concentration in fungal culture aliquots. Since yeasts or bacteria do not build up a comparably high filter cake, suspensions of these organisms are less likely to be estimated accurately. To our knowledge, practically useful experience has so far been made only with *Penicillium* cultures and in the Pekilo process.

2.3
Software Sensors

Software sensors are virtual sensors which calculate the desired variable or parameter from related physical measurements [58]. In other words, there must always be a model available that relates reliably the measured variable with the target variable or parameter. Normally, measured variables are easily measurable effects that are caused and influenced by the target. The most prominent software sensor is the respiratory quotient (RQ-value) which characterizes the physiological state of a culture. However, its determination can be tricky (see Sect. 5).

Generally, software sensors are typical solutions of so-called inverse problems. A so-called forward problem is one in which the parameters and starting conditions of a system, and the kinetic or other equations which govern its behavior, are known. In a complex biological system, in particular, the things which are normally easiest to measure are the variables, not the parameters. In the case of metabolism, the usual parameters of interest are the enzymatic rate and affinity constants, which are difficult to measure accurately in vitro and virtually impossible in vivo [93, 118, 275, 384]. Yet to describe, understand, and simulate the system of interest we need knowledge of the parameters. In other words, one must go backwards from variables such as fluxes and metabolite concentrations, which are relatively easy to measure, to the parameters. Such problems, in which the inputs are the variables and the outputs the parameters, are known as system identification problems or as so-called inverse problems.

However, the situation is even more complicated since things develop in relation to time. The estimation of the physiological state of a culture involves more than one (measurable) variable at a time and the recent history of this set of individual signal trajectories involved. In other words, physiological state estimation requires recognition of complex patterns. Various algorithms used for this purpose, e.g. [7, 134, 220, 246, 290, 426], have in common that it is not the present values alone that are evaluated, there is always the recent history of signal trajectories involved.

In some cases, the data describing the actual state and their recent history are compared with so-called reference patterns: these are data from historical experiments or runs which an expert has associated with a typical physiological state. A physiological state is recognized either if the actual constellation matches any one of the reference sets best – in this case, there is always an identification made – or if the match exceeds a pre-defined degree of certainty, e.g. 60% – then it can happen that no identification or association is possible with a too high limit selected. The direct association with reference data needs normalization (amplitude scaling) and, probably, frequency analysis in order to eliminate dependencies on (time) shifts, biases or drifts.

In other cases, the data trajectories are translated into trend-qualities via shape descriptors such as: glucose uptake rate is decreasing (concave down) while RQ is increasing (linear up) and ... These combinations of trends of the trajectories of various state variables or derived variables define a certain physiological state; the advantage of this definition is that the association is no longer dependent on time and on the actual numerical values of variables and rates [413].

2.4
Validation

On-line measurements produced with in situ sensors are difficult to validate. The usual procedure for evaluating the quality of a measurement is restricted to calibration/checking prior to and after a cultivation. A few sensors such as the pCO_2- or the Cranfield/GBF-glucose sensor [42] allow removal (at least of measuring buffer and also of the transducer itself) and, therefore, recalibration of the transducer during a cultivation (Fig. 23).

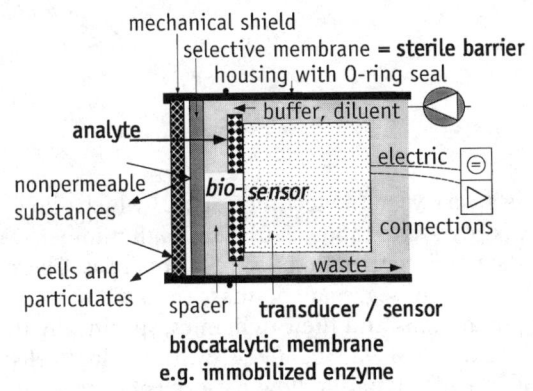

Fig. 23. Schematic design of a biosensor that can be mounted in situ. The biosensor itself sits in a housing and consists of a biocomponent such as one or more immobilized enzymes or cells on top of and in close contact with a suitable type of transducer. A buffer or diluent stream can help to extend the useful dynamic range of the biosensor. The analyte arrives at the biosensor by passing a suitable membrane which enhances selectivity and protects the biosensor. An additional mechanical shield in the form of a mesh, grid or frit may be necessary to assure mechanical stability in the highly turbulent zone

External chromatographs and FIA systems can be regularly recalibrated but the sterile interface cannot. Other sensors such as a pH or a pO_2 probe can be mounted via a retractable housing which allows either sterile exchange or withdrawal for external recalibration during a run. A further possibility to gain some information about the reliability of a measurement is to mount a number (>1) of identical sensors in easy to compare positions and to check the individual signals for equality. This opportunity has been exploited for many years in other technologies.

It is highly desirable to have alternative principles of measurement at hand which operate simultaneously. For instance, both oxygen and carbon dioxide can be determined in the exhaust gas using either the classical analyzers sensitive to infrared absorption (for CO_2) and paramagnetic properties (for O_2), respectively, or using a mass spectrometer. Stability of these instruments is reasonable and interference from other components is not very likely. The partial pressures of both gases can be monitored with membrane-covered probes. The membranes are the weakest elements in this measuring chain and are exactly those that are the most difficult to validate experimentally. The electrodes or the mass spectrometer behind the membrane are just transducing elements and quite reliable. The measurements obtained from the gas and from the liquid phase quantify, of course, different state variables but they are tightly linked via gas-liquid mass transfer. Pattern recognition can help to generally identify malfunctioning sensors in some cases [247].

It is, however, interesting to realize that in the last few years more and more industrialists are convinced that the risk of implementing on-line measurements coupled to controllers, for instance, for feed, pays back due to reproducible and highly productive biotransformations with constantly high quality; for example, Rohner and Meyer [353] have implemented on-line FIA, HPLC and photometry with the respective controllers on the 15 m^3 production scale.

3
Off-Line Analyses

3.1
Flow Cytometry

Flow cytometry is a very versatile technique [223] which allows the analysis of more than 10^4 cells per second [369, 370]. This high number results in statistically significant data and distributions of cell properties. Therefore, flow cytometry is a key technique to segregate biomass (into distinct cell classes) and to study microbial populations and their dynamics, specifically the cell cycle [76, 87, 116, 200, 214, 221, 295, 329, 330, 409, 418]. Individual cells are aligned by means of controlled hydrodynamic flow patterns and pass the measuring cell one by one. One or more light sources, typically laser(s), are focused onto the stream of cells and a detection unit(s) measure(s) the scattered and/or fluorescent light (Fig. 24). Properties of whole cells such as size and shape can be

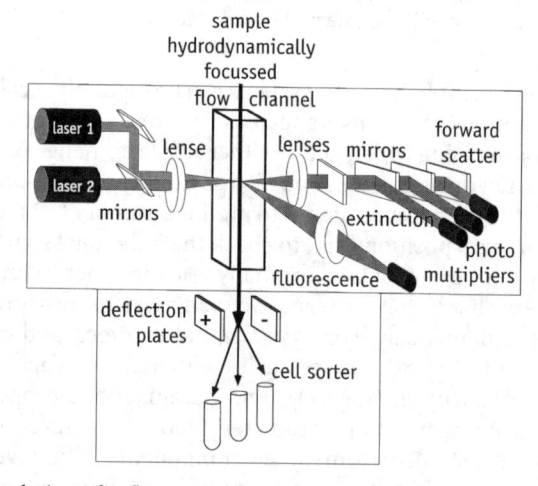

Fig. 24. Schematic design of a flow cytometer. The exciting light is created by one or more laser sources which is focused by means of mirrors and lenses to a small measuring space (typically in the order of the particles to analyze) in the flow channel. Whenever a particle passes this space, extinction, forward or side scatter of light, or emission of fluorescent light occurs. These different light qualities are separated by lenses and mirrors and quantified by detectors (usually photomultipliers) mounted in appropriate positions. The particles are hydrodynamically aligned in the thin flow channel so that only individual single particles pass the measuring zone sequentially. If necessary, a cell sorter can be appended: the on-line executed data evaluation algorithms must classify every measured event in a short time so that an appropriate voltage can be applied to the deflection plates as soon as the respective particle arrives in this space. A tiny fraction of the liquid stream, most probably containing the particle of interest, can be deflected into one container. This allows particles to be sorted according to individual properties determined by the flow cytometer. Since individual properties can be quantitatively determined for each quality (e.g. forward scatter reflects the size, various fluorescences reflect different intra-particle components), population distributions can be analyzed

estimated as well as distinct cellular components. The latter requires specific staining procedures which normally do not allow this technique to be simply used on-line. So far, there has only been one report [480] of an on-line application.

Among the items that have been measured are: vitality, intracellular pH, DNA and RNA content, and specific plasmids [77, 408]. Besides nucleic acids [204], other intracellular components can also be analyzed, e.g. storage materials [2, 82, 294], enzymes and protein content [6, 338], or the cell size [60, 61]. The physiological state can also be rapidly assessed [331]. Furthermore, this technique allows the separation of certain cells using a cell sorter, e.g. for strain improvement [28]. The flow cytometry technique has also been used in connection with molecular probes for identification and viability determination of microbial communities [98]. This application of viability estimation is becoming increasingly important [63, 136, 188, 454]. Unfortunately, the equipment is expensive and most of the measurements are tricky and laborious and not well designed for on-line application.

Jayat and Ratinaud [190] discuss the advantages of multi-parametric analyses taking DNA and other cellular components simultaneously into account; see also, e.g. [73, 261]. Flow cytometry is also useful for identification of (new) species [78] or for biomass estimation [299, 349].

Hewitt et al. [167] have exploited flow cytometry to quantify the impact of fluid mechanical stress on bacterial cultures. A modified technique, called the slit-scan method, allows the determination of cell shapes and of intracellular location of stained components [34]. Image cytometry and fluorescence microscopy are variants for determination of the volume growth of cells or morphology changes and seem to have become increasingly important [60, 382, 461].

3.2
Nuclear Magnetic Resonance (NMR) Spectroscopy

Responses of atomic nuclei with net magnetic moments that are exposed to a magnetic field and irradiated by electromagnetic energy reveal a variety of information. Absorption of energy, i.e. resonance of exciting and nuclear frequency, is typical for a nucleus in a certain molecular-electronic environment. The resonance frequency is shifted according to the shielding effects of the physicochemical environment (other nuclei in the molecular vicinity) on the local magnetic field (the so-called chemical shift which is measured; Fig. 25).

NMR spectroscopy has so far been suited for non-invasive investigation of biochemical structures, fluxes through pathways, distribution of marker-nuclei among various cellular components and enzymatic mechanisms rather than for quantitative determination of small molecules. Biochemical applications have involved NMR spectroscopy mainly for structural determination of complex molecules, e.g. [27, 180], as well as inside the cells, i.e. in vivo [184, 189]. In biotechnology, the potential of determining intracellular components without cell disruption is increasingly used for in vivo studies of metabolism, e.g. [15, 55, 88, 121, 146, 197, 250–252, 271, 335], and effectors [419].

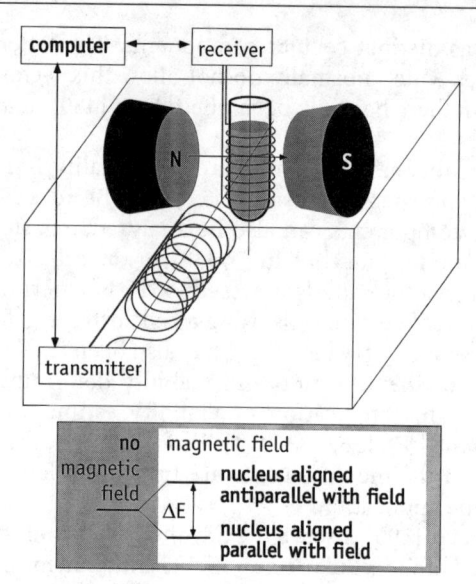

Fig. 25. Schematic design of a nuclear magnetic resonance (NMR) spectrometer in which a monoseptic bioreaction takes place. Nuclei with a net magnetic moment align in a strong static magnetic field. An additional electromagnetic field can flip the orientation of the alignment provided the frequency (energy) is appropriate; this additional field is created by the transmitter coil. The corresponding nuclear magnetic resonance that occurs is associated with an energy uptake from the high frequency field which is detected by the receiver coil. The sample is usually kept homogeneous by spinning the sample vial. In on-line NMR spectrometers, the homogeneity of the sample is assured by vigorous flow-exchange with the bulk liquid, usually above the measuring zone. For aerobic cultivations, the measuring volume represents the bottom tip of a cyclone reactor which allows excellent gas transfer in its top part and separates (gas and liquid) phases towards its bottom part. This design assures a minimal interference of gas bubbles with the measurement

In more recent years, NMR spectroscopy has been developed as an in situ technique [55, 100, 203, 365, 419]. This is (indirectly) used for metabolic flux determination [39] (see also the contribution by Nielsen, this volume) and may require specifically designed reactors [151]. Scanning times lie at least in the range of minutes to hours and cell concentrations needed are above 10^{10} cells l^{-1}. Schuppenhauer et al. [381] have determined spatial gradients in a fluidized bed reactor of some relevant metabolites in a high density animal cell culture using localized ^{31}P-NMR spectroscopy. Mass transfer resistance can also be directly observed using NMR spectroscopy [427].

NMR can help to monitor energization, see, e.g. [121, 273], especially the levels of ^{31}P-containing metabolites, e.g. [45, 366], enzyme kinetics, compartmentalized intracellular ion activities, the fate of ^{3}H-, ^{2}H-, ^{13}C-, ^{15}N-, or ^{19}F-labeled tracers, e.g. [108, 109], O_2 tension, compartmentalized redox potential, membrane potential, cell number or cell volume, see [133], and even pH. Major drawbacks are the cost of the equipment, the low intrinsic sensitivity and the interpretation of spectra [430].

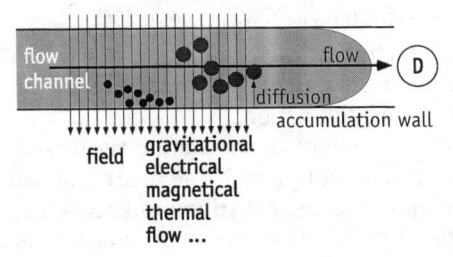

Fig. 26. Schematic design of field flow fractionation (FFF) analysis. A sample is transported along the flow channels by a carrier stream after injection and focusing into the injector zone. Depending on the type and strength of the perpendicular field, a separation of molecules or particles takes place: the field drives the sample components towards the so-called accumulation wall. Diffusive forces counteract this field resulting in discrete layers of analyte components while the parabolic flow profile in the flow channels elutes the various analyte components according to their mean distance from the accumulation wall. This is called "normal mode". Particles larger than approximately 1 μm elute in inverse order: hydrodynamic lift forces induce steric effects: the larger particles cannot get sufficiently close to the accumulation wall and, therefore, elute quicker than smaller ones; this is called "steric mode". In asymmetrical-flow FFF, the accumulation wall is a mechanically supported frit or filter which lets the solvent pass; the carrier stream separates asymmetrically into the eluting flow and the permeate flow which creates the (asymmetrical) flow field

3.3
Field Flow Fractionation (FFF)

Field flow fractionation (FFF) is an elution technique suitable for molecules with a molecular weight > 1000 up to a particle size of some 100 μm. Separating, driving, external field forces are applied perpendicular to a liquid carrier flow, causing different species to be placed in different stream lines (Fig. 26). Useful fields are gravity, temperature, cross flow, electrical charge, and others [128–131].

The range of the (molecular) size of the analytes usually exceeds that which can be determined by classical laboratory analytical methods such as size exclusion chromatography, etc. [351]. Reports on investigated substances are widespread and cover applications such as the separation and characterization of proteins [450] and enzymes [240, 241], of viruses [132], the separation of human and animal cells [50, 51], the isolation of plasmid DNA [367], and the molecular weight and particle size distribution of polymers [216, 217]. The approach is relatively new in biotechnology; therefore, practical experiences are not yet abundant. Langwost et al. [229] have provided a comprehensive survey of various applications in bio-monitoring.

3.4
Biomass

The classical way to determine biomass concentration is typically an off-line method, namely to harvest a known aliquot of the culture suspension, separate cells by centrifugation, wash the cells and dry them to constant weight at a few degrees above the boiling point of the solvent (i.e. aqueous medium, usually

105 °C). After gravimetric determination of the dry mass in the vial or on a filter, the mass concentration in the original sample volume can easily be calculated if there are neither particulate components in the fresh medium nor precipitates formed during cultivation.

What does the cell dry weight concentration mean and teach us about the progress and behavior of a bioprocess? The analytical value reflects only the cellular mass and does not allow a distinction between structural or storage materials and biocatalytically active components of the cells on the one hand, or between live and dead cellular material on the other.

Cell populations are normally viewed as unstructured and unsegregated; their only property is to have mass. We also abbreviate biomass with X – the great unknown. Substantial losses occur in the bioprocess industry each year due to this unknown factor, specifically, the variability in inoculum cultures and, hence, in yields and productivities of production cultures [455].

There are alternatives for this measure, such as total and/or viable cell number concentration; this is determined by counting cells in a given culture volume aliquot. Its determination is quite laborious and error-prone. The result is available more rapidly, but this is the only obvious advantage. However, both measures for cells are not equivalent under transient culture conditions. So-called standard cells are cells with a constant individual cell mass but these are the exception rather than the rule, see Rodin et al. [350] and references cited therein, as well as Schuster in this volume.

Both methods (mass and number concentration estimations) cannot reflect the biocatalytic activity sufficiently. This derives from the fact that (microbial) biomass is not a homogeneous continuum but a very large, however finite, population of individual cells with individual properties. These may change significantly, e.g. depending on the microenvironmental conditions, on the history of a single cell proliferation, on the position in the cell cycle, or on the interaction with other cells or surfaces. The only good reason to neglect the individual character of cells is the statistically relevant high number of individuals in a process population. This is also why strongly simplifying growth models, such as the Monod model with the biomass concentration as the only state variable besides substrate concentration, have been so successful.

The classical ways to quantify a population density by either a mass or a number concentration can be sufficiently accurate and precise but, then, they are accordingly laborious and time-consuming. This does not hinder good scientific research (although not convenient) but is a severe obstacle to process control: the signal must be generated on-line and should not be prone to delays. The signal may be discontinuous but the discrete time intervals between updates must be sufficiently small. On-line estimation of the biomass concentration is no longer a matter of comfort, it is essential for any functional controller.

The latter requirement and the concomitant improvement of comfort led to the exploitation of alternative methods to estimate the biomass concentration. All of them have in common: (1) that they are indirect measures, and (2) that models are mandatory to relate these measures to biomass concentration. The models are of the descriptive rather than the mechanistic type which means that: (1) they must be experimentally verified for each single application, i.e. separately calibrated for each biological and reaction system, and (2) they must not be misused for extrapolation outside the data space that serves for calibration [215, 394]. Most of the indirect methods cannot be used generally; application is subject to several constraints typical for the cells (growing singly or in filaments or aggregated) and the culture liquid (color, particulate substances other than cells, viscosity or another immiscible liquid phase).

3.4.1
Cell Mass Concentration

Off-line determination of biomass concentration by classical gravimetric methods requires cell separation, washing steps and drying to constant weight. The separation of cells can be made either by centrifugation or by filtration.

The first method is cheaper with respect to consumables but has some general disadvantages. (1) The required volume is high: a significant amount of biomass must be accumulated because the dead weight of the centrifugation vial is relatively high; since dry weight is calculated from the difference of the vials with and without cells, error propagation relative to this (small) difference increases with decreasing cell mass. (2) Conditioning of the vials prior to weighing is decisive since vial surfaces are generally large. A considerable amount of water (from ambient air) can adsorb to the surface especially when glass vials are used. Plastic vials can lose considerable amounts of plasticizers. This increases the error. (3) Inactivation of cellular activities is not performed during centrifugation without further precautions. This can result in continuation of mass increase (growth) or decrease (due to mobilization of storage materials) of cells and, consequently, non-representative values. Although these errors are not expected to exceed the standard error of the gravimetric method by much, it might be decisive in scientific research, see e.g. Kätterer [212]. (4) The washing step requires skilled personnel in order to avoid losses of cells during withdrawal of supernatant and the washing step(s).

In contrast, the filtration method is more expensive (filters cannot be reused) but works with smaller net weights (i.e. sample volumes) and is usually more rapid, provided the cells are not slimy and do not clog the filter. Nylon or PVDF filters withstand the pre-drying procedure excellently whereas modified cellulose filters lose their mechanical robustness.

In any case, a strict standard operating procedure (SOP) must be followed to assure at least reproducibility. Figure 27 shows a useful SOP which might require some modifications depending on the characteristic properties of cell suspensions. Wet weight determinations are by far less accurate since a defined water content is not easy to arrange reproducibly.

3.4.2
Cell Number Concentration

Determination of the cell number concentration (cell counting) requires that cells are suspended singly. If this is not the case in the culture, an additional step

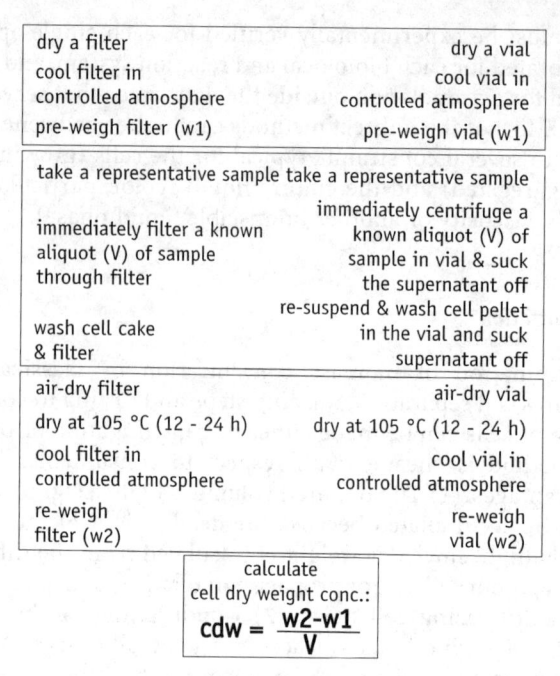

dry a filter cool filter in controlled atmosphere pre-weigh filter (w1)	dry a vial cool vial in controlled atmosphere pre-weigh vial (w1)
take a representative sample	take a representative sample
immediately filter a known aliquot (V) of sample through filter wash cell cake & filter	immediately centrifuge a known aliquot (V) of sample in vial & suck the supernatant off re-suspend & wash cell pellet in the vial and suck supernatant off
air-dry filter dry at 105 °C (12 - 24 h) cool filter in controlled atmosphere re-weigh filter (w2)	air-dry vial dry at 105 °C (12 - 24 h) cool vial in controlled atmosphere re-weigh vial (w2)

$$\text{calculate cell dry weight conc.:} \quad cdw = \frac{w2-w1}{V}$$

Fig. 27. Standard operating procedure (SOP) for determination of cell dry weight

of disintegration of aggregates (e.g. by sonification) or of suspending attached cells (e.g. by trypsinization) is necessary [117, 165]. A further dilution step may be required to achieve the necessary resolution of single cells in either the microscopic picture or in a Coulter counter, e.g. [224], or in a flow cytometer [153]. Inactivation is not a decisive factor unless the population to be quantified is highly synchronized and close to the cell cycle phase of cell separation.

In a few cases, biomass is estimated as colony forming units (cfu's); however, this method is employed nearly exclusively in hygienic analyses of water, waste water, soil and sludge samples. An appropriately diluted, concentrated or suspended homogeneous aliquot of the sample is plated on solid medium and incubated for a predetermined time. Then, the visible colonies are counted and a potential viable cell concentration is calculated based on the assumption that any one colony has proliferated from a single viable cell or spore originally present in the plated volume. This naturally results in a notorious underestimation of cells in the sample because the assumption seldom holds 100% true and a fraction of the live cells may not proliferate to a visible colony within the incubation period. On the other hand, an overestimation of active cells must occur when the organisms are sporulated in the original sample and the incubation time is sufficient to allow germination and sufficient growth.

3.4.3
Viability

Viability – widely used to characterize animal cell cultures – is estimated according to the ability of individual cells to catalyze a biochemical reaction, e.g. the reduction of Methylene Blue or Trypan Blue to the respective leuko-form [65, 332]; after staining, blue cells are distinguished microscopically from non-colored cells. This method may give a significant overestimation of cell viability because lysed or lysing dead cells are not accounted for. Flow cytometric approaches are mentioned above (see Sect. 3.1).

LDH activity in the supernatant can be an excellent indicator for dead or non-viable cells as demonstrated for CHO cells [343]. This method is much faster than all of the counting techniques (and is again a solution of an inverse problem).

3.4.4
Cellular Components or Activities

The trick for analyzing a biocatalytic activity must also be exploited whenever the medium contains particulate matter because this, of course, would falsify any direct gravimetric determination. Herbert [165] also gives an overview of many different indirect methods. One such often-used method is the measurement of dehydrogenase activity, for example, INT dehydrogenase [117, 253, 254]. An extension is the determination of substances that are typical for either vegetative cells or spores such as ATP [92, 173, 277, 393], NAD(P)H, DNA [236, 277, 434], phospholipids or dipicolinic acid [176, 456]; see [37, 400] or [328] for application even in sewage sludge characterization. These methods have in common that they are very laborious, time-consuming, difficult to calibrate and questionable to validate. Some are not precise which is either caused by the chemical instability of the extracts and/or the low relaxation times of the intracellular compounds in comparison with the required sample preparation time. One must also bear in mind that the macromolecular composition of cells may change during a cultivation [71] and even organelles underlie certain dynamics [391]. Physiological state based on the cellular level is extensively treated by Schuster in this volume.

Properties such as cell volume, morphology or cell size are nowadays determined using flow cytometers or image analysis, e.g. [35, 95, 181, 440], or by in situ microscopy, e.g. [422]. Biomass quantification by image analysis is treated extensively by Pons and Vivier in this volume.

3.5
Substrates, Products, Intermediates and Effectors

These important substances do deserve individual consideration but, from an analytical point of view, have already been covered in previous sections under the various techniques discussed.

4
Real Time Considerations

A notorious underestimation of the dynamic properties of microbial and cellular populations results from matching the duration of the respective batch cultivations to the relevant time constant of the biosystem under investigation. However, metabolic regulation of enzyme activities and fluxes often takes place on a time scale of seconds rather than days although the latter may also be true. It is therefore in the scope of promoting biotechnological research to adopt and develop appropriate experimental concepts, methodologies and equipment [397].

What is fast? What is slow? What is relevant? The relaxation time concept of Harder and Roels [148] (Fig. 28) maps typical time constants of microbial and cellular control on the level of modification of enzymes (activation, inhibition, dis/association of subunits, covalent modification or digestion) to the range of ms to s, on the level of regulation of gene expression (induction, repression or derepression of transcription) to min, on the level of population selection and evolution to days and larger units. The examples discussed below will clear up how bioengineering is facing the individual time constants.

A typical bioprocess, if operated in batch mode, extends over several hours or a few days. If operated in continuous mode, it is not reasonable to accept operating times of less than a month (see, e.g. Heijnen et al. [156]). If an organism has special physiological features, such as baker's yeast, a transition from one to the other domain (e.g. from low to high dilution rates or, in other words, from purely oxidative to oxido-reductive growth) may also result in considerable changes in the time required to approach a new steady state. Axelsson et al. [14] reported a rough estimate for this case: the time constant for experiments at low dilution rates ($D < D_R$, where $D_R = D$ at which the regulatory switch between oxidative and oxido-reductive metabolism occurs) is, as expected, in the order of the mean residence time ($\tau = D^{-1}$) but, above D_R, the time

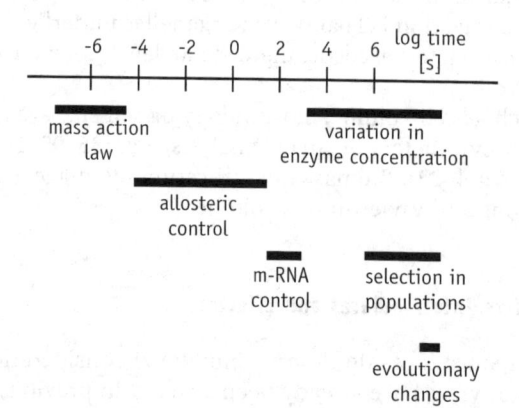

Fig. 28. Concept of relevant relaxation times (according to [148]). Note that the time scale is logarithmic

constant was predicted to be at least one order of magnitude greater. In experiments specifically designed to verify this, we found either even greater time constants or no unique stable steady states at all [404].

Within a given bioprocess, specific fluxes of substrates into the cell (q_S, qO_2, ...) and of products out of the cell (q_P, qCO_2, ...) are determined either by the capacity of the cell (usually the maximal velocity in a spontaneous process, so-called balanced growth) or by an operating parameter such as feed or dilution rate (sub-maximal in controlled processes such as fed batch or chemostat, so-called limited growth). Changes of fluxes due to changes of these rather steady operating conditions are generally subject to metabolic control.

If excess carbon and energy source is pulsed to a C- and energy-limited culture, the intracellular ATP concentration initially drops due to the phosphorylation sink (glucokinase and/or hexokinases). Only later, during catabolism, can the energy provided by this extra substrate be liberated in terms of new ATP. The duration of the ATP sink was predicted to be in the order of a few tens of seconds by Nielsen et al. [306] (for an experimental confirmation, see Sect. 4.1). Obviously, such rapid reactions are under kinetic control and the necessary enzyme activities are present in sufficient quantities. It is not clear, at present, whether the fluxes attain their organism-typical maximal values immediately or whether some fine tuning of enzymatic control precedes this event. It is likely that, in this case, a regulation of (intracellular) enzyme activity, if at all necessary, takes place on the level of enzyme modification (e.g. phosphorylation) rather than on the level of de novo production (i.e. control on the transcriptional level) because the latter would require much more time (see [405]).

4.1
Dynamics of Biosystems

The dynamics of microbial cultures have an important impact on the characteristics of measurement and process control. The "typical time constant" in a bioprocess is often erroneously anticipated to be equivalent to the entire duration of a cultivation.

The quantitative investigation of substrate uptake requires a time resolution of a few 100 ms; otherwise, artifacts must result [84, 442, 443]. This statement was evidenced only after a suitable technique had been established: glucose metabolism was stopped within 100 ms by spraying the cell suspension from the over-pressurized bioreactor into 60% methanol which was pre-chilled to $-40\,°C$. This procedure does not damage the integrity of the cells and the pretreated sample remains liquid which is advantageous for further processing. Although no report has yet appeared, it is quite realistic to imagine that this technique is easy to automate.

The relevant relaxation times of a culture system are determined by the actual cell density and the specific conversion rate (capacity) of the culture, i.e. by one or more operational and state variables (for instance feed rate and the concentrations or activities of cell mass and of effectors, if relevant) and inherent characteristic properties of the biosystem which are parameters. There are metabolites with a long lifetime and other (key) metabolites with very short lifetimes, e.g. molecules representing the energy currency of cells such as ATP and other nucleotides. Rizzi et al. [348] and Theobald et al. [433] have shown that the energy charge response to a pulse challenge of a yeast culture is a matter of a few seconds only. However, the responses can differ considerably when pulses or shifts are applied to cultures of different recent history [399, 404, 405]. Larsson [230] and Neubauer et al. [301] found that substrate oscillations greatly affected the growth performance of *Escherichia coli*. Short-term, i.e. less than 2 min, glucose excess and starvation were investigated in a looped system of a stirred tank and a plug flow reactor (a so-called compartmented reactor). Similar observations have also been made for yeast [115, 124].

Two other paradigms demonstrate that the band width of relaxation times is extremely broad: the time required to achieve a new steady state in a chemostat culture is approximately determined by $(\mu_{max} - D)^{-1}$. The time required by a culture to consume a considerable fraction of small amounts of residual

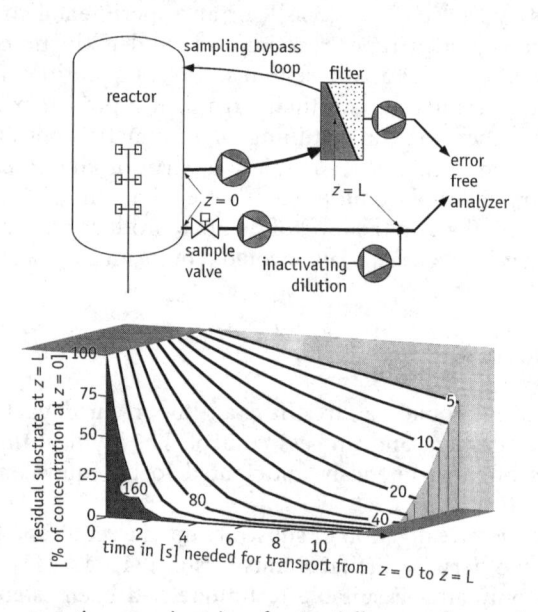

Fig. 29. Origin of systematic errors in spite of potentially error-free analysis. On-line sampling setups (*top*) and time trajectories of limiting substrate concentration during sample preparation in the two paradigmatic setups depending on the actual culture density (*bottom*). Either a filter in bypass loop is used for the preparation of cell-free supernatant (*upper part in top insert*) or an aliquot of the entire culture is removed using an automatic sampler valve and a sample bus for further inactivation and transport of the samples taken (*lower part*). Both methods require some finite time for sample transportation from the reactor outlet (at $z = 0$) to the location where separation of cells from supernatant or inactivation by adding appropriate inactivators (at $z = L$) takes place. During transport from $z = 0$ to $z = L$, the cells do not stop consuming substrate. A low substrate concentration in the reactor (namely $s \sim K_s$) and a maximal specific substrate consumption rate of 3 g g^{-1} h^{-1} were assumed in the simulation example to reflect the situation of either a fed-batch or a continuous culture of an industrially relevant organism such as yeast. The actual culture density (in g l^{-1}) marks some trajectories in the mesh plot. Note that the time scale is in seconds

substrate during sampling – and thus systematically falsify the analytical results if no appropriate inactivation takes place – depends, among others, on the cell density and can also be in the order of a few seconds only (Fig. 29).

Even simple static models are very valuable for compensation of systematic errors built into automated analytical procedures. One important example is when sampling requires a well-known but non-negligible time, e.g. when performed using a filter separator operated in bypass or using a sample bus system. A bypass behaves as a plug flow type reactor fraction where flow-dependent spatial gradients develop and where no inactivation can take place because the bulk of the bypassed aliquots is returned to the reactor. The cells continue to consume substrate while they are being transported from the exit of the reactor to the filter. The permeate recovered there is representative for the filter site but not for the reactor. Knowing the transport time and some basic kinetic parameters one can easily compensate on-line for such errors provided that a useful estimate of the actual biomass concentration is available. Even though a bypass can be tuned to be operated at a mean residence time of 5 s or less, this can be enough for a significant decrease in substrate concentration in high density cultures (see Fig. 29). Sample buses need a minimal (dead) time for transportation of the sample and in situ filters tend to fail in high density cultures due to rapid fouling. Hence, the problem is real and must not be ignored.

4.2
Continuous Signals and Frequency of Discrete Analyses

Sensors based on either electrical, optical or electromagnetic principles normally deliver a continuous signal which is very useful. The dynamics of the analyzed system can be resolved according to the time constants of the respective electronic equipment. In the general case, however, these data will nowadays be digitized but the information loss can normally be neglected because 12- or 16-bit converters are state of performance today.

Further, data are stored in a computer in distinct time intervals resulting in discretion with respect to time and concomitantly in a possible loss of information. A data reduction algorithm should therefore be applied which must account for this fact: raw data should be scanned with high frequency and the essential data – i.e. only recording when important changes take place – may be stored with the necessary frequency, i.e. variably or not equidistant with respect to time.

Analytical instruments based on either physicochemical separation methods or relying on (bio)chemical reactions require a finite time to run; these instruments are usually operated in repetitive, non-overlapping batch mode and deliver results with a certain non-negligible dead time. Generally, data density is low, for instance, in the order of 1 min^{-1} for FIA or 2 h^{-1} for HPLC.

5
Relevant Pitfalls

Sampling and sample preparation usually impose systematic errors on the results of analytical procedures. It is, therefore, important to be able to reproduce these errors and to determine them as accurately as possible, as discussed above, in order to compensate by appropriate calculations. Good sampling techniques, such as rapid automatic sampling, fast cooling, heating or immediate chemical inactivation, help to minimize these errors. One needs to be aware of the possible artifacts inherent to individual measuring techniques and underlying assumptions; some general reflections and simple examples can be found in Locher et al. [244, 245]. In the following sections, three randomly selected examples are discussed in more detail.

5.1
α,β-D-Glucose Analyzed with Glucose Oxidase

A generally ignored pitfall is associated with some glucose analyses, specifically with those methods using glucose oxidase. Such methods are very common since the enzyme is relatively inexpensive and very stable. This enzyme is known to react with β-D-glucose but not with α-D-glucose (Merck Index).

In aqueous solution, a mutarotational equilibrium of the two anomers is reached spontaneously – but not instantaneously – in which the ratio α/β is 36:64 at a temperature of approximately 30 °C. In water, the rate is much lower than in buffer(ed medium) [354]. The enzyme mutarotase accelerates mutarotation considerably. The rate at which the equilibrium is reached spontaneously depends greatly on pH, temperature and other solution components.

The numerical values of the time constants summarized in Table 1 confirm that there is no particular time for significant mutarotation to take place in rapid on-line analyses. If glucose uptake is anomerically specific, as has been shown by Benthin et al. [23], analytical results obtained with a glucose oxidase method must be corrected accordingly. The same holds true for analyses of rapid biological transients and for any spiking method (if the spiking solution happens to be freshly prepared).

Table 1. Values for k_{mut}, the kinetic parameter describing mutarotational equilibration of α,β-D-glucose

Solvent	Not specified	Pure water	PO$_4$-buffer
k_{mut} (h^{-1})	2.8 – 11.4	1.2 ± 0.02	12 ± 1
Ref	[23]	[354]	[354]

5.2
CO$_2$ Equilibrium with HCO$_3^-$

Even error-free determination of CO_2 in the exhaust gas does not necessarily assure correct determination of CO_2 production (rate) or carbon recovery if the solubility and reactivity of CO_2 at neutral or alkaline pH is not taken into consideration [311]. CO_2 dissolves in water to hydrated CO_2, namely $H_2O.CO_2$,

Table 2. Parameters describing the CO_2-bicarbonate equilibrium

$k_1 (h^{-1})$	$k_2 (h^{-1})$	$K_b (M\ l^{-1})$	Ref
209	2160	6.8×10^{-4}	[356]
134	134	4.7×10^{-7}	[311]
–	–	4.26×10^{-7}	a

[a] Handbook of chemistry and physics.

which is in equilibrium with the unstable H_2CO_3. This, in turn, is in practically instantaneous equilibrium with HCO_3^-, determined by the pH of the solution:

$$CO_2 + H_2O \underset{k_2}{\overset{k_1}{\rightleftarrows}} H_2CO_3 \xleftarrow{spontaneous} HCO_3^- + H^+ \tag{3}$$

with the following kinetics:

$$r_{c>b} = k_1 \cdot c - k_2 \cdot [H_2CO_3] = k_1 \cdot c - k_2 \cdot \frac{b \cdot 10^{-pH}}{K_b} \tag{4}$$

where $r_{c>b}$ is the net reaction rate, c is the concentration of dissolved CO_2, b is the concentration of bicarbonate (HCO_3^-) and K_b is the dissociation parameter of carbonic acid to bicarbonate. k_1 and k_2 are rate parameters for which quite different sets of numerical values are found in the literature (Table 2). This is most probably associated with the different acidity of $H_2O.CO_2$ and H_2CO_3; the latter is the stronger acid.

Important is the conclusion that a significant amount of CO_2 may not be found in the exhaust gas but rather trapped in the culture liquid in the form of bicarbonate or even carbonate (in alkalophilic cultures). For example, Ponti [327] estimated that up to 25% of total carbon mineralized in sewage sludge treatment was captured as bicarbonate and not released via the gas phase. Disregarding this fact must result in systematically erroneous determinations of respiratory quotient (RQ), carbon dioxide production rate (CPR) and carbon recovery. However, taking all these factors into account allows the determination of the RQ-value, even in mammalian cell cultures, with reasonable accuracy even though bicarbonate and low gas turnover rates hamper the measurements [38].

5.3
Some Remarks on Error Propagation

Elemental balancing permits the determination of (other) metabolic rates provided the stoichiometry is known. Carbon balances are the most useful but the carbon lost via the exhaust gas (as CO_2) and culture liquid (as HCO_3^-) must be measured. Heinzle et al. [161] determined that the state predictions based on experiments with a small quadrupole mass spectrometer were not useful due to unacceptable error propagation; for instance, a 1% relative offset calibration error could result in a prediction error for an intracellular storage material (PHB) of >50%. Using a highly accurate and precise instrument (absolute errors <0.02% gas composition) together with automatic, repetitive recalibration resulted, however, in reasonable estimation of substrates, PHB and biomass. These findings have also been experienced by others, e.g. [245].

The physiological states expressed in yeast fed-batch cultures could be recognized on-line by fuzzy inference based on error vectors. The error vector has been newly defined in a macroscopic elemental balance equation. The physiological states for cell growth and ethanol production can be characterized by error vectors using many experimental data and fuzzy membership functions, constructed from the frequency distributions of the error vectors, allow state recognition by fuzzy inference. In particular, unusual physiological states during yeast cultivations can be recognized accurately [383].

The variability of replicate measurements of soluble protein or enzyme activities was investigated by Dehghani et al. [83]. The variability of the assays depended on the measured concentrations and the standard deviation was found to be roughly proportional to the actual values.

5.4
The Importance of Selecting Data To Keep

Automated measurement and control of bioprocesses, presently an art but routine in the near future, generates a tremendous amount of data [249]. This requires judgment of the importance of these data for documentation and to reduce data effectively without loss of valuable information. The reduction algorithm must not be based on an intention to make one or the other variable a parameter, it must rather keep a true image of the real data. We adopted a simple algorithm to achieve this goal (Fig. 30): all data – independent of whether measured or calculated – are treated as variables and kept in a circular buffer (holding some 50 h) in the frequency with which they were generated. This volatile data base serves as a rapid graphical review of the short history. However, only values (data points) of variables that change significantly with time are written into the archive. The significance of a change is judged by a reasonably defined window for each variable – including all intended culture parameters – the width of which is usually determined by the noise on the respective signal. In any instance, every "first" data point (of an experiment) is archived, together with a time stamp. The next entry into the archive is made only if the variable moves outside the respective window which has been centered around the last archived value; concomitantly, the window is re-centered around the new entry, and so on. This technique assures that all relevant changes – including those not intended – of any considered variable are trapped and that the dynamics of all signal trajectories is fully documented. The only drawbacks to this data treatment are that there is a need to time stamp every data entry individually, resulting in non-equidistant data vectors. The benefits are a data-to-archive reduction of usually between 10^{-2} and 10^{-4} and the assurance that no important data are lost.

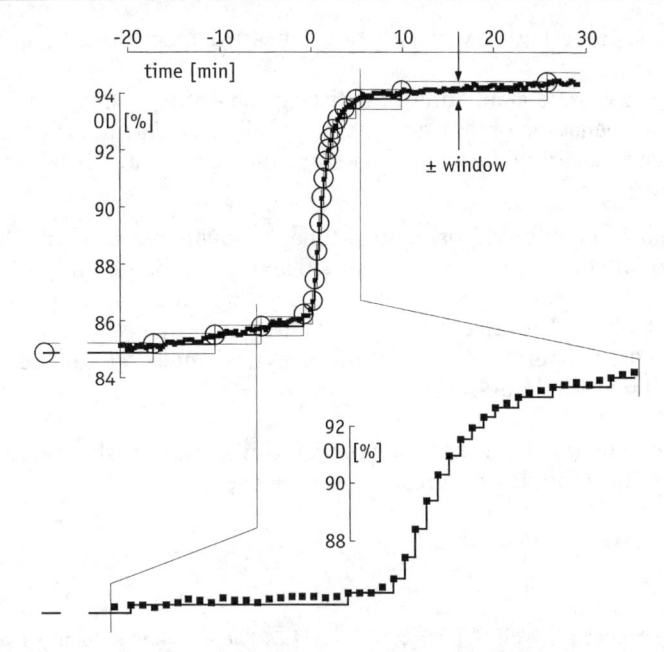

Fig. 30. Schematic design of a simple but very useful and efficient data reduction algorithm. Data representing the time trajectory of an individual variable are only kept (= recorded, stored) when the value leaves a permissive window which is centered around the last stored value. If this happens, the new value is appended to the data matrix and the window is re-centered around this value. This creates a two-column matrix for each individual variable with the typical time stamps in the first column and the measured (or calculated) values in the second column. In addition, the window width must be stored since it is typical for an individual variable. This algorithm assures that no storage space is wasted whenever the variable behaves as a parameter (i.e. does not change significantly with time, is almost constant) but also assures that any rapid and/or singular dynamic behavior is fully documented. No important information is then lost

6
Conclusions

The undelayed evaluation of state of a culture by using software sensors and computers, based on the quantitative analytical information provided by hardware sensors and intelligent analytical subsystems, constitutes an excellent basis for targeted process control. Experts – either human or computer – have the data and the deterministic knowledge to trace observed behavior back to the physical, chemical and physiological roots thereby gaining a qualitative improvement of bioprocess control, a quantum leap: process control can act on the causes of effects rather than just cure symptoms. A simple standard operating procedure [398] has proven useful, namely:

- measure everything that can be measured at the very beginning of process development,
- decide whether or not a variable is relevant,

- determine the relevant variables to be measured, controlled and/or documented, and
- collect all raw data at any time and distinguish on-line between variable and parameter behavior, organize an archive of all these data accordingly and do not discard seemingly useless data since they contribute to the treasure of experience.

If it is, as some people say, correct that today's bioengineering with all its tools and methodologies is too slow and not efficient enough, then it is all the more urgent to improve the performance of the methods, tools and equipment currently available and to invent new and better ones. In essence, techniques of instrumentation, operation, and causal-analytical interpretation of measurements need massive impulses.

Acknowledgements. Financial support of this work by the Swiss Priority Program in Biotechnology is greatly acknowledged.

References

1. Aberl F, Modrow S, Wolf H, Koch S, Woias P (1992) Sensors and Actuators B 6:186
2. Ackermann JU, Muller S, Losche A, Bley T, Babel W (1995) J Biotechnol 39:9
3. Acuna G, Latrille E, Beal C, Corrieu G, Cheruy A (1994) Biotechnol Bioeng 44:1168
4. Agayn VI, Walt DR (1993) Bio/Technology 11:726
5. Ahlmann N, Niehoff A, Rinas U, Scheper T, Schügerl K (1986) Anal Chim Acta 190:221
6. Alberghina L, Porro D (1993) Yeast 9:815
7. Albiol J, Campmajo C, Casas C, Poch M (1995) Biotechnol Prog 11:88
8. Andrew KN, Blundell NJ, Price D, Worsfold PJ (1994) Anal Chem 66:A916
9. Arnold MA, Meyerhoff ME (1988) CRC Crit Rev Anal Chem 20:149
10. Arnold MA, Ostler TJ (1986) Anal Chem 58:1137
11. Asami K, Yonezawa T (1995) BBA-Gen Subjects 1245:99
12. Attridge JW, Leaver KD, Cozens JR (1987) J Phys E Sci Instrum 20:548
13. Auberson LCM, Kanbier T, Stockar U von (1993) J Biotechnol 29:205
14. Axelsson JP, Münch T, Sonnleitner B (1992) In: Karim N (ed) Computer applications in fermentation technology – modelling and control of bioprocesses. Pergamon, p 383
15. Bailey JE, Shanks JV (1991) Bioprocess Eng 6:273
16. Barnbot SB, Lakowicz JR, Rao G (1995) TIBTECH 13:106
17. Bataillard P, Steffgen E, Haemmerli S, Manz A, Widmer HM (1993) Biosens Bioelectron 8:89
18. Bavouzet JM, Lafforguedelorme C, Fonade C, Goma G (1995) Enzyme Microb Technol 17:712
19. Baxter PJ, Christian GD, Ruzicka J (1995) Chem Anal 40:455
20. Beck HP, Wiegand C (1995) Fresenius J Anal Chem 351:701
21. Bedard C, Jolicoeur M, Jardin B, Tom R, Perret S, Kamen A (1994) Biotechnol Tech 8:605
22. Bennetto HP, Box J, Delaney GM, Mason JR, Roller SD, Stirling JL, Thurston CF (1987) Redox-mediated electrochemistry of whole microorganisms: from fuel cells to biosensors. Oxford Scientific Publishers, Oxford, p 291
23. Benthin S, Nielsen J, Villadsen J (1992) Biotechnol Bioeng 40:137
24. Berger A, Blum LJ (1994) Enzyme Microb Technol 16:979
25. Bergveld P (1989) J Phys E Sci Instr 22:678
26. Berovic M (1987) In: Otocec YU (ed) Bioreactor engineering course. Boris Kidric and Slov Chemical Society, p 327

27. Berrada R, Dauphin G, David L (1987) J Org Chem 52:2388
28. Betz JW, Aretz W, Härtel W (1984) Cytometry 5:145
29. Beyeler W, Einsele A, Fiechter A (1981) Eur J Appl Microbiol Biotechnol 13:10
30. Birnbaum S, Bülow L, Hardy K, Danielsson B, Mosbach K (1986) Anal Biochem 158:12
31. Birou B, Marison IW, Stockar U von (1987) Biotechnol Bioeng 30:650
32. Birou B, Marison IW, Stockar U von (1987) Biotechnol Bioeng 30:650
33. Birou B, Stockar U von (1989) Enzyme Microb Technol 11:12
34. Block DE, Eitzman PD, Wangensteen JD, Srienc F (1990) Biotechnol Progr 6:504
35. Bloem J, Veninga M, Shepherd J (1995) Appl Environ Microbiol 61:926
36. Boe I, Lovrien R (1990) Biotechnol Bioeng 35:1
37. Bomio M, Sonnleitner B, Fiechter A (1989) Appl Microbiol Biotechnol 32:356
38. Bonarius HPJ, Degooijer CD, Tramper J, Schmid G (1995) Biotechnol Bioeng 45:524
39. Bonarius HPJ, Timmerarends B, deGooijer CD, Tramper J (1998) Biotechnol Bioeng 58:258
40. Bormann EJ (1988) J Theor Biol 133:215
41. Botrè C, Botrè F (1990) Anal Biochem 185:254
42. Bradley J, Turner APF, Schmid RD (1989) GBF Monographs 13:85
43. Brandt J, Hitzmann B (1993) Am Biotechnol Lab 11:78
44. Brewster JD, Maxwell RJ, Hampson JW (1993) Anal Chem 65:2137
45. Briasco CA, Ross DA, Robertson CR (1990) Biotechnol Bioeng 36:879
46. Brookman JSG (1969) Biotechnol Bioeng 6:323
47. Bruinenberg PM (1986) Antonie v Leeuwenhoek 52:411
48. Busch M, Schmidt J, Rothen SA, Leist C, Sonnleitner B, Verpoorte S (1996) Proc 2nd Int Symp Miniaturized Total Analysis Systems (TAS96), 19–22 Nov 1996, Basel, (Analytical Methods and Instrumentation, Special Issue 1996, Widmer IIM, Verpoorte E, Barnard S (eds), p 120
49. Caldarone EM, Buckley LJ (1991) Anal Biochem 199:137
50. Caldwell KD (1986) Abstr Pap Am Chem Soc 192 Meet ISEC 46
51. Caldwell KD, Cheng ZQ, Hradecky P, Giddings JC (1984) Cell Biophys 6:233
52. Camelbeeck JP, Comberbach DM, Goossens J, Roelants P (1988) Biotechnol Tech 2:183
53. Campmajo C, Cairo JJ, Sanfeliu A, Martinez E, Alegret S, Godia F (1994) Cytotechnology 14:177
54. Carlsen M, Johansen C, Min RW, Nielsen J, Meier H, Lantreibecq F (1993) Anal Chim Acta 279:51
55. Castro CD, Meehan AJ, Koretsky AP, Domach MM (1995) Appl Environ Microbiol 61:4448
56. Chauvatcharin S, Konstantinov KB, Fujiyama K, Seki T, Yoshida T (1995) J Ferment Bioeng 79:465
57. Chauvatcharin S, Seki T, Fujiyama K, Yoshida T (1995) J Ferment Bioeng 79:264
58. Chéruy A (1997) J Biotechnol 52:193
59. Chong K, Loughlin T, Moeder C, Perpall HJ, Thompson R, Grinberg N, Smith GB, Bhupathy M, Bicker G (1996) J Pharm Biomed Anal 15:111
60. Christensen H, Bakken LR, Olsen RA (1993) FEMS Microbiol Ecol 102:129
61. Christensen H, Olsen RA, Bakken LR (1995) Microbiol Ecol 29:49
62. Christensen LH, Mandrup G, Nielsen J, Villadsen J (1994) Anal Chim Acta 296:51
63. Comas J, VivesRego J (1998) J Microbiol Meth 32:45
64. Comberbach DM, Bu'lock JD (1983) Biotechnol Bioeng 25:2503
65. Combrier E, Metezeau P, Ronot X, Gachelin H, Adolphe M (1989) Cytotechnology 2:27
66. ComTec GmbH, Technologiezentrum Jülich (1997) BioWorld 2/97:22 and 3/97:39
67. Connolly P, Lewis SJ, Corry JEL (1988) Int J Food Microbiol 7:31
68. Cook PD, Gao YP, Smith R, Toube TP, Utley JHP (1995) J Mater Chem 5:413
69. Copella SJ, Wang DIC (1990) Biotechnol Tech 4:161
70. Cordier JL, Butsch BM, Birou B, Stockar U von (1987) Appl Microbiol Biotechnol 25:305
71. Cortassa S, Aon JC, Aon MA (1995) Biotechnol Bioeng 47:193
72. Cox RP (1987) Mass Spectrom Biotechnol Process Anal Contr 63

73. Crissman HA, Steinkamp JA (1993) Europ J Histochem 37:129
74. Danigel H. (1995) Opt Eng 34:2665
75. DaPra E, Schneider K, Bachofen R (1989) Experientia 45:1024
76. Darzynkiewicz Z (1994) In: Celis JE (ed) Cell biology. A laboratory handbook, vol 1. Academic Press, San Diego, p 261
77. Darzynkiewicz Z. (1994) In: Darzynkiewicz Z, Robinson JP, Crissman HA (eds) Methods in cell biology, vol 41. Academic Press
78. Davey HM, Jones A, Shaw AD, Kell DB (1999) Cytometry 35:162
79. Deboux BJC, Lewis E, Scully PJ, Edwards R (1994) Optical Fibre Sensors Conference, Glasgow, p 6
80. Decastro MDL, Valcarcel M (1995) In: Hurst WJ (ed) Automation in the laboratory. VCH
81. Decristoforo G (1988) Methods Enzymol 137:197
82. Degelau A, Scheper T, Bailey JE, Guske C (1995) App Microbiol Biotechnol 42:653
83. Dehghani M, Bulmer M, Gregory ME, Thornhill NF (1995) Bioprocess Eng 13:239
84. de Koning W, vanDam K (1992) Anal Biochem 204:118
85. Delaguardia M (1995) Mikrochim Acta 120:243
86. Desgranges C, Georges M, Vergoignan C, Durand A (1991) Appl Microbiol Biotechnol 35:206
87. Dien BS, Srienc F (1991) Biotechnol Prog 7:291
88. Dijkema C, Vries SC de, Booij H, Schaafsma TJ, Kammen A van (1988) Plant Physiol 88:1332
89. Ding T, Schmid RD (1990) Anal Chim Acta 234:247
90. Doherty SJ, Winniford WL (1994) LC-GC12:846
91. Donaldson TL, Palmer HJ (1979) AIChE 25:143
92. Droste RL, Sanchez WA (1983) Water Res 17:975
93. Duggleby RG (1991) TIBS FEB 91:51
94. Dullau T, Reinhardt B, Schügerl K (1989) Anal Chim Acta 225:253
95. Durant G, Cox PW, Formisyn P, Thomas CR (1994) Biotechnol Tech 8:759
96. Ebina Y, Ekida M, Hashimoto H (1989) Biotechnol Bioeng 33:1290
97. Edwards AG, Ho CS (1988) Biotechnol Bioeng 32:1
98. Edwards C, Diaper J, Porter J, Deere D, Pickup R (1994) In: Ritz K, Dighton J, Giller KE (eds) Beyond the biomass. Wiley, Chichester, p 57
99. Effenhauser CS, Manz A, Widmer HM (1993) Anal Chem 65:2637
100. Eggeling L, Graaf A de, Sahm H, Eikmanns B, Marx A, Wiechert W (1995) ECB7, Nice, France, MAC66
101. Englbrecht U, Schmidt HL (1992) J Chem Technol Biotechnol 53:397
102. Englbrecht U, Schmidt HL (1995) J Chem Technol Biotechnol 62:68
103. Eyer K, Oeggerli A, Heinzle E (1995) Biotechnol Bioeng 45:54
104. Fang LL, Zhang R, Williams ER, Zare RN (1994) Anal Chem 66:3696
105. Fardeau ML, Plasse F, Belaich JP (1980) Eur J Appl Microbiol Biotechnol 10:133
106. Fatt I (1976) Polarographic oxygen sensors. CRC Press Inc, Cleveland, OH, USA
107. Fehrenbach R, Comberbach M, Pêtre JO (1992) J Biotechnol 23:303
108. Fernandez EJ, Clark DS (1987) Enzyme Microb Technol 9:259
109. Fernandez EJ, Mancuso A, Clark DS (1988) Biotechnol Prog 4:173
110. Fiechter A, Sonnleitner B (1994) Adv Microb Physiol 36:145
111. Filippini C, Moser JU, Sonnleitner B, Fiechter A (1991) Anal Chim Acta 255:91
112. Filippini C, Sonnleitner B, Fiechter A (1992) Anal Chim Acta 265:63
113. Filippini C, Sonnleitner B, Fiechter A, Bradley J, Schmid RD (1991) J Biotechnol 18:153
114. Forster RJ, Diamond D (1992) Anal Chem 64:1721
115. Frandsen S, Nielsen J, Villadsen J (1994) In: Alberghina L, Frontali L, Sensi P (eds) ECB6: Proceedings of the 6th European Congress on Biotechnology. p 887
116. Fredrickson AG, Hatzis C, Srienc F (1992) Cytometry 13:423
117. Fry JC (1990) Methods Microbiol 22:41
118. Fuhrmann GF, Völker B (1992) J Biotechnol 27:1

119. Furukawa K, Heinzle E, Dunn IJ (1983) Biotechnol Bioeng 25:2293
120. Galaev IY, Mattiasson B (1993) Enzyme Microb Technol 15:354
121. Galazzo JL, Shanks JV, Bailey JE (1990) Biotechnol Bioeng 35:1164
122. Garn M, Gisin M, Thommen C, Cevey P (1989) Biotechnol Bioeng 34:423
123. Gary K, Meier P, Ludwig K (1988) Canbiocon 1988 Biotechnol Res Appl:155
124. George S, Larsson G, Enfors SO (1994) In: Alberghina L, Frontali L, Sensi P (eds) ECB6: Proceedings of the 6th European Congress on Biotechnology, vol 2, p 883
125. Geppert G, Thielemann H (1984) Acta Biotechnol 4:361
126. Geppert G, Thielemann H, Langkopf G (1989) Acta Biotechnol 9:541
127. Gere DR, Knipe CR, Castelli P, Hedrick J, Frank LGR, Schulenbergschell H, Schuster R, Doherty L, Orolin J, Lee HB (1993) J Chromatogr Sci 31:246
128. Giddings JC (1995) Anal Chem A 67:592
129. Giddings JC (1989) J Chromatogr 470:327
130. Giddings JC (1993) Science 260:1456
131. Giddings JC, Moon MH (1991) Anal Chem 63:2869
132. Giddings JC, Yang FJ, Myers MN (1977) J Virol 21:131
133. Gillies RJ, MacKenzie NE, Dale BE (1989) Bio/Technology 7:50
134. Gollmer K, Posten C (1995) In: Munack A, Schügerl K (eds) CAB6 preprints, Garmisch-Partenkirchen, D:41
135. Goodacre R, Trew S, Wrigleyjones C, Saunders G, Neal MJ, Porter N, Kell DB (1995) Anal Chim Acta 313:25
136. Gorczyca W, Melamed MR, Darzynkiewicz Z (1998) In: Jaroszeski MJ, Heller R (eds) Flow cytometry protocols, vol 91. p 217
137. Gorlach E, Richmond R, Lewis I (1998) Anal Chem 70:3227
138. Griot M, Heinzle E, Dunn IJ, Bourne JR (1987) Mass Spectrom Biotechnol Process Anal Contr 75
139. Groboillot A, Pons MN, Engasser JM (1989) Appl Microbiol Biotechnol 32:37
140. Gschwend K, Beyeler W, Fiechter A (1983) Biotechnol Bioeng 25:2789
141. Guilbault GG, Luong JH (1988) J Biotechnol 9:1
142. Guilbault GG, Luong JHT (1989) Selective Electrode Rev 11:3
143. Gurev IA, Zyuzina LF, Lazareva OP (1995) J Anal Chem Engl Tr 50:765
144. Haemmerli S, Schaeffler A, Manz A, Widmer HM (1992) Sens Actuators B 7:404
145. Hall EAH (1986) Enzyme Microb Technol 8:651
146. Hammer BE, Heath CA, Mirer SD, Belfort G (1990) Bio/Technol 8:327
147. Hansen EH (1995) Anal Chim Acta 308:3
148. Harder A, Roels JA (1982) Adv Biochem Eng 21:56
149. Harris CM, Kell DB (1985) Biosensors 1:17
150. Harris CM, Todd RW, Bungard SJ, Lovitt RW, Morris JG, Kell DB (1987) Enzyme Microb Technol 9:181
151. Hartbrich A, Schmitz G, Weusterbotz D, Degraaf AA, Wandrey C (1996) Biotechnol Bioeng 51:624
152. Hatch RT, Veilleux BG (1995) Biotechnol Bioeng 46:371
153. Hatch RT, Wilder C, Cadman TW (1979) Biotechnol Bioeng Symp 9:25
154. Hauser PC, Rupasinghe TWT, Cates NE (1995) Talanta 42:605
155. Hayward MJ, Kotiaho T, Lister AK, Cooks RG, Austin GD, Narayan R (1990) Anal Chem 62:1798
156. Heijnen JJ, Terwisscha van Scheltinga AH, Straathof AJ (1992) J Biotechnol 22:3
157. Heinzle E (1987) Adv Biochem Eng/Biotechnol 35:1
158. Heinzle E (1992) J Biotechnol 25:81
159. Heinzle E, Kramer H, Dunn IJ (1985) Biotechnol Bioeng 27:238
160. Heinzle E, Moes J, Griot M, Sandmeier E, Dunn IJ, Bucher R (1986) Ann NY Acad Sci 469:178
161. Heinzle E, Oeggerli A, Dettwiler B (1990) Anal Chim Acta 238:101
162. Heinzle E, Reuss M (eds) (1987) Mass spectroscopy in biotechnological analysis and control

163. Hemert P van, Kilburn DG, Righelato RC, Wezel AL van (1969) Biotechnol Bioeng 6:549
164. Henschke PA, Thomas DS (1988) J Appl Bacteriol 64:123
165. Herbert RA (1990) Methods Microbiol 22:1
166. Hernandez O, Jimenez AI, Jimenez F, Arias JJ (1995) Anal Chim Acta 310:53
167. Hewitt CJ, Boon LA, McFarlane CM, Nienow AW (1998) Biotechnol Bioeng 59:612
168. Hibino W, Kadotani Y, Kominami M, Yamane T (1993) J Ferment Bioeng 75:
169. Higgins IJ, Cardosi MF, Turner APF (1988) NATO ASI SerA 128, Perspectives Biotechnol p 55
170. Hirose Y (1986) Prog Ind Microbiol 24:67
171. Hitzmann B, Lammers F, Weigel B, Putten A van (1993) BioForum 16:450
172. Hitzmann B, Lohn A, Reinecke M, Schulze B, Scheper T (1995) Anal Chim Acta 313:55
173. Holm-Hansen O (1973) Bull Ecol Res Comm (Stockholm) 17:215
174. Holzhauer-Rieger K, Zhou W, Schügerl K (1990) J Chromatogr 499:609
175. Hong K, Tanner RD, Malaney GW, Wilson DJ (1987) Proc Biochem 22:149
176. Hooijmans CM, Abdin TA, Alaerts GJ (1995) Appl Microbiol Biotechnol 43:781
177. Hooker AD, Goldman MH, Markham NH, James DC, Ison AP, Bull AT, Strange PG, Salmon I, Baines AJ, Jenkins N (1995) Biotechnol Bioeng 48:639
178. Hopkins D, Hatch RT (1990) Abstr Pap Am Chem Soc 199 Meet, part 1, BIOT115
179. Horvath JJ, Enriquez-Ortiz AB, Semerjian HG (1989) Abstr Pap Am Chem Soc 198 Meet, MBTD65
180. Huang Z, Poulter CD, Wolf FR, Somers TC, White JD (1988) J Am Chem Soc 110:3959
181. Huller R, Glossner E, Schaub S, Weingartner J, Kachel V (1994) Cytometry 17:109
182. Humphrey AE (1988) Aust J Biotechnol 2:141
183. Iijima S, Yamashita S, Matsunaga K, Miura H, Morikawa M, Shimizu K, Matsubara M, Kobayashi T (1987) J Chem Technol Biotechnol 40:203
184. Inoue Y, Sano F, Nakamura K, Yoshie N, Saito Y, Satoh H, Mino T, Matsuo T, Doi Y (1996) Polym Int 39:183
185. Isaacs SH, Soeberg H, Christensen LH, Villadsen J (1992) Chem Eng Sci 47:1591
186. Isenschmid A, Marison IW, Stockar U von (1995) J Biotechnol 39:229
187. Iversen JJL, Thomsen JK, Cox RP (1994) Appl Microbiol Biotechnol 42:256
188. Jacobsen CN, Fremming C, Jakobsen M (1997) J Microbiol Meth 31:75
189. Jan S, Roblot C, Courtois J, Courtois B, Barbotin JN, Seguin JP (1996) Enzyme Microb Technol 18:195
190. Jayat C, Ratinaud MH (1993) Biology of the Cell 78:15
191. Jee HS, Nishio N, Nagai S (1987) J Gen Appl Microbiol 33:401
192. Jobst G, Urban G, Jachimowicz A, Kohl F, Tilado O, Lettenbichler I, Nauer G (1993) Biosens Bioelectron 8:123
193. Jolicoeur C, To TC, Beaubien A, Samson R (1988) Anal Chim Acta 213:165
194. Jones JG, Zhou DM (1994) Biotechnol Adv 12:693
195. Jones RP, Greenfield PF (1982) Enzyme Microb Technol 4:210
196. Jorgensen H, Nielsen J, Villadsen J, Mollgaard H (1995) Biotechnol Bioeng 46:117
197. Joy RW, Mcintyre DD, Vogel HJ, Thorpe TA (1996) Physiol Plant 97:149
198. Ju LK, Ho CS (1988) Biotechnol Bioeng 31:995
199. Ju LK, Yang X, Lee JF, Armiger WB (1995) Biotechnol Progr 11:545
200. Juan G, Darzynkiewicz Z (1998) In: Celis JE (ed) Cell biology. A laboratory handbook. Academic Press, San Diego, p 261
201. Junker BH, Wang DIC, Hatton TA (1988) Biotechnol Bioeng 32:55
202. Kadotani Y, Miyamoto K, Mishima N, Kominami M, Yamane T (1995) J Ferment Bioeng 80:63
203. Kalnenieks U, Degraaf AA, Bringermeyer S, Sahm H (1993) Arch Microbiol 160:74
204. Kapuscinski J. (1995) Biotech Histochem 70:220
205. Karube I, Nomura Y, Arikawa Y (1995) TRAC 14:295
206. Karube I, Sode K, Tamiya E (1989) Swiss Biotech 7:25
207. Karube I, Sode K, Tamiya E, Gotoh M, Kitagawa Y, Suzuki H (1988) 8th Int Biotechnol Symp, part 1, p 537

208. Karube I, Tamiya E, Sode K, Yokoyama K, Kitagawa Y, Suzuki H (1988) Anal Chim Acta 213:69
209. Karube I, Yokoyama K, Sode K, Tamiya E (1989) Anal Lett 22:791
210. Kasthurikrishnan N, Cooks RG (1995) Talanta 42:1325
211. Kato C, Sato T, Horikoshi K (1995) Biodiversity and Conservation 4:1
212. Kätterer L (1983) PhD thesis 7172, ETH Zürich
213. Kell D, Markx GH, Davey CL, Todd RW (1990) TRAC 9:190
214. Kell DB, Ryder HM, Kaprelyants AS, Westerhoff HV (1991) Antonie v Leeuwenhoek Int J Gen Molec Microbiol 60:145
215. Kell DB, Sonnleitner B (1995) TIBTECH 13:481
216. Kirkland JJ, Rementer SW, Yau WW (1988) Anal Chem 60:610
217. Kirkland JJ, Yau WW (1991) J Chromatogr 550:799
218. Kjaergaard L (1977) Adv Biochem Eng 7:131
219. Konstantinov K, Chuppa S, Sajan E, Tsai Y, Yoon S, Golini F (1994) TIBTECH 12:324
220. Konstantinov KB, Yoshida T (1992) Biotechnol Bioeng 39:479
221. Kromenaker SJ, Srienc F (1991) Biotechnol Bioeng 38:665
222. Kroner KH (1988) Fresenius Z Anal Chem 329:718
223. Kruth HS (1982) Anal Biochem 125:225
224. Kubitschek HE (1969) Methods Microbiol 1:593
225. Kullick T, Beyer M, Henning J, Lerch T, Quack R, Zeitz A, Hitzmann B, Scheper T, Schügerl K (1994) Anal Chim Acta 296:263
226. Kuriyama H, Mahakarnchanakul W, Matsui S, Kobayashi H (1993) Biotechnol Lett 15:189
227. Kwong SCW, Randers L, Rao G (1993) App Environ Microbiol 59:604
228. Lacoursiere A, Thompson BG, Kole MM, Ward D, Gerson DF (1986) Appl Microbiol Biotechnol 23:404
229. Langwost B, Kresbach GM, Scheper T, Ehrat M, Widmer HM (1995) ECB7, Nice, France, poster
230. Larsson G (1990) PhD thesis, Royal Institute of Technology, Stockholm
231. Larsson C, Stockar U von, Marison I, Gustafsson L (1995) Thermochim Acta 251:99
232. Lee C, Lim H (1980) Biotechnol Bioeng 22:639
233. Lenz R, Boelcke C, Peckmann U, Reuss M (1986) In: Johnson, A (ed) Proc 1st Modelling Contr Biotechnol Process, Helsinki, Pergamon Press, Oxford, p 85
234. Li J, Gomez P, Humphrey A (1990) Biotechnol Tech 4:293
235. Liden G, Jacobsson V, Niklasson C (1993) App Biochem Biotechnol 38:27
236. Liebeskind M, Dohmann M (1994) Water Sci Technol 29:7
237. Lindahl B, Gullberg B (1991) Anticancer Research 11:397
238. Lindberg W, Ruzicka J, Christian GD (1993) Cytometry 14:230
239. L'Italien Y, Thibault J, LeDuy A (1989) Biotechnol Bioeng 33:471
240. Litzen A, Garn MB, Widmer HM (1994) J Biotechnol 37:291
241. Litzen A, Wahlund KG (1991) J Chromatogr 548:393
242. Ljunggren E, Karlberg B (1995) J Autom Chem 17:105
243. Lloyd D, Whitmore TN (1988) Lett Appl Microbiol 6:5
244. Locher G, Hahnemann U, Sonnleitner B, Fiechter A (1993) J Biotechnol 29:57
245. Locher G, Hahnemann U, Sonnleitner B, Fiechter A (1993) J Biotechnol 29:75
246. Locher G, Sonnleitner B, Fiechter A (1990) Bioproc Eng 5:181
247. Locher G, Sonnleitner B, Fiechter A (1992) J Biotechnol 25:55
248. Locher G, Sonnleitner B, Fiechter A (1992) J Biotechnol 25:23
249. Locher G, Sonnleitner B, Fiechter A (1992) Proc Contr Qual 2:257
250. Lohmeier-Vogel EM, Hahn-Hägerdahl B, Vogel HJ (1986) Appl Microbiol Biotechnol 25:43
251. Lohmeier-Vogel EM, Hahn-Hägerdal B, Vogel HJ (1995) App Environ Microbiol 61:1414
252. Lohmeier-Vogel EM, Skoog K, Vogel H, Hahn-Hägerdahl B (1989) Appl Environ Microbiol 55:1974
253. Lopez JM, Koopman B, Bitton G (1986) Biotechnol Bioeng 28:1080

254. Lopez JM, Koopman B, Bitton G (1986) J Environ Eng 109:915
255. Lovrien RE, Williams KK, Ferrey ML, Ammend DA (1987) Appl Environ Microbiol 53:2935
256. Lüdi H, Garn MB, Bataillard P, Widmer HM (1990) J Biotechnol 14:71
257. Luong JHT, Carrier DJ (1986) Appl Microbiol Biotechnol 24:65
258. Luong JHT, Volesky B (1982) Eur J Appl Microbiol Biotechnol 16:28
259. Maclaurin P, Worsfold PJ, Norman P, Crane M (1993) Analyst 118:617
260. Maddox J (1994) Nature 368:95
261. Maftah A, Huet O, Gallet PF, Ratinaud MH (1993) Biology of the Cell 78:85
262. Mailinger W, Schütz M, Reuss M (1993) BIOforum 16:336
263. Manz A, Verpoorte E, Effenhauser CS, Burggraf N, Raymond DE, Harrison DJ, Widmer HM (1993) HRC – J High Resolut Chromatogr 16:433
264. Marison I, Stockar U von (1989) Adv Biochem Eng/Biotechnol 40:93
265. Markx GH, Davey CL, Kell DB (1991) J Gen Microbiol 137:735
266. Markx GH, Kell DB (1995) Biotechnol Prog 11:64
267. Matanguihan RM, Konstantinov KB, Yoshida T (1994) Bioprocess Eng 11:213
268. Mattiasson B, Hakanson H (1993) TIBTECH 11:136
269. Mcgown LB, Hemmingsen SL, Shaver JM, Geng L (1995) Appl Spectrosc 49:60
270. McLaughlin JK, Meyer CL, Papoutsakis ET (1985) Biotechnol Bioeng 27:1246
271. Meehan AJ, Eseky CJ, Koretsky AP, Domach MM (1992) Biotechnol Bioeng 40:1359
272. Meierschneiders M, Grosshans U, Busch C, Eigenberger G (1995) Appl Microbiol Biotechnol 43:
273. Melvin BK, Shanks JV (1996) Biotechnol Progr 12:257
274. Memmert K, Wandrey C (1987) Ann NY Acad Sci 506:631
275. Mendes P, Kell DB, Welch GR (1995) In: Brindle K (ed) Enzymology in vivo (Advanced molecular cell biology), vol 11. JAI Press, London, p 1
276. Merchuk JC, Yona S, Siegel MH, Zvi AB (1990) Biotechnol Bioeng 35:1161
277. Merten OW, Palfi GE, Steiner J (1986) Adv Biotechnol Processes 6:111
278. Mesaros JM, Gavin PF, Ewing AG (1996) Anal Chem 68:3441
279. Meschke J, Bennemann H, Herbst H, Dormeier S, Hempel DC (1988) Bioprocess Eng 3:151
280. Meyer C, Beyeler W (1984) Biotechnol Bioeng 26:916
281. Meyer HP, Beyeler W, Fiechter A (1984) J Biotechnol 1:341
282. Meyerjens T, Matz G, Märkl H (1995) App Microbiol Biotechnol 43:341
283. Min RW, Nielsen J, Villadsen J (1995) Anal Chim Acta 312:149
284. Mizutani F, Yabuki S (1994) Biosens Bioelectro 9:411
285. Mizutani F, Yabuki S, Katsura T (1993) Sens Actuators B 13/14:574
286. Moes J, Griot M, Keller J, Heinzle E, Dunn IJ, Bourne JR (1985) Biotechnol Bioeng 27:482
287. Molin G (1983) Eur J Appl Microbiol Biotechnol 18:214
288. Möller J, Hiddessen R, Niehoff J, Schügerl K (1986) Anal Chim Acta 190:195
289. Molt K (1992) GIT 36:107
290. Montague G, Morris AJ (1994) Trends Biotechnol 12:312
291. Montesinos JL, Campmajo C, Iza J, Valero F, Lafuente J, Sola C (1993) Biotechnol Tech 7:429
292. Monzambe KM, Naveau HP, Nyns EJ, Bogaert N, Bühler H (1988) Biotechnol Bioeng 31:659
293. Moskvin LN, Simon J (1994) Talanta 41:1765
294. Muller S, Losche A, Bley T, Scheper T (1995) App Microbiol Biotechnol 43:93
295. Münch T, Sonnleitner B, Fiechter A (1992) J Biotechnol 24:299
296. Munkholm C, Walt DR, Milanovich FP (1988) Talanta 35:109
297. Nagata R, Clark SA, Yokoyama K, Tamiya E, Karube I (1995) Anal Chim Acta 304:157
298. Nagata R, Yokoyama K, Clark SA, Karube I (1995) Biosens Bioelectron 10:261
299. Natarajan A, Boxrud D, Dunny G, Srienc F (1999) J Microbiol Meth 34:223
300. Nestaas E, Wang DIC (1983) Biotechnol Bioeng 25:1981
301. Neubauer P, Haggstrom L, Enfors SO (1995) Biotechnol Bioeng 47:139

302. Nielsen J (1992) Proc Contr Qual 2:371
303. Nielsen J, Johansen CL, Villadsen J (1994) J Biotechnol 38:51
304. Nielsen J, Jorgensen HS (1995) Biotechnol Prog 11:299
305. Nielsen J, Nikolajsen K, Benthin S, Villadsen J (1990) Anal Chim Acta 237:165
306. Nielsen J, Nikolajsen K, Villadsen J (1991) Biotechnol Bioeng 38:1
307. Nielsen J, Villadsen J (1993) In: Rehm HJ, Reed G (eds) Biotechnology, 2nd completely revised edition, vol 3, Bioprocessing. Stephanolpoulos G (vol ed), VCH, Weinheim, p 78
308. Nilsson M, Vijayakumar AR, Holst O, Schornack C, Hakanson H, Mattiasson B (1994) J Ferment Bioeng 78:356
309. Nipkow A, Andretta C, Käppeli O (1990) Chem Ing Tech 62:1052
310. Nipkow A, Sonnleitner B, Fiechter A (1985) Appl Microbiol Biotechnol 21:287
311. Noorman HJ, Luijx GCA, Luyben KCAM, Heijnen JJ (1992) Biotechnol Bioeng 39:1069
312. Oeggerli A, Eyer K, Heinzle E (1995) Biotechnol Bioeng 45:42
313. Oeggerli A, Heinzle E (1994) Biotechnol Prog 10:284
314. Ohashi E, Karube I (1995) J Biotechnol 40:13
315. Ohashi M, Watabe T, Ishikawa T, Watanabe Y, Miwa K, Shode M, Ishikawa Y, Ando T, Shibata T, Kitsunai T, Kamiyama N, Oikawa Y (1979) Biotechnol Bioeng Symp 9:103
316. Oosterkamp AJ, Irth H, Tjaden UR, Vandergreef J (1994) Anal Chem 66:4295
317. Owen VM, Turner APF (1987) Endeavour 11:100
318. Owens JD, Thomas DS, Thompson PS, Timmermann JW (1989) Lett Appl Microbiol 9:245
319. Park JK, Shin MC, Lee SG, Kim HS (1995) Biotechnol Prog 11:58
320. Park SH, Hong KT, Lee JH, Bae JC (1983) Eur J Appl Microbiol 17:168
321. Pasquini C, Defaria LC (1991) J Automatic Chem 13:143
322. Peng WF, Li IIM, Wang E (1994) J Electroanal Chem 375:185
323. Peterson JI, Fitzgerald RV, Buckhold DK (1984) Anal Chem 56:62
324. Pieles U, Zurcher W, Schar M, Moser HE (1993) Nucl Acids Res 21:3191
325. Pih N, Bernardez E de, Dhurjati P (1988) Biotechnol Bioeng 31:311
326. Pohlmann A, Stamm WW, Kusakabe H, Kula MR (1990) Anal Chim Acta 235:329
327. Ponti C (1994) Dissertation, ETH Zürich, No 10504
328. Ponti C, Sonnleitner B, Fiechter A (1995) J Biotechnol 38:173
329. Porro D, Ranzi BM, Smeraldi C, Martegani E, Alberghina L (1995) Yeast 11:1157
330. Porro D, Srienc F (1995) Biotechnol Prog 11:342
331. Porter J, Edwards C, Pickup RW (1995) J Appl Bacteriol 79:399
332. Postgate JR (1969) Methods Microbiol 1:611
333. Preuschoff F, Spohn U, Janasek D, Weber E (1994) Biosens Bioelectron 9:543
334. Puhar E, Einsele A, Bühler H, Ingold W (1980) Biotechnol Bioeng 22:2411
335. Pullen FS, Swanson AG, Newman MJ, Richards DS (1995) Rapid Commun Mass Spectrom 9:1003
336. Pungor E Jr, Schaefer EJ, Cooney CL, Weaver JC (1983) Eur J Appl Microbiol Biotechnol 18:135
337. Rank M, Gram J, Nielsen KS, Danielsson B (1995) App Microbiol Biotechnol 42:813
338. Ranzi BM, Compagno C, Martegeni E (1986) Biotechnol Bioeng 28:185
339. Rao G, Mutharasan R (1989) Appl Microbiol Biotechnol 30:59
340. Reardon KF, Scheper T, Bailey JE (1987) Biotechnol Progr 3:153
341. Reardon KF, Scheper T, Bailey JE (1987) Chem Ing Tech 59:600
342. Rebelo MJF, Compagnone D, Guilbault GG, Lubrano GJ (1994) Anal Lett 27:3027
343. Renner WA, Jordan M, Eppenberger HM, Leist C (1993) Biotechnol Bioeng 41:188
344. Reuss M, Boelcke C, Lenz R, Peckmann U (1987) BTF – Biotech Forum 4:3
345. Ries P (1983) PhD thesis 7214, ETH Zürich
346. Riesenberg D (1991) Curr Opin Biotechnol 2:380
347. Riesenberg D, Schulz V, Knorre WA, Pohl HD, Korz D, Sanders EA, Ross A, Deckwer WD (1991) J Biotechnol 20:17
348. Rizzi M, Theobald U, Baltes M, Reuss M (1993) In: Nienow AW (ed) Bioreactor and bioprocess fluid dynamics. Mechanical Engineering Publishers Ltd, London, p 401

349. Robertson BR, Button DK, Koch AL (1998) Appl Environ Microbiol 64:3900
350. Rodin JB, Lyberatos GK, Svoronos SA (1991) Biotechnol Bioeng 37:127
351. Roessner D, Kulicke WM (1994) J Chromatogr A 687:249
352. Rohner M, Locher G, Sonnleitner B, Fiechter A (1988) J Biotechnol 9:11
353. Rohner M, Meyer HP (1995) Bioprocess Eng 13:69
354. Rothen SA, Saner M, Meenakshisundaram S, Sonnleitner B, Fiechter A (1996) J Biotechnol 50:1
355. Rothen SA, Sauer M, Sonnleitner B, Witholt B (1998) Biotechnol Bioeng 58:92
356. Royce PN (1992) Biotechnol Bioeng 40:1129
357. Rui CS, Sonomoto K, Ogawa HI, Kato Y (1993) Anal Biochem 210:163
358. Ruzicka J (1994) Analyst 119:1925
359. Ruzicka J, Hansen EH (1981) Flow injection analysis. Wiley, New York
360. Ruzicka J, Lindberg W (1992) Anal Chem 64:A537
361. Sakato K, Tanaka H, Samejima H (1981) Ann NY Acad Sci 369:321
362. Salmon JM (1987) Biotechnol Tech 1:7
363. Samson R, Beaumier D, Beaulieu C (1987) J Biotechnol 6:175
364. Sandmeier EP, Dunn IJ, Bourne JR, Heinzle E (1987) Eur Congr Biotechnol 3:137
365. Santos H (1995) ECB7, Nice, France, MAC65
366. Sattur AP, Karanth NG, Divakar S (1988) Biotechnol Tech 2:73
367. Schallinger LE, Gray JE, Wagner LW, Knowlton S, Kirkland JJ (1985) J Chromatogr 342:67
368. Scheper T (1991) Bioanalytik. Vieweg, Braunschweig
369. Scheper T, Hitzmann B, Rinas U, Schügerl K (1987) J Biotechnol 5:139
370. Scheper T, Hoffmann H, Schügerl K (1987) Enzyme Microb Technol 9:399
371. Scheper T, Lorenz T, Schmidt W, Schügerl K (1987) Ann NY Acad Sci 506:431
372. Scheper T, Schügerl K (1986) J Biotechnol 3:221
373. Schill N, Stockar U von (1995) Thermochim Acta 251:71
374. Schmid RD, Künnecke W (1990) J Biotechnol 14:3
375. Schmidt E (1988) Appl Microbiol Biotechnol 27:347
376. Schmitz A, Eberhardt R, Spohn U, Weuster-Botz D, Wandrey C (1992) DECHEMA Biotechnology Conferences 5:1117
377. Schneider M, Marison IW, Stockar U von (1994) Enzyme Microb Technol 16:957
378. Schügerl K (1991) Analytische Methoden in der Biotechnologie. Vieweg, Braunschweig
379. Schügerl K, Hitzmann B, Jurgens H, Kullick T, Ulber R, Weigal B (1996) Trends Biotech 14:21
380. Schulze U, Larsen ME, Villadsen J (1995) Anal Biochem 228:143
381. Schuppenhauer MR, Kühne G, Tiefenauer L, Smala A, Dunn IJ (1994) 13th Meeting European Society Animal Cell Technology (ESACT), Veldhoven, NL, p 5
382. Scudder KM, Christian GD, Ruzicka J (1993) Exp Cell Res 205:197
383. Shimizu H, Miura K, Shioya S, Suga K (1995) Biotechnol Bioeng 47:165
384. Shiraishi F (1994) Enzyme Microb Technol 16:349
385. Shu HC, Hakanson H, Mattiasson B (1995) Anal Chim Acta 300:277
386. Shulga AA, Gibson TD (1994) Anal Chim Acta 296:163
387. Silman RW (1988) Abstr Pap Am Chem Soc 196 Meeting, MBTD81
388. Simutis R, Havlik I, Dors M, Lübbert A (1993) Proc Contr Qual 4:211
389. Simutis R, Havlik I, Lübbert A (1993) J Biotechnol 27:203
390. Simutis R, Havlik I, Schneider F, Dors M, Lübbert A (1995) CAB6, Garmisch-Partenkirchen May 1995, preprints, p 59
391. Smeraldi C, Berardi E, Porro D (1994) Microbiology-UK 140, part 11, p 3161
392. Smolenski W, Suflita JM (1987) J Microbiol Methods 6:71
393. Sonderhoff SA, Kilburn DG, Piret JM (1992) Biotechnol Bioeng 39:859
394. Sonnleitner B (1989) BTF – Biotech Forum 6:156
395. Sonnleitner B (1991) Bioprocess Eng 6:187
396. Sonnleitner B (1991) Antonie v Leeuwenhoek Int J Gen Molec Microbiol 60:133

397. Sonnleitner B (1993) In: Mortensen U, Noorman HJ (eds) Bioreactor performance. Biotechnology Research Foundation, Lund, Sweden, p 143
398. Sonnleitner B (1996) Adv Biochem Eng/Biotechnol 54:155
399. Sonnleitner B (1998) J Biotechnol 65:47
400. Sonnleitner B, Bomio M (1990) Biodegradation 1:133
401. Sonnleitner B, Fiechter A (1988) Anal Chim Acta 213:199
402. Sonnleitner B, Fiechter A (1992) Adv Biochem Eng/Biotechnol 46:143
403. Sonnleitner B, Fiechter A, Giovannini F (1984) Appl Microbiol Biotechnol 19:326
404. Sonnleitner B, Hahnemann U (1994) J Biotechnol 38:63
405. Sonnleitner B, Rothen SA, Kuriyama H (1997) Prog Biotechnol 13:8
406. Soper SA, Legendre BL, Williams DC (1995) Anal Chem 67:4358
407. Spinnler HE, Bouillanne C, Desmazeaud MJ, Corrieu G (1987) Appl Microbiol Biotechnol 25:464
408. Srienc F, Campbell JL, Bailey JE (1986) Cytometry 7:132
409. Srienc F, Dien BS (1992) Biochemical Engineering VII Ann of NY Acad Sci 665:59
410. Srinivas SP, Mutharasan R (1987) Biotechnol Letters 9:139
411. Srinivas SP, Mutharasan R (1987) Biotechnol Bioeng 30:769
412. Srinivasan N, Kasthurikrishnan N, Cooks RG, Krishnan MS, Tsao GT (1995) Anal Chim Acta 316:269
413. Stephanopoulos G, Locher G, Duff M (1995) In: Munack A, Schügerl K (eds) CAB6 preprints. Garmisch-Partenkirchen, D:195
414. Stockar U von, Birou B (1989) Biotechnol Bioeng 34:86
415. Stockar U von, Marison IW (1989) Adv Biochem Eng/Biotechnol 40:93
416. Stockar U von, Marison IW, Birou B (1988) 1st Swiss-Japanese joint meeting on bioprocess development, Interlaken, Switzerland
417. Stöcklein W, Schmid RD (1990) Anal Chim Acta 234:83
418. Strässle C, Sonnleitner B, Fiechter A (1989) J Biotechnol 9:191
419. Strohhacker J, Degraaf AA, Schoberth SM, Wittig RM, Sahm H (1993) Arch Microbiol 159:484
420. Suda M, Sakuhara T, Karube I (1993) Appl Biochem Biotechnol 41:3
421. Suda M, Sakuhara T, Murakami Y, Karube I (1993) Appl Biochem Biotechnol 41:11
422. Suhr H, Wehnert G, Schneider K, Bittner C, Scholz T, Geissler P, Jahne B, Scheper T (1995) Biotechnol Bioeng 47:106
423. Suleiman AA, Villarta RL, Guilbault GG (1993) Anal Lett 26:1493
424. Suzuki T, Yamane T, Shimizu S (1986) J Ferment Technol 64:317
425. Svendsen CN (1993) Analyst 118:123
426. Takiguchi N, Shimizu H, Shioya S (1997) Biotechnol Bioeng 55:170
427. Tallarek U, vanDusschoten D, VanAs H, Guiochon G, Bayer E (1998) Angew Chem Int Ed Engl 37:1882
428. Tanaka H, Aoyagi H, Jitsufuchi T (1992) J Ferment Bioeng 73:130
429. Taya M, Hegglin M, Prenosil JE, Bourne JR (1989) Enzyme Microb Technol 11:170
430. Tellier C, Guillou-Charpin M, Grenier P, Botlan D le (1989) J Agric Food Chem 37:988
431. Thatipamala R, Rohani S, Hill GA (1994) J Biotechnol 38:33
432. Thatipamala R, Rohani S, Hill GA (1991) Biotechnol Bioeng 38:1007
433. Theobald U, Mailinger W, Reuss M, Rizzi M (1993) Anal Biochem 214:31
434. Thomanetz E (1982) Stuttgarter Berichte zur Siedlungswasserwirtschaft 74
435. Thomas DC, Chittur VK, Cagney JW, Lim HC (1985) Biotechnol Bioeng 27:729
436. Tservistas M, Weigel B, Schügerl K (1995) Anal Chim Acta 316:117
437. Ukeda H, Wagner G, Bilitewski U, Schmid RD (1992) J Agric Food Chem 40:2324
438. Urban G, Jobst G, Aschauer E, Tilado O, Svasek P, Varahram M, Ritter C, Riegebauer J (1994) Sens Actuators B 18–19:592
439. Uttamlal M, Walt DR (1995) BioTechnology 13:597
440. Vaija J, Lagaude A, Ghommidh C (1995) Antonie v Leeuwenhoek Int J Gen Molec Microbiol 67:139
441. Valero F, Lafuente J, Poch M, Sola C (1990) App Biochem Biotechnol 24/25:591

442. vanDam K, Jansen N (1991) Antonie v Leeuwenhoek Int J Gen Molec Microbiol 60:209
443. vanDam K, Vandervlag J, Kholodenko BN, Westerhoff HV (1993) Eur J Biochem 212:791
444. vandeMerbel NC, Kool IM, Lingeman H, Brinkman UAT, Kolhorn A, Derijke LC (1992) Chromatographia 33:525
445. van de Merbel NC, Zuur P, Frijlink M, Holthuis JJM, Lingeman H, Brinkman UAT (1995) Anal Chim Acta 303:175
446. van der Pol JJ, Spohn U, Eberhardt R, Gaetgens J, Biselli M, Wandrey C, Tramper J (1994) J Biotechnol 37:253
447. van Kleeff BHA (1995) Thermochimica Acta 251:111
448. van Kleeff BHA, Kuenen JG, Heijnen JJ (1993) Biotechnol Bioeng 41:541
449. van Staden JF (1995) Fresenius J Anal Chem 352:271
450. Wahlund KG, Litzen A (1989) J Chromatogr 461:73
451. Wang HY, Li XM (1989) Biosensors 4:273
452. Wang J, Chen L, Chicharro M (1996) Anal Chim Acta 319:347
453. Wang NS, Simmons MB (1987) Abstr Pap Am Chem Soc 194 Meet MBTD84
454. Weaver JL (1998) In: Jaroszeski MJ, Heller R (eds) Flow cytometry protocols, vol 91. p 77
455. Webb C, Kamat SP (1993) World J Microbiol Biotechnol 9:308
456. White DC, Bobbie RJ, Herron JS, King JD, Morrison SJ (1979) In: Costerton JW, Colwell RR (eds) Native aquatic bacteria, enumeration, activity, and ecology, ASTM STP 695. American Society for Testing and Materials, Philadelphia, p 69
457. Whitmore TN, Jones G, Lazzari M, Lloyd D (1987) Mass Spectrom Biotechnol Process Anal Contr:143
458. Wilson DF, Westerhoff HV van (1982) TIBS 7:275
459. Wilson PDG (1987) Biotechnol Tech 1:151
460. Winter EL, Rao G, Cadman TW (1988) Biotechnol Tech 2:233
461. Woldringh CL, Huls PG, Vischer NOE (1993) J Bacteriol 175:3174
462. Wolfbeis OS (1987) GBF Monograph 10, Biosensors Int Workshop, p 197
463. Wolfbeis OS, Li H (1993) Biosen Bioelectron 8:161
464. Worsfold PJ (1994) J Autom Chem 16:153
465. Wu P, Ozturk SS, Blackie JD, Thrift JC, Figueroa C, Naveh D (1995) Biotechnol Bioeng 45:495
466. Wu X, Bellgardt KH (1998) J Biotechnol 62:11
467. Wu XA, Bellgardt KH (1995) Anal Chim Acta 313:161
468. Wyatt PJ (1973) Methods Microbiol 8:183
469. Xie B, Mecklenburg M, Danielsson B, Ohman O, Norlin P, Winquist F (1995) Analyst 120:155
470. Xie XF, Suleiman AA, Guilbault GG (1992) Biotechnol Bioeng 39:1147
471. Yamane T (1993) Biotechnol Prog 9:81
472. Yamane T, Hibino W, Ishihara K, Kadotani Y, Kominami M (1992) Biotechnol Bioeng 39:550
473. Yang JD, Wang NS (1993) Appl Biochem Biotechnol 42:53
474. Yano Y, Nakayama A, Yoshida K (1995) Appl Environ Microbiol 61:4480
475. Yao T, Matsumoto Y, Wasa T (1990) J Biotechnol 14:89
476. Ye BC, Li QS, Li YR, Li XB, Yu JT (1995) J Biotechnol 42:45
477. Yegneswaran PK, Gray MR, Thompson BG (1990) Biotechnol Bioeng 36:92
478. Zabriskie DW, Humphrey AE (1978) Enzyme Microb Technol 35:337
479. Zhang F, Scully PJ, Lewis E (1994) Institute of Physics Applied Optics and Optoelectronics Conference, York, AS13.1, p 173
480. Zhao R, Natarajan A, Srienc F (1999) Biotechnol Bioeng 62:609
481. Zhi ZL, Rios A, Valcarcel M (1996) Anal Chim Acta 318:187
482. Zumbusch PV, Meyerjens T, Brunner G, Märkl H (1994) Appl Microbiol Biotechnol 42:140

Electronic Noses for Bioreactor Monitoring

Carl-Fredrik Mandenius

Department of Physics and Measurement Technology; Linköping University,
S-581 83 Linköping, Sweden
E-mail: cfm@ifm.liu.se

Electronic noses provide new possibilities for monitor the state of a cultivation non-invasively in real-time. The electronic nose uses an array of chemical gas sensors that monitors the off-gas from the bioreactor. By taking advantage of the off-gas components' different affinities towards the sensors in the array it is possible with the help of pattern recognition methods to extract valuable information from the culture in a way similar to the human nose. For example, with artificial neural networks, metabolite and biomass concentration can be predicted, the fermentability of a medium before starting the fermentation estimated, and the growth and production stages of the culture visualized. In this review these and other recent results with electronic noses from monitoring microbial and cell cultures in bioreactors are described.

Keywords. Artificial nose, Chemical multisensor array, Artificial neural network, Bioprocess monitoring, Bioprocess control

1 Introduction 66

2 Methodology 67

2.1 Sensor Types 67
2.2 Sensor Array Systems 69
2.3 Bioreactor Interface 69
2.4 Evaluation Methods 71

3 Application to Bioreactor Monitoring 73

3.1 Prediction of Metabolites and Other State Variables 74
3.2 Prediction of Media Quality 77
3.3 Visualization of Bioreactor State 77

4 Conclusion 81

References 81

List of Symbols and Abbreviations

ANN   artificial neural network
CHO   Chinese hamster ovary
CP   conductive polymer

hGH	human growth hormone
hCAII	human carboanhydrase II
hFVIII	human blood coagulation factor VIII
MOS	metal oxide semiconductor gas sensor
MOSFET	metal oxide semiconductor field effect transistor
PCA	principal component analysis
PLS	partial least square
RMSE	root mean square error
QCM	quartz crystal microbalance

1
Introduction

The quality of bioprocess products could be improved substantially by using non-invasive real-time monitoring methods that could provide additional and new information on the state of the culture in the bioreactor. The electronic nose technique provides such an opportunity. The concept of electronic nose was introduced in 1982 by Persaud and Dodd [1]. They demonstrated how an array of chemical gas sensors could be used to improve the capacity of analysis by taking advantage of the redundancy of information that is enclosed in the responses from several sensors, all slightly different in affinity to the analyte(s). Furthermore, they analyzed the responses by using the powerful artificial neural networks (ANN) method allowing non-linear modeling of the complex response pattern of the sensor array with training from known analyte concentrations. In this way their system operated similarly to the human olfactory sensory organs with their molecular receptors linked to neurons that interpret the signals on higher perception level, thereby justifying the term "electronic nose".

Subsequently, other researchers developed the electronic nose idea with a variety of chemical gas sensor arrays using different pattern recognition techniques for improving the interpretation of responses [2–5].

The electronic nose arrays have today successfully been used in a vast number of applications. The electronic noses are particularly appealing in food analysis since they resemble the traditional way of controlling the quality of foodstuffs. The electronic nose is already applied as a complement to sensory test panels in the food industry for product quality classification. Examples of applications are classification of grains [6] and beer [7]. Examples of other consumables tested are tobacco [8] and perfumes [9]. Applications in other areas such as environmental control and pulp and paper quality are also reported in over five hundred references currently found in the literature on electronic noses.

Although it would seem reasonable to apply the electronic noses for measuring biological variables such as metabolic products and microbial activity, only a limited number of studies have been described. Of these, the following can be mentioned: detection of infection bacteria activity in ulcers [10], microbial contamination in meat [11], classification of microbial strains [12, 13], and monitoring of bioreactors [14].

In this review, it is described how the electronic noses are applied to on-line bioreactor monitoring for meeting the requirement of non-invasive real-time measurement and facilitating rapid and safe data generation from microbial and cell cultivations.

2
Methodology

Particularly when applied to bioreactor monitoring the electronic noses require high quality sensors with long and stable operation, efficient signal processing and evaluation methods, and practical interfacing with the bioreactor. The currently used sensor types and evaluation methods are summarized below.

2.1
Sensor Types

A variety of chemical gas sensors are or could be used in electronic nose instruments. So far, successful results have been reached with conductive polymer (CP) sensors, metal oxide semiconductor (MOS) sensors, metal oxide semiconductor field effect transistor (MOSFET) sensors, quartz crystal microbalance (QCM) sensors, and infrared sensors.

The MOS sensors [15] are based on reactions taking place on electrochemically active metal oxide surfaces (e.g. tin oxide, copper oxide). The selectivity and sensitivity of the surface are modulated by oxide composition, amounts of trace elements, e.g. palladium, gold or rhodium, and operating temperatures, normally in the 100–400 °C range. The MOS sensors exhibit varying sensitivity towards organic compounds and the analyte affinity to the surface is more or less affected by its chemical properties. Examples of typical analytes to which selectivity can be attained are hydrogen, carbon monoxide, ammonia, hydrogen sulfide, nitrogen oxide, sulfurous compounds, alcohols, and hydrocarbons. Sensitivity is in the 10–10,000 ppm range depending on the analyte. The MOS sensors are manufactured by several commercial vendors for diverse practical purposes such as smoke alarms, methane detection in mines, and combustion control (e.g. Figaro Inc., Japan; Capture Ltd., UK; F I S Inc., Japan).

By covering a silicon oxide field effect transistor with a thin film of a catalytic metal, analytical properties of the transistor device can be obtained [16]. A number of organic and inorganic compounds can be catalytically decomposed on such a surface resulting in dipoles that cause shifts in the capacitance of the semiconductor device. The selectivity of the MOSFET sensors is modulated by using different catalytic metals, e.g. palladium, iridium, platinum, thereby rendering them sensitive to hydrogen, ammonia, alcohols, aldehydes, sulfides, and other degradable molecules. Typical range of sensitivity is 1–1000 ppm.

CP sensors [17] exploit the electrochemical properties of compounds such as polypyrrole and polyindole. Films of the polymers are deposited on electrical conductor components. When analyte molecules are absorbed into the film the conductivity changes. The activity of CP films is fine-tuned by derivatizing the polymer with different functional groups thereby rendering the CP sensors selective to numerous chemical compounds. Films of CP such as poly(N-methylpyrrole), poly(5-carboxyindole), and poly(3-methylthiophene) have been shown to discriminate between electron-rich and electron-deficient species. Hundreds of different CPs have been synthesized and used in electronic noses.

The QCM sensors [5] have gained popularity because they are operated at room temperature and allow combinations of selective layers different to those of MOS, MOSFET and PC sensors. The QCM measures physical mass of the analyte by recording the change in frequency of a quartz crystal when the analyte binds to it. Layers of gas chromatographic stationary phases and natural or synthetic lipids can discriminate between alcoholic drinks, perfume and flavor odorants.

In a few cases, small optical infrared absorption monitors have been integrated into electronic nose sensors [11], mostly for detection of the carbon dioxide evolution from cells. The 3000–4000 nm filters are normally used.

Figure 1 depicts the frequently used sensors in electronic noses.

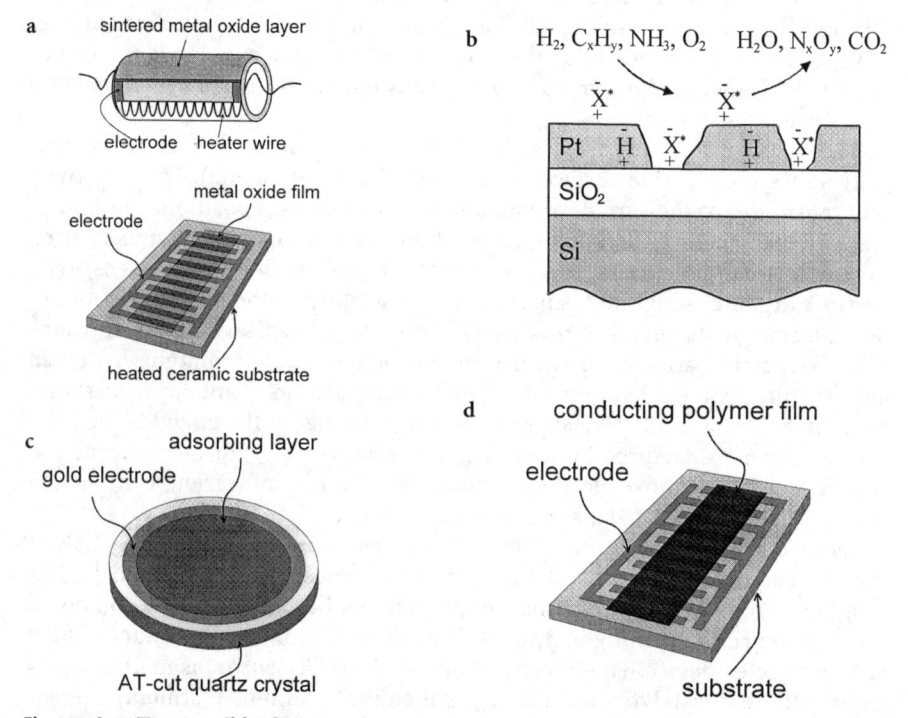

Fig. 1a–d. a Two possible designs of metal oxide semiconductor (MOS) sensors where the sensitive metal oxide is either applied on a heated wire or on a heated ceramic substrate; b a typical design of a MOSFET sensor with a silicon substrate covered by a thin layer of silicon oxide and on top of that a thin layer of catalytic metal, e.g. palladium; c a quartz crystal microbalance (QCM) sensor design with a 10 mm quartz oblate to which a gold electrode is attached that is coated with a selective adsorbing layer; d the design of a conducting polymer (CP) sensor where a integrated circuit with electrode components is coated with a thin polymer film

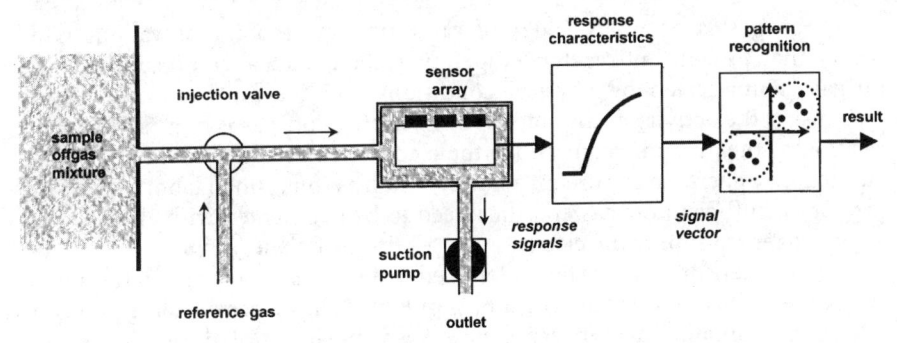

Fig. 2. The principle configuration of an electronic nose system where the analyte mixture is contacted with a chemical sensor array that produces raw data which subsequently are treated with a pattern recognition algorithm that delivers the predicted result

2.2
Sensor Array Systems

The sensors of the electronic nose are assembled in an array. The array is normally a small electronic unit that integrates the different sensors into a practical circuit card or another appropriate system that is easy to insert into the electronic nose instrument. If the array is to be used in a flow injection setup the unit also comprises a flow cell compartment with minimal volume. The system depicted in Fig. 2 shows how MOS and MOSFET arrays are integrated in a flow injection system [11]. Larger arrays can be integrated into silicon chips, as described for CP sensors where, for example an ASIC chip with 32 sensors has been fabricated with BiCMOS technology and having an area of 7×7 mm [18]. If the array is be inserted in the headspace volume of a bioreactor, the technical solution is a remote array probe that can be placed in a gas sample container [19].

An advantage of the array arrangement is that all sensors can be in operation during monitoring and subsequently be selected at the mathematical evaluation based on selection methods described below. If a signal is not useful for the pattern recognition procedure, it is simply disregarded in the calculation.

Today, several companies sell commercial electronic nose systems with their favorite sensor configurations [20-22]. The commercial systems have the drawback that the types of sensors used in the array cannot be changed. If these configurations are not the appropriate ones for the analytes to be measured, it becomes necessary to combine different commercial instruments. Alternatively, a research instrument may be used.

2.3
Bioreactor Interface

Liquid samples can be collected from the bioreactor sampling port and introduced into the electronic nose instrument manually. More appealing in bioprocessing is to sample on-line. An electronic nose system monitors non-invasively by sampling from the off-gas port of the bioreactor [23]. The humidity of the off-gas is normally 100% of saturation in aerobic cultivations. This means that, at constant reactor temperature, the chemical composition of the off-gas is unaffected by physical conditions and the only cause of sensor response is the activity of the culture. The supply of off-gas is quite sufficient in relation to the amount required for the electronic nose analysis in an aerobic culture. An anaerobic culture may however, at least on a small laboratory scale, produce insufficient off-gas and thus need to be augmented with an inert gas, e.g. nitrogen, to supply the electronic nose with sufficient sample.

A typical sensor array system interfaced to a bioreactor representative of the studies described in this paper is shown in Fig. 3. The bioreactor off-gas is conducted to a container by its overpressure. At sample injection the gas in the container is withdrawn by a suction pump placed in the electronic nose instrument. The sample gas passes the sensor array, which is distributed over three serially coupled units. In the example in Fig. 3 the first unit contains ten MOSFET sensors, the second six MOS sensors and the third unit an infrared sensor. After injection, a valve is switched to a reference carrier gas, taken from the ambient air or from a gas flask with a controlled composition.

The sample gas is injected into the reference, or carrier, gas flow. The reference needs to have the same composition as the sample with regard to humidity and chemical background in order to give responses which only reflect the bioreactor composition. Thus, the reference gas has to be humidified

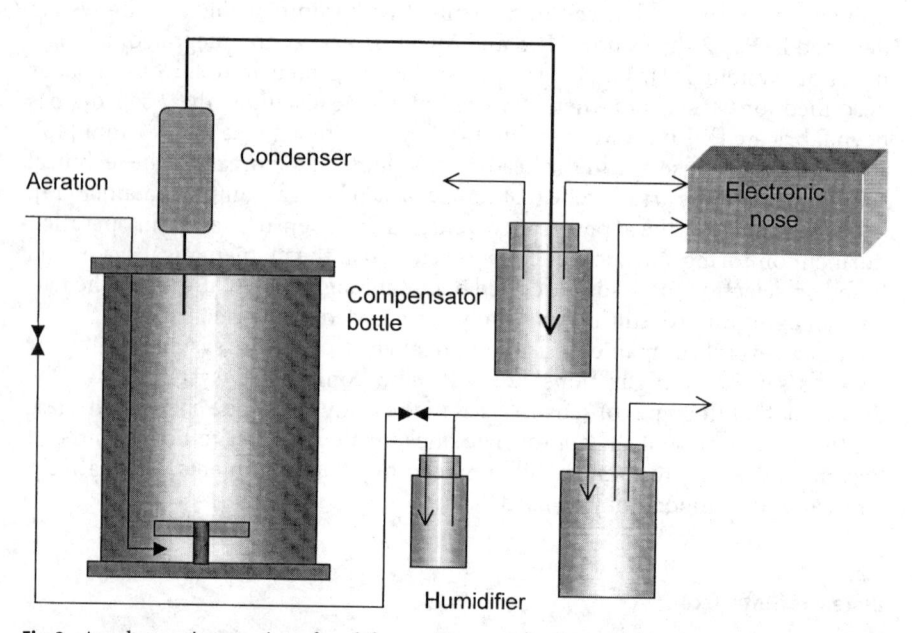

Fig. 3. An electronic nose interfaced for on-line monitoring of a bioreactor. The electronic nose unit is preceded by a compensator bottle from which the necessary analyte gas volume is sampled. The bottle serves as a compensator for minor variation in the gas composition and a liquid trap. Reference gas in sampled from the aeration inlet to the bioreactor with the same type of compensator bottle before the nose unit

before being mixed with the sample gas. This is done with a humidifier as shown in Fig. 3.

During bioreactor monitoring regular recalibration or checking of performance is necessary.

2.4
Evaluation Methods

The response signals from the array when arranged in a flow injection system have each a general curve shape as shown in Fig. 4. The signals' rise and fall time and their plateau values contain information specific to the analyte or analyte mixture. Integration and derivation of the curve add more information. Thus, one injection produces at least five parameters that can be used for the prediction. An array with, for example twenty sensors, consequently generates one hundred response parameters.

One of the best methods of analyzing the responses from the array is ANN [24]. The common back-propagation ANN with one input per sensor variable, one hidden layer and one or more output nodes for the predicted parameter(s) are mostly used. A sigmoidal transfer function is applied on the input responses:

$$y_i = -\left[1 + \exp\left(\sum_{i=1}^{n} x_i w_{ij} + \theta_j\right)\right]^{-1} \qquad (1)$$

where y_i is the output signal, x_i an input signal, w_{ij} a weight coefficient and θ a

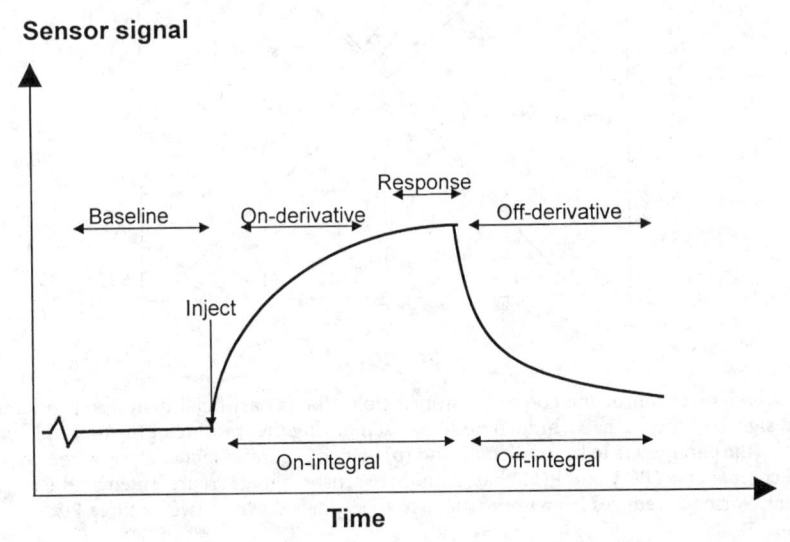

Fig. 4. Electronic nose signal in flow injection instrumentation. The *arrow* (inject) indicates when the sample is injected into the array. The on-integral and off-integral represent chosen times over which the rise of the signal is integrated, and the on-derivative and off-derivative chosen times where the signal rise and fall are derivated. The response is the plateau value of the response. All of the signal values are mean values of at least 20 data points

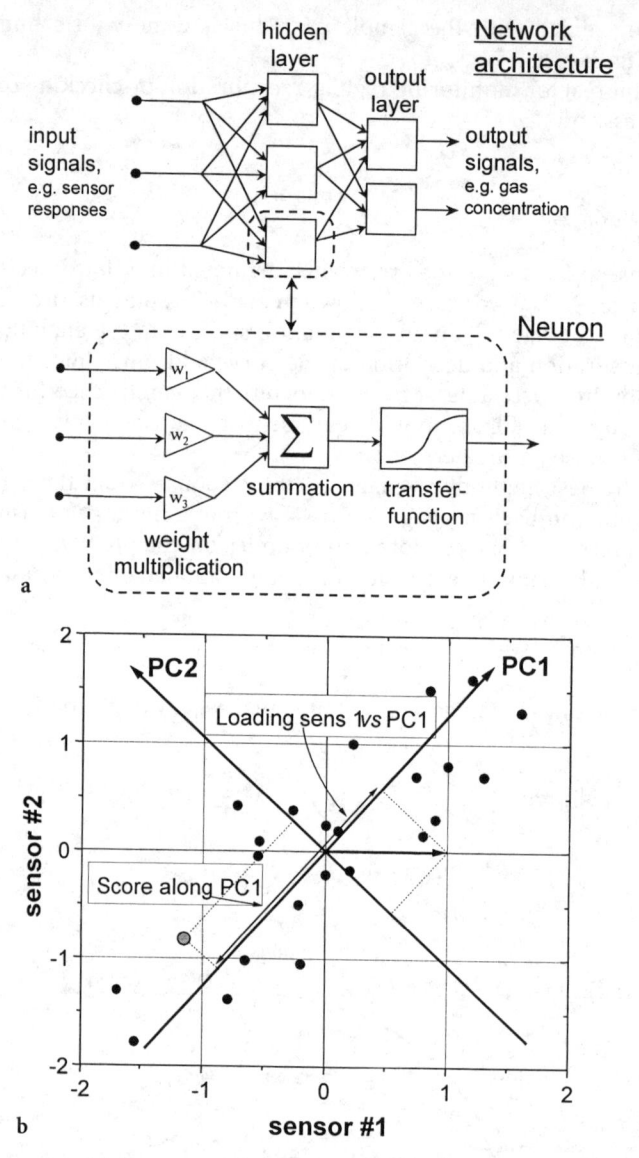

Fig. 5a, b. Illustration of the computation principles for (a) artificial neural networks with input signals (array signals), hidden nodes chosen during the training of the net, and output signals (the parameters to be predicted); and (b) principal component analysis with two principal components (PC 1 and PC 2) based on three sensor signals (represented by the x, y and z axes). Normally reduces from approximately 100 signals down to two or three PCs

threshold value of neuron j. The topology of the back-propagation network is based on a multilayer feed forward network, where the input layer receives information from the sensor array, the output layer provides information to the user and one or more hidden layers connect the input and output layers with different weights (Fig. 4). In the training (or learning calibration) process weights are changed according to an error feedback method which first updates the values of all neurons corresponding to input data based on current weights and then determines new weights my minimizing the output error. An update algorithm, such as Levenberg-Marquardt, is useful:

$$w_{new} = w_{old} - (Z^T Z + \lambda I)^{-1} Z^T \varepsilon(w_{old}) \qquad (2)$$

where λ governs the step size, ε is an output error vector and Z the Jacobean matrix of derivatives of each error with respect to each weight w.

Also very useful for evaluating the electronic nose response pattern is PCA [25]. PCA is a linear supervised pattern recognition technique that can be used to reduce the dimensionality of the array data. The calculation results in principal components, where the nth principal component describes the direction of the nth largest variation in the array data set. The result is often presented in a plot where the two or three most important principal components are plotted against each other, either as a score plot describing how the samples relate to each other or as a loading-plot describing the relation between the variables. In addition, PCA can be used for regression analysis of the array data. Standard algorithms for PCA are included in many software packages, for example the MATLAB chemometric toolbox.

Partial least square (PLS) regression has been used with some success for identifying electronic nose responses [26]. Comparison between the above methods seems to favor ANN but is case dependent.

In many cases, the number of response variables from the array needs to be reduced. There can be several reasons for making a variable selection. Elimination of irrelevant variables can improve the prediction performance, fewer variables give statistically better prediction models, and some of the variables might have a high correlation which can cause problems in the model calculation. Depending on the problem and the prediction model type, different types of variable selection schemes can be applied. One method that has been successfully applied to several gas sensing problems is the forward selection procedure, which is especially well suited for a feed-forward ANN [27]. In order to find the best sensors to use for a particular problem, variables are added one at a time based on their contribution to the error from a multi-linear regression model [27]. The method can also be applied to select suitable variables to use in conjunction with other data processing tools, e.g. PCA or PLS [25].

3
Application to Bioreactor Monitoring

How can the off-gas emission from a culture in a bioreactor provide any useful information about culture state? Under favorable conditions, an experienced practitioner can identify the type and stage as well as detect contamination with his own nose from the odor of a culture. The electronic nose could thus be expected to provide the same type of information if its sensors are at least as sensitive as the human nose and the integrated array system has the same ability to learn by training. By applying pattern recognition methods in a smart way to the responses even more information can be extracted. Either the electronic nose can (1) identify defined chemical components contained in the off-gas, such as ethanol, acetic acid or volatile fatty acids, using ANN prediction or other pattern recognition techniques; or (2) relate soluble components in the medium, for example biomass, to volatiles in the off-gas that are in proportion to the component to be predicted; or (3) predict variables where the volatile compounds in the off-gas in some more indirect way can be modeled by ANN or other methods. In addition, the pattern recognition methods can be used for identifying stages or conditions in the cultivation irrespectively of considering the components per se. In the following paragraphs examples of these types of application are given.

3.1
Prediction of Metabolites and Other State Variables

The ethanol concentration in the medium of a *Saccharomyces cerevisiae* cultivation can be monitored from its content in the gas phase by directly recording the current from a chemical MOS sensor [28]. The accuracy of such a measurement was significantly improved by using an electronic nose with five sensors in the array and recognizing the response pattern with ANN [29, 30]. The sensors were a combination of MOS and MOSFET sensors selected from a PCA loading plot. Data sets from three cultivations were used to train the ANN. When the trained net was applied on new cultivations the ethanol was predicted with a mean square error (RMSE) of 4.6% compared to the off-line determined ethanol (Fig. 6). With only one sensor the RMSE was 18%.

In a subsequent study an extended electronic nose with twenty sensors in the array (MOS, MOSFET, IR) monitored a 34-h batch cultivation of *S. cerevisiae* in a semisynthetic medium [31, 32]. The bioreactor was also monitored with an on-line high performance liquid chromatography (HPLC) system measuring the content of glucose, acetate, acetaldehyde and glycerol. Biomass was measured off-line with dry-weight sampling. The responses from the electronic nose from repeated cultivations were used for training ANNs versus the HPLC and dry weight biomass values. Each component required a separate ANN for attaining the best accuracy. Up to six training cultivations were used in the training and subsequently validated on one or two new cultivations. The RMSE for ethanol and biomass was 2.7–4%. These components appeared in the cultivation at relatively high concentrations and where smelled either directly (ethanol) or as a result of by-products (biomass). The non-odorant glucose was however also predicted with good accuracy (1.7–3.9%) probably because it is directly related with the biomass and ethanol emission. Components present at much lower concentrations (approximately twenty-five-fold less) were predicted as well. Acetaldehyde was predicted with an accuracy of 3.5% and acetate and glycerol with 8.3–10% (Table 1).

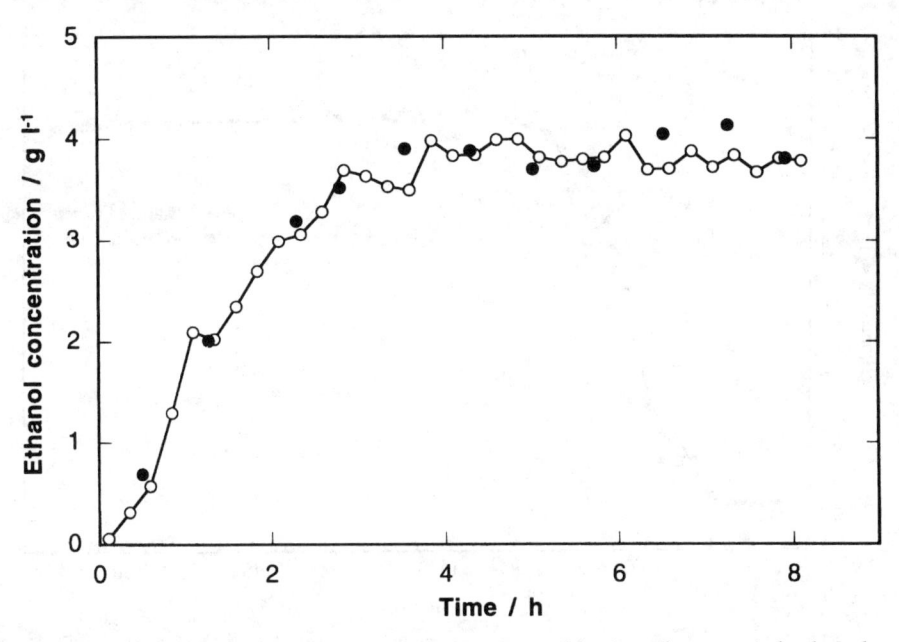

Fig. 6. Prediction of the ethanol concentration in a yeast cultivation. The *open circles* show the predicted ethanol concentration using a three layer ANN. The *filled circles* show off-line analyses of ethanol using a reference method (from [29] with permission from Elsevier)

Table 1. Prediction of components in a *S. cerevisiae* culture using ANN (from [31] with permission)

State variable	Accuracy [%]
Ethanol	2.7–5.2
Biomass	1.5–3.8
Glucose	1.7–3.9
Glycerol	10
Acetate	8.3
Acetaldehyde	3.5

A generically different microorganism, an *Escherichia coli* producing recombinant human carboanhydrase II (hCAII), was monitored with the same electronic nose as above [23]. The biomass concentration was predicted from nine of the sensors after training in four batch cultivations. The induction of hCAII did not affect the result of the training. The validation was exhibiting an RMSE of 1.46% compared to the dry weight measurements of cell mass (Fig. 7). The odor from *E. coli* was easily identified by the experimenter although an accurate estimation of the quantitative value cannot be done. The specific growth rate was also predicted but from another selection of sensor responses in the same cultivations. The prediction was slightly less accurate with 2.8%.

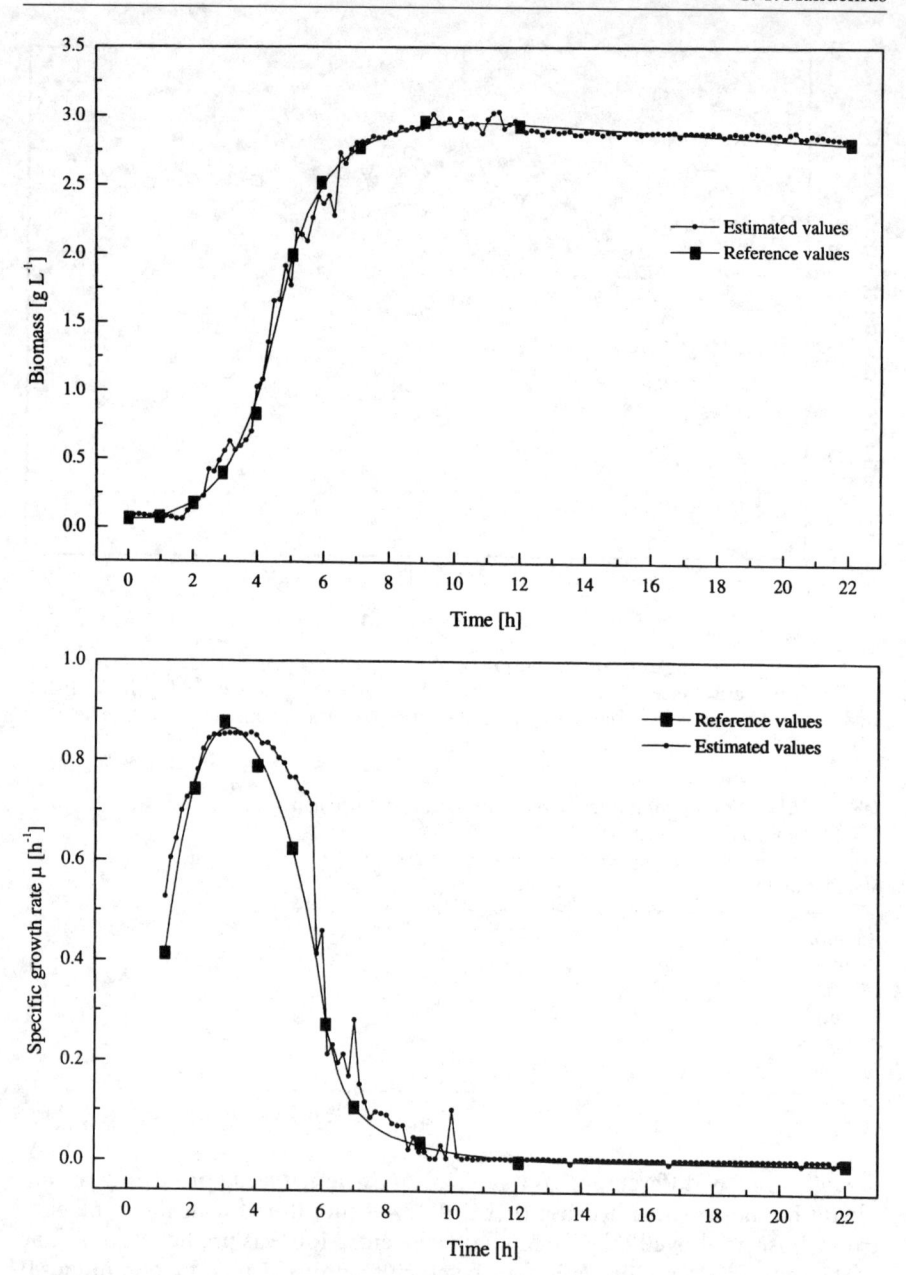

Fig. 7. Prediction of biomass and specific growth rate in a recombinant *E. coli* cultivation producing CAII. The *upper diagram* shows prediction of the biomass value predicted with a 9-8-1 ANN compared to dry weight values. The *lower diagram* shows prediction of the specific growth rate with a 12-7-1 ANN compared with calculated values from the dry weight values (from [23] with permission from Elsevier)

Monitoring of large-scale fed-batch manufacture of baker's yeast was also possible with the electronic nose [33]. The cultivation took place in a 200-m^3 bubble-column reactor. The monitoring procedure is complicated by the large phase variation and circulation times in the bioreactor. On the 200-m^3 scale, ethanol and biomass were predicted but with lower accuracy than in the laboratory (10%). The data was compensated for increasing reactor liquid volume and aeration rate during the fed-batch cycle, simply by including these variables in the inputs to the ANN.

3.2
Prediction of Media Quality

The quality of a complex growth medium is decisive for high growth rate and product yield of a cultivation. A common ingredient in complex recombinant *E. coli* media is casein hydrolysate, whose quality is known to influence the performance of the cultivation considerably. This natural source shows considerable lot-to-lot variability due to source, treatment and sterilization. The electronic nose has been successful in discriminating casein hydrolysate lots with favorable quality for growth of a recombinant *E. coli* from hydrolysate lots with poor quality [34]. The variability was detected using an electronic nose with twelve CP sensors and where the response profiles were analyzed with two-dimensional PCA resulting in distinct classification of the good and bad quality hydrolysate lots.

The electronic nose has also been applied to prediction of the quality of lignocellulose hydrolysates from pine, spruce, birch, and aspen for production of ethanol with *S. cerevisiae* [35]. During acid hydrolysis of wood, substances, such as furfural, are formed which inhibit the yeast cells' production capacity of ethanol. The lot-to-lot variability of the wood as well as temperature changes during hydrolysis have profound effects on the fermentability of the hydrolysate. With a combination of two electronic nose systems containing arrays with CP, MOS and MOSFET sensors it was possible to predict the outcome of the anaerobic yeast fermentation by analyzing the hydrolysates before starting the fermentation. Ethanol productivity as well as yield of ethanol and furfurals could be predicted with at least 15% accuracy. The sensor signals were selected with the forward selection procedure. Different lignocellulose hydrolysates of different origin were distinguished in a two-dimensional PCA diagram (Fig. 8).

3.3
Visualization of Cultivation State

In a number of different bioreactor cultivations with bacteria, yeasts, molds and mammalian cells, it was shown how the electronic nose can serve to visualize the course of the processes. The pattern recognition method that best manages to mirror the complex sensor array responses during extended cultivations is two- or three-dimensional PCA. Examples from such electronic nose applications are given below.

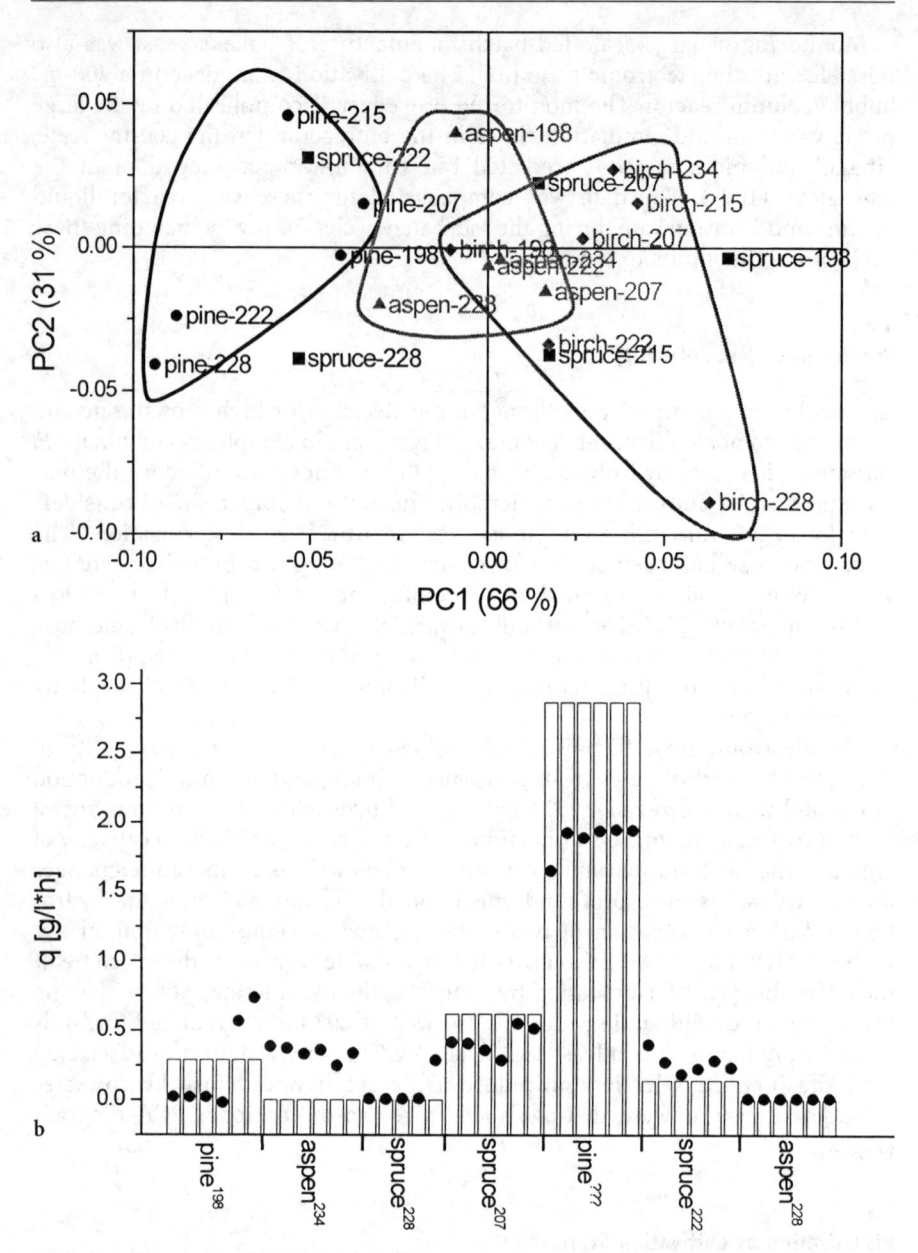

Fig. 8a, b. a PCA classification of lignocellulose hydrolysates from pine, spruce, aspen and birch using a combination of MOS, MOSFET and CP sensors. b Prediction of the fermentability of the same hydrolysates expressed as specific ethanol production rate using ANNs with topologies adapted to the sensor array (from [34] with permission of ACS)

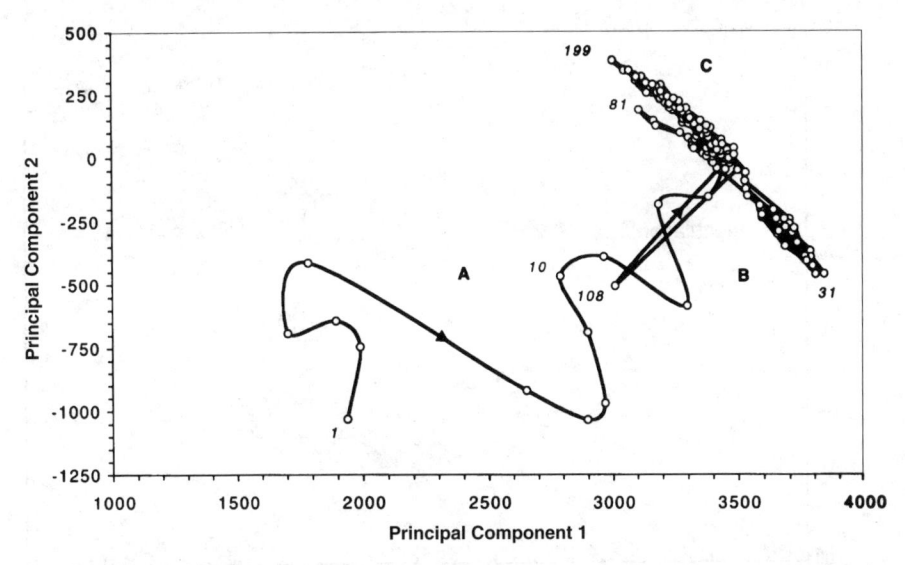

Fig. 9. PCA trajectory of a fed-batch *E. coli* cultivation producing recombinant hGH. The trajectory mirrors the lag phase directly after inoculum (*A*), the exponential growth phase (*B*) and the stationary hGH production phase (*C*) (from [36] with permission from Elsevier)

The idea was first demonstrated on a 14-L fed-batch cultivation producing recombinant human growth hormone (hGH) with *E. coli* in a complex medium [36]. The cultivation lasted for 33-h and had typical lag, exponential and stationary phases. The expression of the hGH occurred in the stationary phase. The PCA plot from the electronic nose responses separated the phases clearly in a two-dimensional PCA from MOS, MOSFET, and infrared signals as well as in a PCA where also the dissolved oxygen signal was included (Fig. 9). Fusion of the dissolved oxygen signal to the electronic nose signals altered the shape of the PCA plot but gave in principle equivalent information.

The methodology was applied to fed-batch baker's yeast production on a 200-m^3 scale [33]. The typical phases in a baker's yeast cultivation were visualized including lag phase, formation and consumption of ethanol and increase and decrease of cell mass. Fusion of signals from external sensors for volume, aeration flow rate and dissolved ethanol resulted in different character of the trajectory in the PCA but with the same principal information.

The visualization method also worked with a 500-L perfusion reactor system for production of recombinant human coagulation factor VIII (hFVIII) in Chinese hamster ovary (CHO) cells [36, 37]. Despite the diluted concentration of CHO cells and low titer of hFVIII in the medium, the nose could differentiate between the batch phase, medium replacement phase, and the high and low productivity phases during the five-week long cultivations (Fig. 10). The low concentration of hFVIII makes it credible to believe that there are other components associated with the product formation that the electronic nose responds to.

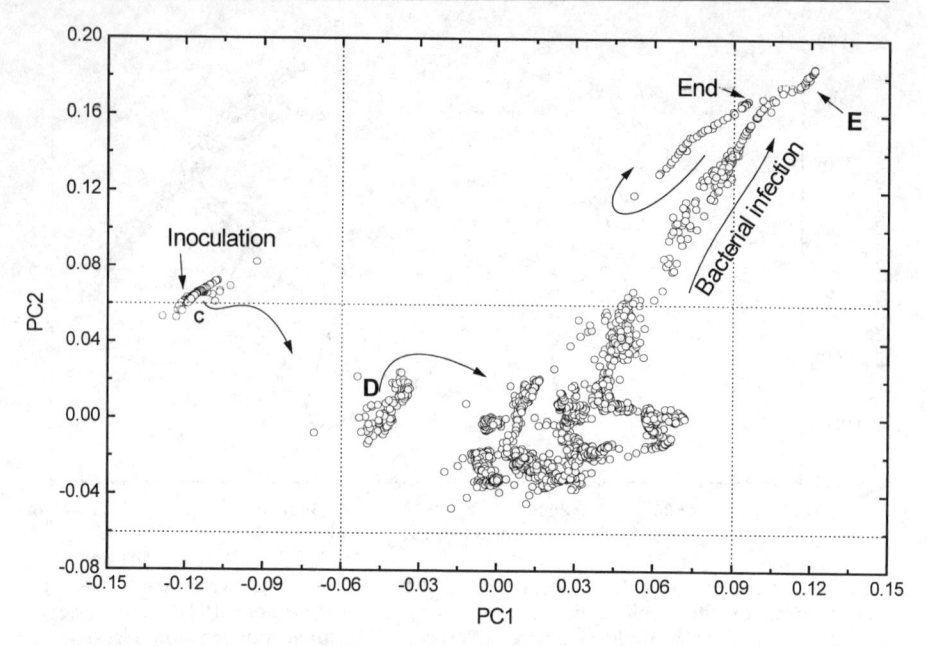

Fig. 10. PCA plot of a perfusion culture with CHO cells producing recombinant hFVIII. The data points in the plot mirror inoculum of the bioreactor and a phase of batch growth (c), transfer from growth to production medium (D), production phase of rFVIII (demarcated with dotted lines) and the deviation of a bacterial infection of the cell culture (E) (from Bachinger et al.[37], with permission)

Of particular interest to note is that a bacterial infection of the CHO culture was clearly seen in the PCA plot (Fig. 10). The PCA revealed the infection at least one day before the in-process analysis. From the above experience of the nose in *E. coli*, it is reasonable to believe that the culture with the two organisms produces a different response pattern that the pure CHO culture.

The same effect of bacterial contaminations was observed with an electronic nose equipped with CP sensors in an antibiotic fermentation with *Micromonospora carbonacea* [34]. Infections of *E. coli* and Gram-positive bacteria can be discriminated in the plot. The same culture also exhibited characteristic response patterns at the different fermentation stages over a nine-day period. The character of the trajectory in the PCA mirrored the growth phase, the antibiotic synthesis phase as well as the declination phase. The starting point of the trajectory almost coincided with the end point.

It is tempting to speculate if metabolic shifts and shifts in the physiological state of the culture can be more precisely monitored with the electronic nose. Physiologically important odorant metabolites in antibiotic strains such as geosmin and oxolone have been analyzed with conventional methods in the fermentation off-gas [38]. Recently, responses from electronic noses have been possible to relate to plasmid copy number [39]. Other types of intracellular shifts, such as viral infection, inclusion body formation or increase in mRNA level, could challenge the capacity of electronic noses.

4
Conclusion

Electronic noses allow non-invasive, real-time monitoring of bioreactors. There are two principally different ways of using them in bioreactor monitoring: (1) to predict defined state variables in the bioreactor such as metabolite concentrations or other components in the culture as well as typical bioengineering variables such as production rates, or (2) to visualize the course of a cultivation by identifying the response pattern in relation to cultivation stages or other conditions in the culture. The limitation of solely monitoring the off-gas from the cultivation could be circumvented by fusing the electronic nose responses with conventional on-line bioreactor variables such dissolved oxygen or reactor volume. With the electronic nose tool, process development could be enhanced, quality control and safety of bioreactor production improved and process control realized more efficiently. Ongoing development of solid state sensors and signal processing methods could, in combination with other new non-invasive chemical sensors, extend further the usefulness of electronic noses for bioreactor monitoring.

Acknowledgements. Thanks are due to Thomas Bachinger, Helena Lidén, Tomas Eklöv, Hans Sundgren, Per Mårtensson, Fredrik Winquist, and Ingemar Lundström for their valuable contributions to the development of the electronic nose for bioreactor monitoring. The financial support by the Swedish National Board for Technical and Industrial Development is gratefully acknowledged.

References

1. Persaud K, Dodd GH (1982) Nature 299:352
2. Gardner JW, Bartlett PN (1992) In: Gardner JW, Bartlett PN (eds) Sensors and sensory systems for electronic nose. Kluwer Academic Publisher, London, p 161
3. Slater JM, Watt EJ (1991) Analyst 116:1125
4. Gardner JW (1991) Sens Actuators B 4:109
5. Kress-Rogers E (1997) Handbook of biosensors and electronic noses, medicine, food and environment, CRC Press, New York
6. Börjesson T, Eklöv T, Jonsson A, Sundgren H, Schnurer J (1996) Cereal Chem 73:457
7. Pearce TC, Gardner JW, Friel S, Bartlett PN, Blair N (1993) Analyst 118:371
8. Shurmer HV, Gardner JW, Chan HT (1989) Sens Actuators 18:361
9. Nakamoto T, Fukuda A, Moriizumi T (1993) Sens Actuators B 10:85
10. Parry AD, Chadwick PR, Simon D, Oppenheimer BA, McCollum CN (1995) J Wound Care 4:404
11. Winquist F, Hörnsten G, Sundgren H, Lundström I (1993) Meas Sci Technol 4:1493
12. Gibson TD, Prosser O, Hulbert JN, Marshall RW, Corcoran P, Lowery P, Ruck-Keene EA, Heron S (1997) Sens Actuators B 44:413
13. Gardner JW, Craven M, Dow C, Hines EL (1998) Meas Sci Technol 9:120
14. Mandenius CF, Lundström I, Bachinger T (1996) 1st Eur Symp Biochem Eng Sci, p 104
15. Yamazoe N, Miura N (1992) In: Sberveglieri G (ed) Gas sensors. Kluwer, Andrecht, p 1
16. Spetz A, Winquist F, Sundgren H, Lundström I (1992) In: Sberveglieri G (ed) Gas sensors. Kluwer Academic Publishers, Andrecht, p 219

17. Pelosi P, Persaud, KC (1988) In: Dario P (ed) Sensors and sensory systems for advanced robots, NATO ASI Series F Computer and system science. Springer, New York, p 361
18. Hatfield JV, Neaves P, Hicks PJ, Persaud K, Travers P (1994) Sens Actuators B 18–19:221
19. Shurmer HV, Gardner JW, Corcoran P (1990) Sens Actuators B 1:256
20. Nordic Sensor Technologies AB, Linköping, Sweden
21. AromaScan plc, Crewe, UK
22. AlphaMos SA, Toulouse, France
23. Bachinger T, Mårtensson P, Mandenius CF (1998) J Biotechnol 60:55
24. Bishop CM (1995) Neural networks for pattern recognition. Oxford University Press
25. Jolliffe IT (1986) Principal component analysis. Springer, New York
26. Sundgren H, Winqusit F, Lukkari I, Lundström I (1991) Meas Sci Technol 2:464
27. Eklöv T, Mårtensson P, Lundström I (1999) Anal Chim Acta 381:221
28. Bach HP, Woehrer W, Roehr M (1978) Biotechnol Bioeng 20:799
29. Lidén H, Mandenius, CF, Gorton L, Meinander N, Lundström I, Winquist F (1998) Anal Chim Acta 361:223
30. Lidén H (1998) PhD thesis, Lund University, Sweden
31. Bachinger T, Lidén H, Mårtensson P, Mandenius CF (1998) Seminars Food Analysis 3:85
32. Lidén H, Bachinger T, Gorton L, Mandenius CF (1998) Submitted to Anal Chem
33. Mandenius CF, Eklöv T, Lundström I (1997) Biotechnol Bioeng 55:427
34. Namdev PK, Alroy Y, Singh V (1998) Biotechnol Prog 14:75
35. Mandenius CF, Lidén H, Eklöv T, Taherzadeh M, Lidén G (1999) Biotechnol Prog (in press)
36. Mandenius CF, Hagman A, Dunås F, Sundgren H, Lundström I (1998) Biosens Bioelectron 13:193
37. Bachinger T, Riese U, Eriksson R, Mandenius CF (1999) J Biotechnol (accepted for publication)
38. Rezanka T, Libalova D, Votruba J, Viden I (1994) Biotechnol Lett 16:75
39. Bachinger T, Bayer K, Mandenius CF (1999) (to be submitted)

Rapid Analysis of High-Dimensional Bioprocesses Using Multivariate Spectroscopies and Advanced Chemometrics

A. D. Shaw[1], M. K. Winson, A. M. Woodward, A. C. McGovern, H. M. Davey, N. Kaderbhai, D. Broadhurst, R. J. Gilbert, J. Taylor, É. M. Timmins, R. Goodacre, D. B. Kell[2]

Institute of Biological Sciences, University of Wales, Aberystwyth, Ceredigion SY23 3DD, UK
[1] E-mail: ais@aber.ac.uk, [2] E-mail: dbk@aber.ac.uk

B. K. Alsberg, J. J. Rowland
Dept. of Computer Science, University of Wales, Aberystwyth, Ceredigion SY23 3DD, UK

There are an increasing number of instrumental methods for obtaining data from biochemical processes, many of which now provide information on many (indeed many *hundreds*) of variables *simultaneously*. The wealth of data that these methods provide, however, is useless without the means to extract the required information. As instruments advance, and the quantity of data produced increases, the fields of bioinformatics and chemometrics have consequently grown greatly in importance.

The chemometric methods nowadays available are both powerful and dangerous, and there are many issues to be considered when using statistical analyses on data for which there are numerous measurements (which often exceed the number of samples). It is not difficult to carry out statistical analysis on multivariate data in such a way that the results appear much more impressive than they really are.

The authors present some of the methods that we have developed and exploited in Aberystwyth for gathering highly multivariate data from bioprocesses, and some techniques of sound multivariate statistical analyses (and of related methods based on neural and evolutionary computing) which can ensure that the results will stand up to the most rigorous scrutiny.

Keywords. Vibrational spectroscopy, Mass spectrometry, Dielectric spectroscopy, Flow Cytometry, Chemometrics

1	General Introduction – Multivariate Analyses in the Post-Genomic Era	84
2	Mass Spectrometric Measurements on Bioprocesses	85
3	Monitoring Bioprocesses by Vibrational Spectroscopies	87
3.1	Infrared Analysis	87
3.1.1	Advantages of NIR Application to Bioprocess Monitoring	87
3.1.2	Instrumentation and Standardisation	88
3.1.3	Interpreting Spectra in Quantitative Terms	88
3.1.4	Applications	89
3.2	MIR Analysis	90
3.3	Monitoring Bioprocesses Using Raman Vibrational Spectroscopy	92

4	**Measurement of Biomass**	94
4.1	Dielectrics of Biological Samples – Linear or Nonlinear?	95
4.1.1	The Nonlinear Dielectric Spectrometer	96
4.1.2	Nonlinear Dielectrics of Yeast Suspensions	98
4.1.3	Multivariate Analysis	98
4.1.4	Electrode Polarisation and Fouling	100
4.1.5	Electrode Coating	101
4.1.6	Genetic Programming	102
4.1.7	Other Microbial Systems	103
5	**Flow Cytometry**	103
6	**Data Analysis**	104
6.1	Data Pre-processing	104
6.2	Model Simplification	105
6.3	Data Partitioning	106
6.3.1	Training and Testing	107
6.3.2	The Extrapolation Problem	107
7	**Concluding Remarks**	108
References		108

1
General Introduction – Multivariate Analyses in the Post-Genomic Era

"But one thing is certain: to understand the whole you must look at the whole" – Kacser H (1986). On parts and wholes in metabolism. In: Welch GR, Clegg JS (eds) The organisation of cell metabolism, Plenum Press, New York, p 327

As we enter the post-genomic era [1, 2], there is a growing realisation that the search for gene function in complex organisms is likely to require analyses not just of one or two genes or other variables in which an experimenter happens to have an interest but of everything that is going on inside a cell and its surroundings. Such analyses are now occurring at the level of the transcriptome (e.g. [3, 4]), the proteome (e.g. [5–7]) and the metabolome [2], to define, respectively the expressed performance of the genome at the level of transcription, translation and small molecule transactions. However, the present level of analysis of such data is comparatively rudimentary [8].

The bioprocess analyst has long realised that the more (useful) measurements we can make the more likely are we to understand our bioprocesses, and we ourselves have long sought to increase the number of non-invasive, on-line probes available [9, 10]. Classical methods, monitoring factors such as pH, dissolved oxygen tension, and so on, however, are in essence *univariate* methods, and only give information on individual determinands.

The strategy that we have therefore sought to follow is to exploit *multivariate* methods which can measure many variables *simultaneously*. The resulting data floods necessitate the use of robust, multivariate chemometric methods. These too are now available in many flavours, with different strengths and weaknesses.

The purpose of the present review, then, as requested by the Editor, is to review some of the types of method we have developed and exploited in Aberystwyth for the rapid, precise, quantitative, and – where possible – non-invasive measurement of bioprocesses. Our website http://gepasi.dbs.aber.ac.uk may also be consulted. We start with mass spectrometry.

2
Mass Spectrometric Measurements on Bioprocesses

Whilst on-line desorption chemical ionisation mass spectrometry (MS) has been used to analyse fermentation biosuspensions for flavones [11], the majority of MS applications during fermentations have been for the analysis of gases and volatiles produced over the reactor [12–15], or by employing a membrane inlet probe for volatile compounds dissolved in the biosuspensions [16–22]. It is obvious that more worthwhile information would be gained by measuring the non-volatile components of fermentation biosuspensions, particularly when the product itself is non-volatile, which is usually the case.

The introduction of non-volatile components into an MS has typically been via the pyrolysis of whole fermentation liquors. Pyrolysis is the thermal degradation of a material in an inert atmosphere or a vacuum. It causes molecules to cleave at their weakest points to produce smaller, volatile fragments called pyrolysate [23]. An MS can then be used to separate the components of the pyrolysate on the basis of their mass-to-charge ratio (m/z) to produce a pyrolysis mass spectrum, which can then be used as a "chemical profile" or fingerprint of the complex material analysed [24].

Figure 1 gives typical pyrolysis mass spectra of *Penicillium chrysogenum* and of penicillin G, indicating the rich structural and process information that is available from highly multivariate methods of this type.

Pyrolysis MS (PyMS) has been applied to the characterisation and identification of a variety of microbial systems over a number of years (for reviews see: [25–27]) and, because of its high discriminatory ability [28–30], presents a powerful fingerprinting technique applicable to any organic material. Whilst the pyrolysis mass spectra of complex organic mixtures may be expressed in the simplest terms as sub-patterns of spectra describing the pure components of the mixtures and their relative concentrations [24], this may not always be true because during pyrolysis intermolecular reactions can take place in the pyrolysate [31–33]. This leads to a lack of superposition of the spectral components and to a possible dependence of the mass spectrum on sample size [31]. However, suitable numerical methods (or chemometrics) can still be employed to measure the concentrations of biochemical components from pyrolysis mass spectra of complex mixtures.

Heinzle et al. [34] were able to characterise the states of fermentations using off-line PyMS, and this technique was extended to on-line analysis [35].

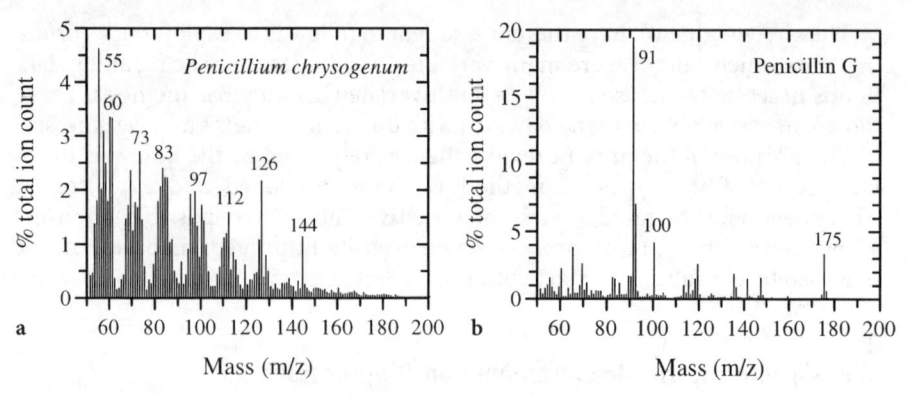

Fig. 1. a Normalised pyrolysis mass spectra of *Penicillium chrysogenum*; this complex 'fingerprint' can be used to type this organism. **b** Normalised pyrolysis mass spectra of 200 µg pure Penicillin G; this somewhat simpler 'biochemical profile' is one of the range of penicillins produced by *Penicillium chrysogenum*

However, they were not very satisfied with their system because there was no suitable data processing for the PyMS spectra. Although Heinzle and colleagues continued to use mass spectrometry for the analysis of volatiles produced during fermentation [13, 36], the analysis of non-volatiles by PyMS has not been investigated further by these authors.

With the advent of user-friendly chemometric software packages, PyMS can now be used for gaining accurate and precise quantitative information about the chemical constituents of microbial (and other) samples [37–39]. Within biotechnology the combination of PyMS with chemometrics has the potential for the screening and analysis of microbial cultures producing recombinant proteins; for instance this technique has permitted the amount of mammalian cytochrome b_5 [40] or α_2-interferon [41] expressed in *E. coli* to be predicted accurately. Chemometrics, and in particular artificial neural networks (ANNs), have also been applied to the quantitative analysis of the pyrolysis mass spectra of whole fermentor biosuspensions [31]. Initially a model system consisting of mixtures of the antibiotic ampicillin with either *Escherichia coli* or *Staphylococcus aureus* (to represent a variable biological background) was studied. It was especially interesting that ANNs trained to predict the amount of ampicillin in *E. coli* (having seen only mixtures of ampicillin and *E. coli*) were able to generalise so as to predict the concentration of ampicillin in an *S. aureus* background to approximately 5%, illustrating the very great robustness of ANNs to rather substantial variations in the biological background. (Genetic algorithms can also be used to simplify analyses of these data [42].) Samples from fermentations of a single organism in a complex production medium were also analysed quantitatively for a drug of commercial interest, and this could be extended to a variety of mutant producing strains cultivated in the same medium, thus effecting a rapid screening for the high-level production of desired substances [31]. In related studies *Penicillium chrysogenum* fermentation biosuspensions were analysed quantitatively for penicillins using PyMS [43] and this approach has also been used successfully to monitor *Gibberella fujikuroi* fermentations producing gibberellic acid [25, 44], to measure clavulanic acid production by *Streptomyces clavuligerus* [45], and to investigate various differentiation states in *Streptomyces albidoflavus* [46].

In conclusion, PyMS is undoubtedly very useful for the discrimination of micro-organisms at the genus, species and subspecies level, and whilst it has relatively low throughput (2 min per sample), which would make it unsuitable for very-high-throughput screening programmes, it does present itself as a suitable method for the rapid, precise and accurate analysis of the biochemical composition of bioprocesses.

3
Monitoring Bioprocesses by Vibrational Spectroscopies

3.1
Infrared Analysis

The measurement of compounds in bioprocesses, including fermentations, using conventional laboratory techniques such as HPLC, TLC or calorimetric assays is often tedious, invasive, requires sample handling and difficult to do in real time. For a bioprocess where it is important to gain information about the reactor status for feedback control, methods enabling rapid and reliable measurement of components are desirable.

Infrared spectroscopy is a powerful alternative analytical technology for process monitoring which has found wide application as an off-line method in the chemical and food industries. The additional advantage over other methods is that in many circumstances it is possible to quantify a number of components *simultaneously*.

The Near-Infrared (NIR) region extends from 780 nm to 2526 nm (12820 to 3959 cm^{-1}), as defined by the American Society for Testing and Materials. Molecules that contain covalent bonds and have a dipole moment absorb IR radiation. The majority of the bands observed in the NIR are due to overtones or combinations of fundamental vibrations occurring in the Mid-IR (MIR) region that extends from 2.5 to 25 µm (4000–400 cm^{-1}) [47]. The light mass of the hydrogen atom and consequently its anharmonic nature means that most of the combination bands in NIR are due to hydrogen-stretching vibrations (3600–2400 cm^{-1}). Consequently, the greatest utility of NIR is in the determination of functional groups that contain unique hydrogen atoms [48].

3.1.1
Advantages of NIR Application to Bioprocess Monitoring

Peaks in the NIR region are not nearly as distinct as those observed in the fingerprint region of the MIR. As the intensity of first overtones are generally an order of magnitude less than the fundamentals, pathlengths are usually much longer in the NIR. The advantages of these lower intensities include the fact that nonlinearities due to strong absorptions are less likely [49]. NIR analysis can be employed as a non-destructive process requiring little or no sample preparation and the sample may be re-introduced into the bioreactor. This is advantageous in a process environment where time is an important factor in the analysis [50].

3.1.2
Instrumentation and Standardisation

Modern NIR equipment is generally robust and precise and can be operated easily by unskilled personnel [51]. Commercial instruments which have been used for bioprocess analyses include the Nicolet 740 Fourier transform infrared spectrometer [52, 53] and NIRSystems, Inc. Biotech System [54, 55]. Off-line bioprocess analysis most often involves manually placing the sample in a cuvette with optical pathlengths of 0.5 mm to 2.0 mm, although automatic sampling and transport to the spectrometer by means of tubing pump has been used (Yano and Harata, 1994). A number of different spectral acquisition methods have been successfully applied, including reflectance [55], absorbance [56], and diffuse transmittance [51].

At-line sampling may involve a flow-through cell in the NIR spectrometer; in one process a glass-lined steel reaction vessel was used in combination with a fibre optic probe for measurements in a full scale chemical plant reactor [57]. Fibre optic bundles can be used to transmit NIR radiation to the reaction matrix and take signal back to the spectrometer. NIR is notoriously sensitive to changes in temperature and methods for keeping the temperature constant must be incorporated into the instrumentation.

3.1.3
Interpreting Spectra in Quantitative Terms

Broad superimposed bands are observed in NIR spectroscopic measurements and in most instances the peaks are not directly proportional to sample concentration. Statistical approaches are therefore required for modelling the behaviour of spectra for quantification. In the application of NIR to real world bioprocess samples, which are highly turbid scattering matrices, quantification of a constituent of interest can be particularly difficult. Vibrations are often observed that are common both to the determinand and the medium and cells in fermentations. Qualitative interpretation, and selection of unique spectral windows for calibration is therefore not always possible. One approach in the determination of wavelengths that can be used to quantify the constituent levels in bioprocess samples is to collect the spectra of raw materials alone and in combination, and then overlay spectra for isolation of unique bands. Second derivative pre-processing of spectral data can enhance spectral features and in addition baseline differences are often eliminated by this calculation; as cell density increases, the effective pathlength traversing through the sample increases because of light scattering by the cells, producing baseline offset [58]. Brimmer and Hall, [55] derived a Multiple Least squares Regression (MLR) equation that compensated for scattering differences attributable to changes in the biomass of the fermentation process. This was accomplished by using a reference wavelength at which the spectral data varies with penetration depth in a reproducible manner. Background information such as that attributable to water or the sample holder may be subtracted or used as a ratio [53, 59], however, in some instances this correction does not appear to affect the modelling ability of the algorithms [56]. NIR can be applied to whole cells, supernatant and aqueous mixtures of constituent samples, which may be also used to form calibration models [60].

Multivariate calibration methods. These are capable of extracting meaningful information from seemingly uninterpretable NIR spectra of bioprocess samples; however for these methods measurements made using other techniques must be available for training. It may be necessary to form a model for different times in the bioprocess e.g. for the start-up period and for later stages when inhibitors are accumulating and substrates are depleting in the fermentation.

Transferability of spectral data and models in NIR spectroscopy. This subject is an issue that is pertinent to the future use of NIR for bioprocess monitoring. Pre-processing to remove baseline shifts and noise in spectra from individual machines or direct standardisation by data transformation with a representative subset can be used to calibrate across instruments [61].

3.1.4
Applications

NIR spectroscopy continues to be applied to on-line fermentation and biotransformation monitoring, for example, of ethanol and biomass in rich medium in a yeast fermentation [62, 63], lactic acid production [64, 65], bioconversion of glycerol to 1,3-dihydroxyacetone [66] and nutrient and product concentrations in commercial antibiotic fermentations [67, 68]. Hall, Macaloney and colleagues [51, 58] reported NIR spectroscopic monitoring of industrial fed-batch *E. coli* fermentation of varying levels of acetate, ammonium, glycerol and biomass which they had previously studied in shake flasks [54], while Yano and colleagues [56] used NIR spectroscopy to determine with good precision the concentrations of ethanol and acetate in rice vinegar fermentations. The spectral signature of biomass with respect to wavelength regions was found to be essentially identical when groups of industrially-important microorganisms [69] were analysed. The concentration of many species may be determined from one spectroscopic measurement, as long as their concentration is 1 mM or greater [59].

New methods of variable selection include evolutionary methods based on Darwinian principles including Genetic Algorithms and Genetic Programming [70] and as such help to deconvolute whole spectral models in terms of which variables are important in the modelling procedure. When applied to a NIR glucose sensor, fewer than 25 variables were selected to produce errors statistically equivalent to those yielded by the full set containing 500 wavelengths and the algorithm correctly chose the glucose absorption peak areas as the information-carrying spectral regions [71], and these approaches, coupled to digital filtering, appear to be the methods of choice [72, 73].

It is important that calibration models are rigorously validated and in the first instance that all variations are accounted for in the model using diverse samples that are expected to be observed in future bioprocess runs. Some investigators attempt to keep process conditions very reproducible but such conditions are uncommon in an industrial environment. In addition, multivariate calibration models will work well if identical media (composition) and process conditions are used on each successive run. Simple modifications such as use of a different media supplier can affect the spectral background. The predictive ability of the models will then be affected as they will be challenged with samples which they have not been trained to recognise [74].

3.2
MIR Analysis

The higher level of spectral resolution in the MIR range often allows peaks to be assigned to specific medium components or chemical entities. Although analysis of bioprocesses in the MIR range would be especially useful for monitoring products of interest because of the feature rich spectra between 4000–200 cm^{-1}, application to on-line aqueous systems at an industrial level is hindered by the broad water absorption across most of the so-called 'fingerprint' spectral range. For off-line analysis this can be overcome simply by drying samples; however, for on-line analysis success with mid-IR monitoring of bioprocesses has been limited to use of transmission cells with extremely short pathlengths or Attenuated Total Reflectance (ATR) spectroscopy. ATR utilises the phenomenon of total internal reflection. ATR can be used essentially as an 'in-line' method, where the sample interface is located in the process line itself, thus eliminating the requirement for an independent sampling system. The sample to be analysed is placed in direct contact with a crystal made from zinc selenide, germanium, thallium/iodide, sapphire, diamond or zirconium. Quantitative monitoring by FT-IR spectroscopy of the enzymatic hydrolysis of penicillin V to 6-aminopenicillanic acid and phenoxyacetic acid using a 25 µl flow through cell with a zinc selenide crystal demonstrated that the IR method allowed better prediction of the process termination time than the standard method based on monitoring the addition of sodium hydroxide [75].

On-line MIR ZnSe ATR analysis of microbial cultures has been used primarily for non-invasive monitoring of alcoholic or lactic fermentations. Alberti et al. [76] reported the use of a ZnSe cylindrical ATR crystal to monitor accurately substrate and product concentrations from a fed-batch fermentation of *Saccharomyces cerevisiae*. Picque et al. [77] also used a ZnSe ATR cell for monitoring fermentations and found that whereas NIR spectra obtained from alcoholic or lactic fermentation samples contained no peaks or zones whose absorbance varied significantly, both transmission and ATR MIR could be used successfully to measure products. Fayolle et al. [78] have employed MIR for on-line analysis of substrate, major metabolites and lactic acid bacteria in a fermentation process (using a germanium window flow-through cell), and studied the effects of temperature on the ability to quantify the substrates (glucose and fructose) and metabolites (glycerol and ethanol) in an alcoholic fermentation using a ZnSe ATR crystal. Hayakawa et al. [79] described the use of a remote ZnSe ATR probe for determining glucose, lactic acid and pH simultaneously in a lactic acid fermentation process using *Lactobacillus casei*. The benefits of the ATR method of analysis are generally those that would be considered advantageous for any on-line system, being non-destructive, requiring no sample preparation or reagents and only a short analysis time, with minimal expertise necessary in the industrial environment. Practical drawbacks for the technique, particularly for microbial fermentations, centre on the need to purge the flow cell or clean the ATR probe to prevent surface contamination through biofilm formation. Some ATR crystal materials are toxic, limiting certain applications to the use of sapphire, diamond or zirconia. Sapphire crystals are non-transmitting below 2000 cm^{-1} which means that the MIR fingerprint region cannot be investigated with this device [80]. Developments in optical fibre design and coupling to spectrometers makes IR analysis a practical consideration for industrial reactors, as the IR spectrometer can be kept remote from the sampling probe, although at present chalcogenide fibres can only be used over short distances.

Off-line analysis of bioprocesses is clearly less desirable for a rapid response. However, MIR analysis of fermentation samples off-line does offer certain advantages over other techniques. A method we have introduced and called DRASTIC (Diffuse Reflectance-Absorbance IR Spectroscopy Taking in Chemometrics) [81] for MIR analysis of bioprocess samples has been successfully applied to the estimation of drug concentrations in biological samples, including fermentations from a microbial strain development programme [82, 83]. In this technique fermentation samples (5 µl) were applied to wells in an aluminium plate or aluminium-coated plastic 384-well microtitre plate, dried, mounted on a motorised mapping stage and analysed by the diffuse reflectance-absorbance method using a Bruker IFS28 FT-IR spectrometer. This allows rapid non-destructive analysis of samples (typically 1 per second) at a high signal to noise ratio. We were thus able to predict concentrations of ampicillin in a biological background of *E. coli* (see Fig. 2 for example spectra) and *Staphylococcus aureus* cells, and we used spectral data obtained from analysis of fermentations of *Streptomyces citricolor* to predict the concentrations of the carbocyclic nucleosides aristeromycin and neplanocin A. PLS routine was used to create a training set using the MIR spectral data and information provided from HPLC analysis of samples. This method can be fully automated and allows for a particularly high sample throughput rate.

The use of multivariate spectral information is particularly advantageous where quantification of a particular metabolite in a complex biological background is being attempted and application of the technique necessitates the use of chemometric processing techniques for quantification of components.

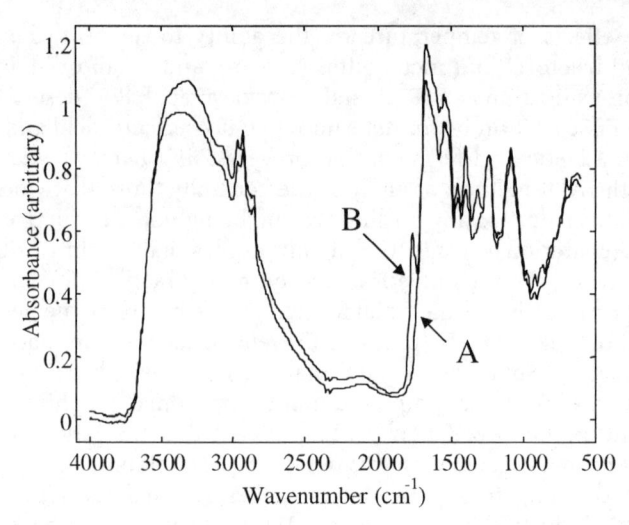

Fig. 2. MIR diffuse reflectance spectra of *Escherichia coli* cells without (*A*) and with (*B*) 20 mM ampicillin

3.3
Monitoring Bioprocesses Using Raman Vibrational Spectroscopy

Recent exploitation of biotechnological processes for pharmaceutical and food industries has necessitated rapid screening and quantitative analysis of the specific components. Therefore, there is continuing need for developing on-line methods for monitoring such biological processes [84–86]. The ideal method [87] would be rapid, non-invasive, reagentless, precise and cheap, although to date, with the possible exception of near-IR spectroscopy almost no such single method has been found. Generally these bioprocesses progress from translucent to increasingly opaque matrices as the microbial cells multiply and become highly light scattering and rich in molecular vibrational information. The use of specific molecular vibrations allowing specific fingerprinting of singular or multi-components for identification and quantification using the vibrational FT-IR and Raman spectroscopies for monitoring these bioprocesses can provide suitable alternatives to the present day process monitoring.

Raman spectroscopy relies on vibrational signals generated by focusing a laser beam onto the sample to be analysed, where most of the incident photons are either transmitted through the sample, absorbed by it, or scattered (elastic scattering). In a very few cases, approximately 1 in 10^9, the vibrations and rotations of the scattering molecules cause energy quanta to be transferred between molecules and photons in the collision process (inelastic scattering). A monochromator and a detector are then used to measure these inelastically scattered photons to give a Raman spectrum.

Raman spectroscopy can be used to analyse aqueous biological and bio-organic samples e.g., bacteria, spores, diseased tissues, neurotransmitters, protein structures, membrane lipids, biochemical assays, drug-nucleotide interactions, constituents of oils, water for toxic analytes and bioprocesses.

During the last few years there has been a renaissance in Raman instrumentation suitable for the analysis of biological systems, initially with the development of Fourier Transform (FT)-Raman instruments in which the wavelength of the exciting laser is in the near-infrared laser (usually a Nd:YAG (neodymium doped yttrium-aluminium garnet) at 1064 nm) rather than in the visible region, an arrangement which therefore avoids the background fluorescence typical of biological samples illuminated in the visible [47, 88–104]. In addition, and at least as importantly, exceptional Rayleigh light rejection has come from the development of holographic notch filters [105–108], and a recent innovation is the use of Hadamard-transform-based spectrometers [109, 110].

Although the FT approach to both infrared and Raman spectroscopy possesses well-known advantages of optical throughput [47, 111], there are still problems for FT-Raman with many aqueous biological samples as water may absorb both the exciting laser radiation at 1064 nm and the Raman scattered light. In addition, it is often necessary to co-add many hundreds of spectra to produce high-quality data from biological systems, and acquisition times are frequently 15–60 min. More recently, therefore, it has been recognised that charge coupled device (CCD) array detectors are ideal elements for use in *dispersive* (non FT) Raman spectroscopy. However, they normally have very low quantum efficiency at 1064 nm photons. Thus holographic notch filters and CCD array detectors have been combined with a dispersive instrument, using diode laser excitation at 780 nm (a wavelength which suppresses fluorescence from most samples but which penetrates water well). The cooled CCD is a multi-channel device which has exceptional sensitivity and very low intrinsic noise (dark current), so that the signal:noise ratio is improved by at least 2 orders of magnitude (compared with an uncooled CCD) and data acquisition is correspondingly fast [89]. These and other major technical advances [112, 113] now make Raman a very promising tool for the rapid, non-invasive and multiparameter analysis of aqueous biological systems, including the estimation of metabolite concentrations in ocular tissue [114, 115].

In 1987, Shope and colleagues [116] used attenuated total reflectance (ATR) Raman spectroscopy for the on-line monitoring of the fermentation by yeast of sucrose to ethanol, using the argon ion laser line at 514.5 nm. Gomy et al. [117, 118] monitored their alcoholic fermentation using the same laser with a fiber optic probe attached to a Raman spectrometer but analysed the ethanol levels only at higher wavenumber (2600–3800 cm^{-1}). This was because the Raman monitoring of these processes using 514.5 nm excitation gave significant fluorescence in the lower wavenumber region, as can be observed in the spectra shown in these papers.

Although fluorescence has been a major hindrance for the use of Raman spectroscopy in biology, Shope and colleagues [116] clearly showed that the narrow Raman peaks were distinct from the broad features of fluorescence and proposed the use of full widths at half-height of the peaks for chemical quantitation from Raman spectra. Shope et al. [116] used a least squares fit to analyse the Raman spectra for quantification of the production of ethanol during the

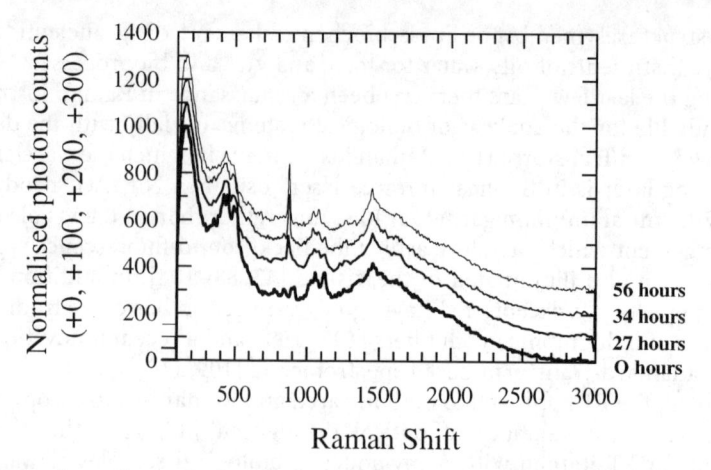

Fig. 3. Comparison of smoothed, normalised spectra from a biotransformation of glucose to ethanol, taken at intervals through the experiment, showing the change in the spectrum over time. Spectra are artificially displaced by 100 photon counts for clarity

yeast fermentation process. Finally, Spiegelman and colleagues [119] have recently shown that the amount of glucose in aqueous solution can be measured using Raman spectroscopy.

4
Measurement of Biomass

This laboratory long ago devised [120] the use of radio-frequency dielectric spectroscopy [121, 122] for the on-line and real-time estimation of microbial and other cellular biomass during laboratory and industrial fermentations. The principle of operation is that only intact cells (see [123] for what is meant in this context by the word 'viable'), and nothing else likely to be in a fermentor, have intact plasma membranes and that the measurement of the electrical properties of these membranes allows the direct estimation of cellular biomass (Fig. 4).

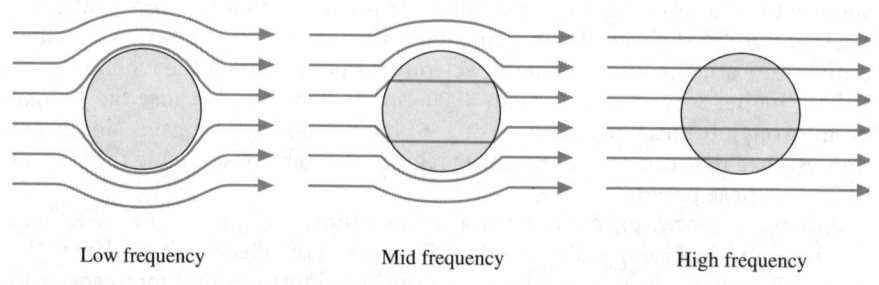

Fig. 4. Fields and cell membranes. At low frequency the field cannot penetrate the cell wall and is dropped almost entirely across the outer membrane such that the membrane amplifies the field across itself by a factor of up to 1000 From *left* to *right* – Low frequency, Mid frequency, High frequency

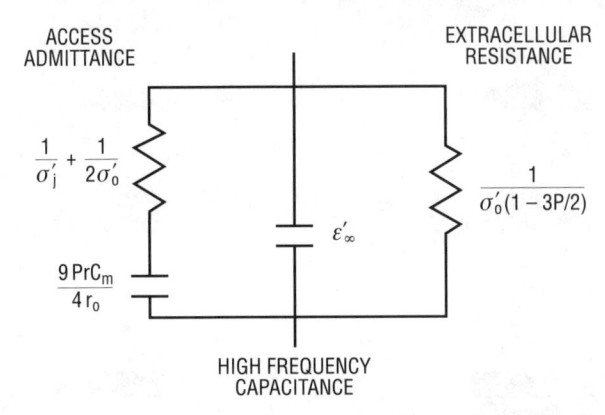

Fig. 5. Standard linear equivalent circuit of an assumed linear dielectric cell membrane can be modelled with simple standard components. This assumption breaks down if the field is amplified across the membrane as in Fig. 1 to a degree sufficient to produce nonlinearity

This situation is modelled as the equivalent circuit of Fig. 5, where all the components are assumed linear.

The probe has been long and successfully commercialised (see http://www.aber-instruments.co.uk) and since we have reviewed this approach on a number of occasions (e.g. Kell et al. 1990, Davey 1993 a, b, Davey et al. 1993 a, b) we will not do so here, save to point out (in the spirit of this review) the trend to the exploitation of *multi-frequency excitation* for acquiring more (and more robust) information on the underlying spectra. [124, 125]. Most recently, we have also devised a number of novel routines for correcting for the electrode polarisation that can occur under certain circumstances [126, 127], and have turned our attention to the *nonlinear* dielectric spectra of biological systems.

4.1
Dielectrics of Biological Samples – Linear or Nonlinear?

The dielectric response of biological tissue has long been assumed linear. Thus an enzyme is treated as a hard sphere which relaxes linearly in an a.c. field at all but high field strengths [128]. In a suspension of cells, the electric field cannot penetrate to the interior of the cell at the low frequencies currently of interest in nonlinear dielectric spectroscopy [129], and is dropped almost entirely across the outer membrane of the cell which is predominantly capacitive at these frequencies, as was shown in Fig. 4.

However, an enzyme which has different dipole moments in different conformations during its operation (Fig. 6) may affect and be affected by electromagnetic fields [130]. Change between states is unlikely to be smoothly or linearly related to the field due to the constraints imposed on the enzyme by its environment in the membrane, so the dielectric response of the material is nonlinear even at low applied fields [131].

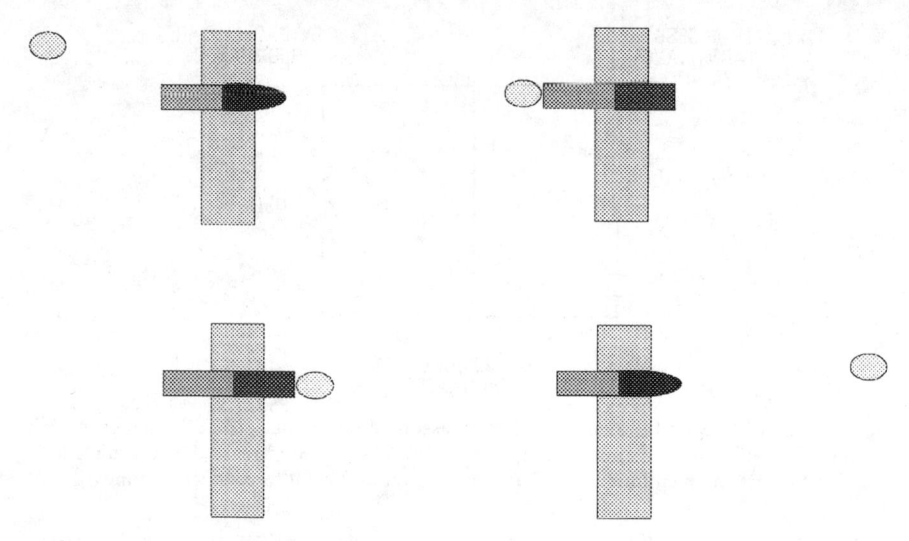

Fig. 6. Enzyme transporting ion across membrane via conformational change. If the different conformations have different dipole moments, the enzyme will be sensitive to electric fields and will be detectable by its effect on these fields

The equivalent circuit of Fig. 5 is no longer very useful since its individual components are no longer linear. This behaviour shows up as the generation by the tissue of harmonics of the applied frequency [129].

A nonlinear dielectric spectrometer has been designed around a standard IBM PC; and realised almost completely in software, with a minimum of extraneous hardware [129].

4.1.1
The Nonlinear Dielectric Spectrometer

A sinusoidal (or otherwise) signal is generated by the PC and applied to the outer terminals of a 4-terminal electrode system. The resulting signal across the inner electrodes is fed back differentially to the PC. This signal is then transformed into its power spectrum and the harmonics studied (Fig. 7).

Of course things are never quite this simple. At the low frequencies (a few Hz to a few kHz) studied so far, there is a strong polarisation layer around the driver electrodes. The i/V relation of this layer is both strongly nonlinear and highly variable with time, and its effects must be removed from the (weak) harmonics generated by the biology, if direct visualisation of the harmonic spectra is needed.

A reference spectrum (dB power spectrum) is taken using the supernatant of the suspension under test. This is the polarisation signature. This is then subtracted from the equivalent spectrum from the whole suspension. This procedure deconvolves the polarisation harmonics from those produced by the tissue nonlinearity (Fig. 8).

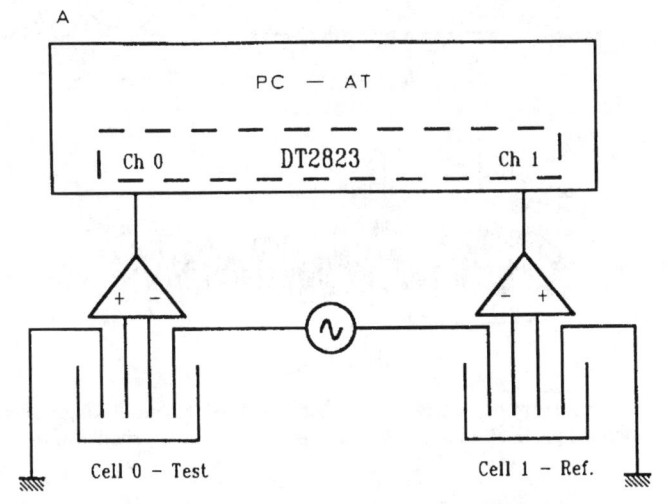

Fig. 7. Dielectric spectrometer schematic: Two standard four-terminal electrode chambers are connected to A/D converters and on into a PC. Fourier analysis is done by the PC to produce the nonlinear dielectric spectra

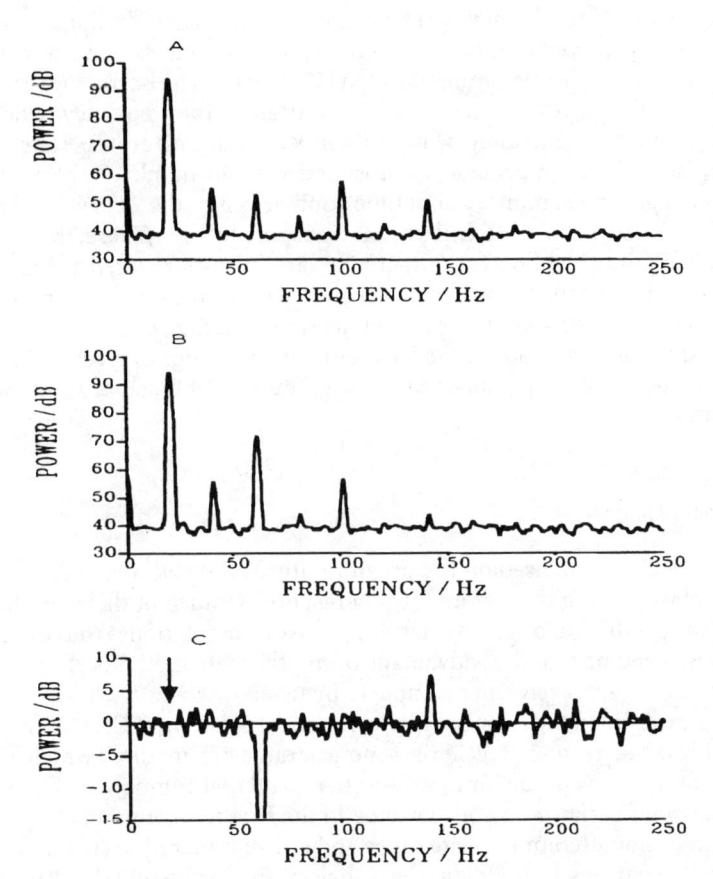

Fig. 8. Reference, suspension, and difference spectra of resting yeast. Predominantly odd harmonics only are produced in this metabolic state signifying a symmetric system in equilibrium

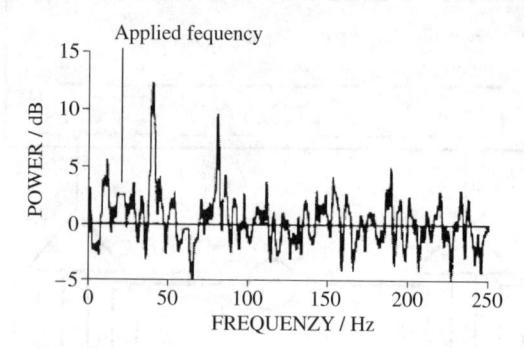

Fig. 9. Difference spectrum of metabolising yeast cells Even harmonics appear under these conditions showing an activation of the ATPase signifying the disturbance of the equilibrium of Fig. 5

4.1.2
Nonlinear Dielectrics of Yeast Suspensions

In a suspension of *Saccharomyces cerevisiae*, an inhibitor study along with use of mutant strains showed that the predominant source of the nonlinear signature in this organism is the membrane-located H+ ATPase. The harmonics are highly voltage- and frequency-windowed, with the peak of the frequency window for the resting enzyme coinciding neatly with its k_{cat} value. In a resting state, at equilibrium, the suspension generates almost entirely odd-numbered harmonics, as in Fig. 8, suggesting symmetry about the equilibrium of the ATPase. If glucose is adddded to the suspension to fuel proton transport by the ATPase, then the shift away from equilibrium breaks the symmetry and even-numbered harmonics appear, giving a measure of the activity or inactivity of this enzyme and the consequent metabolic state of the yeast cells as shown in Fig. 9.

Analysing the behaviour of the harmonics over a range of frequencies/voltages allows the rapid collection of a very large amount of metabolism-dependent information.

4.1.3
Multivariate Analysis

Recently, work has focused on the use of multivariate methods to form models capable of predicting the factors causing responses. Much of this work has centred on the prediction of glucose levels in yeast fermentations from the cellular responses. A major practical advantage of multivariate methods is that there is no requirement for a reference sample to be taken.

Initial experiments used principal component analysis (PCA) to investigate the multivariate response. PCA is a non-parametric method which outputs linear combinations of the input values (the "principal components"), such that the majority of variation is concentrated in the first few components.

PCA does not attempt to relate cause and effect; it merely serves to highlight the larger variations in the data. Nevertheless, the results obtained from PCA

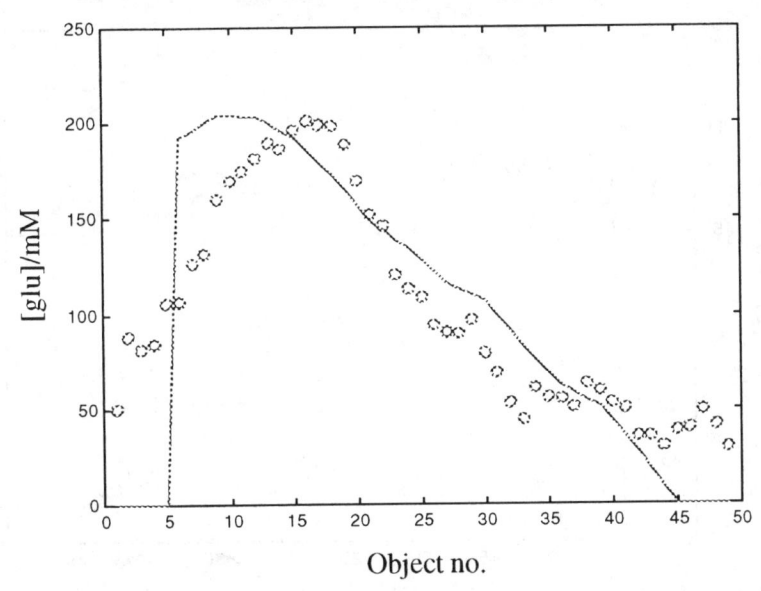

Fig. 10. PLS based prediction of glucose levels in one yeast batch fermentation by a model formed and validated on glucose levels in two other independent fermentations. The rmsep is 41% of the mean value of the data

proved promising, showing large variations which could be due to the cells' activity in response to glucose.

Subsequent work has used partial least squares regression (PLS) to form predictive models of glucose concentration during batch fermentations as shown in Fig. 10 (where object number = sample number and gives a measure of the progress of the fermentation). PLS produces models by projecting the large number of response X-variables (the harmonics in the NLDS spectra) into a smaller number of 'Latent' variables, while retaining as much relevant variability as possible. The variables in this space are then used to form a regression onto the predicted Y-variables (the actual glucose levels measured by a reference method). This "two-way'" modeling tends to form much more accurate models than other simple linear multivariate methods (e.g. principal component regression and multiple linear regression) as it automatically detects relevant X-variables and preferentially forms the model on these. The precision of the prediction is assessed by the commonly used Root Mean Square Error of Prediction (rmsep) [132]. Three independent datasets are required; one to form the model, one to validate the model, and one which the modelling process has not seen to test the model against 'unknown' data

Examination of the "residual" unmodelled variation in these experiments indicates that there is a nonlinearity in the relationship between the X and Y variables. This detracts from the models' accuracy. To this end the inherently nonlinear capabilities of ANNs have been employed with an improved predictive capability resulting in the prediction of Fig. 11.

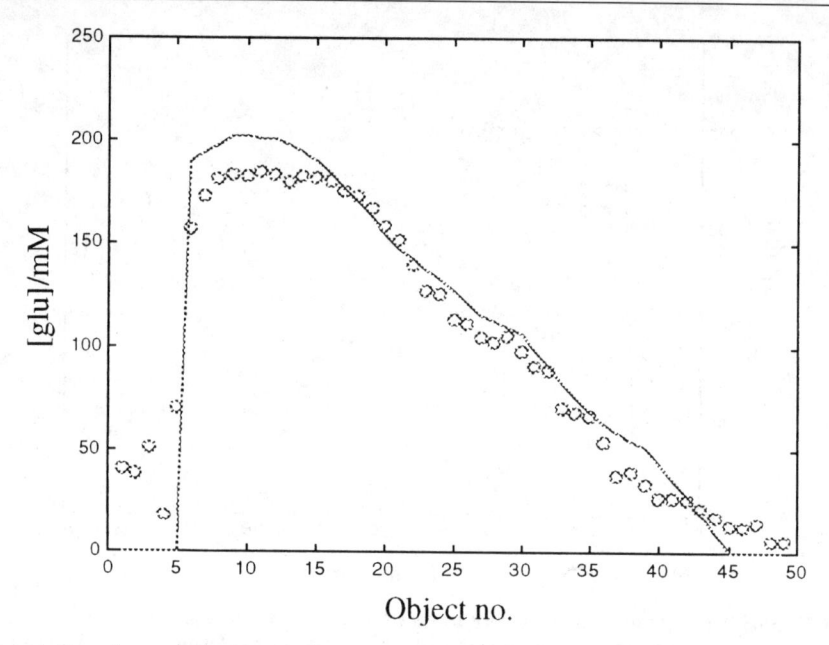

Fig. 11. Neural net prediction of glucose levels in one yeast batch fermentation by a model formed and validated on glucose levels in two other independent fermentations. This experiment uses uncoated gold electrodes. The rmsep is 19%

The current area of interest is in trying to reduce the electrode instabilities which are responsible for large baseline offsets when a model based on one fermentation is used to predict results from another. This can be done in either hardware, by coating the electrode to stabilise the interface [133], or in software, by using more powerful modelling methods such as Genetic Programming (GP) to automatically remove the effect of these instabilities from the model [134].

4.1.4
Electrode Polarisation and Fouling

In biological NLDS work, electrode polarisation is a serious problem at the low frequencies (up to a few tens of Hz) where the biology typically reacts most strongly to the electric field; and its fluctuations can be similar in size to, or bigger than, the small changes due to biological activity (e.g. upon glucose metabolism). It is therefore vital to control electrode polarisation insofar as is possible. To obtain nonlinear electrochemical reproducibility, electrode surfaces must be scrupulously clean, and this is very difficult to achieve. If any contamination is present, the biologically relevant signal may be unstable, distorted or concealed completely [135].

Electrode cleaning to ensure repeatable nonlinear dielectric spectra is a complex and empirical task, due to the lack of knowledge of the exact form of the causative mechanisms operating in the electrode/electrolyte interface. No repeatable and certain ways of obtaining a quiet and repeatable reference signal from an individual electrode surface have been found but simple abrasion works best. Once clean, electrodes may stay stable for days, or become unstable within a few minutes. Continual control readings, performed as indicated above, are vital during any series of experiments to be sure the electrode surface behaviour has not substantially altered during the experiments, in which case the results must be abandoned and the experiments repeated. This Byzantine process can make the process of obtaining a lengthy series of results with continually clean electrodes a nightmare.

4.1.5
Electrode Coating

To prevent a protein from adhering to a metal surface, the surface can be coated with a sheet of poloxamers. These are a triblock copolymer consisting of PEO-PPO-PEO, in which two polyethylene oxide (PEO) chains are attached to a hydrophobic polyproylene oxide (PPO) anchor. This prevents the protein binding by steric repulsion overpowering the attraction between the protein and the coating layer [136]. This coating layer stabilises the electrode interface slightly and prevents protein fouling, allowing the electrodes to be used after a simple cleaning and coating procedure. They then stay useable for a month.

The coating allows three independent datasets leading to the prediction of PLS prediction of Fig. 12 (to be compared with that of Fig. 10) to be obtained

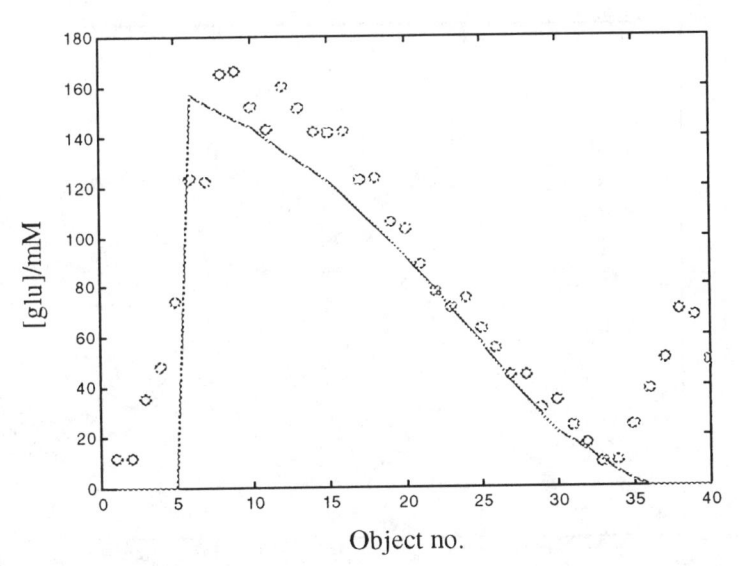

Fig. 12. PLS prediction of glucose levels in one yeast batch fermentation by a model formed and validated on glucose levels in two other independent fermentations. This experiment uses polymer coated electrodes. The rmsep is 35%

rapidly and conveniently, without the prohibitive electrode problems discussed above. It is also found that the coating linearises the data and allows PLS to perform better in relation to nonlinear modelling methods.

4.1.6
Genetic Programming

Genetic programming [137] is an evolutionary technique which uses the concepts of Darwinian selection to generate and optimise a desired computational function or mathematical expression. It has been comprehensively studied theoretically over the past few years, but applications to real laboratory data as a practical modelling tool are still rather rare. Unlike many simpler modelling methods, GP model variations that require the interaction of several measured nonlinear variables, rather than requiring that these variables be orthogonal.

An initial population of individuals, each encoding a potential solution to the optimisation problem, is generated randomly and their ability to reproduce the desired output is assessed. New individuals are generated either by mutation (the introduction of one or more random changes to a single parent individual) or by crossover (randomly re-arranging functional components between two or more parent individuals). The fitness of the new individuals is then assessed, and the fitter individuals from the total population are more likely to become the parents of the next generation. This process is repeated until either the desired result is achieved or the rate of improvement in the population becomes zero. It has been shown [137] that if the parent individuals are chosen according

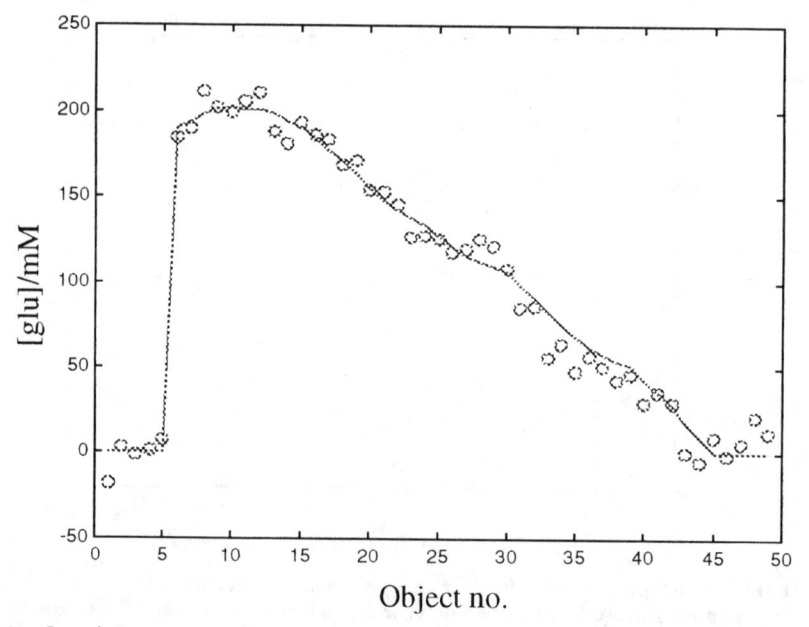

Fig. 13. Genetic Program prediction of the data of Figures 7 and 8. The rmsep is 9%

to their fitness values, the genetic method can approach the theoretical optimum efficiency for a search algorithm.

This technique allows the prediction of Fig. 10 and 11 to be improved to produce Fig. 13.

Given the very heavy computational load of GP, it would not be the method of choice for problems which yield to simpler approaches. However the above data show that it can be very beneficial on problems that have defeated other methods.

4.1.7
Other Microbial Systems

NLDS has also been successfully applied in this laboratory to measurements of photosynthesis in *Rhodobacter capsulatus* [138]; of glucose levels in erythrocytes, both invasively and non-invasively [135]. It has also been used successfully to detect the subtle interaction of weak low-frequency magnetic fields with membrane proteins of aggregating amoebal cells of *Dictyostelium discoideum*. Using PCA, a significant distinction was shown between cells previously exposed to pulsed magnetic fields (PMF) of 0.4 mT and 6 mT and their respective controls. Significant distinction was also shown between cells exposed to 50 Hz sinusoidal magnetic fields of 9 µT and 90 µT and their respective controls. NLDS was able to demonstrate a dose response with respect to both duration of exposure and field strength. In all cases significant changes in intracellular biochemistry had also been shown. There is some evidence to support a hypothesis that voltage gated calcium channels are involved in the response of *Dictyostelium* to PMFs [139, 140].

5
Flow Cytometry

Flow cytometry [141, 142] is a technique that allows the measurement of multiple parameters on individual cells. Cells are introduced in a fluid stream to the measuring point in the apparatus. Here, the cell stream intersects a beam of light (usually from a laser). Light scattered from the beam and/or cell-associated fluorescence are collected for each cell that is analysed. Unlike the majority of spectroscopic or bulk biochemical methods it thus allows quantification of the heterogeneity of the cell sample being studied. This approach offers tremendous advantages for the study of cells in industrial processes, since it not only enables the visualisation of the distribution of a property within the population, but also can be used to determine the relationship *between* properties. As an example, flow cytometry has been used to determine the size, DNA content, and number of bud scars of individual cells in batch and continuous cultures of yeast [143, 144]. This approach can thus provide information on the effect of the cell cycle on observed differences between cells that cannot be readily obtained by any other technique.

Flow cytometry has been applied to the study of the formation of the biopolymer poly-*b*-hydroxybutyrate (PHB). While the formation of the polymer can be detected by changes in the light scattering behaviour of cells [145], its accumulation has also been analysed using the hydrophobic fluorescent dye Nile Red [146]. PHB is produced commercially for use in the manufacture of biodegradable plastic materials and this approach has enabled researchers to determine the effect of changes in nutrient limitation conditions on the production and storage of PHB in individual cells [147].

While these examples illustrate the role of flow cytometry in bioprocess monitoring, the analyses have been conducted off-line thus making their use in bioprocess control impractical. Recently, a portable flow cytometer – the Microcyte – [148] has been described, which due to its small size and lower cost (compared to conventional machines) allows flow cytometry to be used as an at-line technique [149]. Rønning showed that this instrument had a role to play in the determination of viability of starter cultures and during fermentation. The physiological status of each individual cell is likely to be an important factor in the overall productivity of the culture and is therefore a key parameter in optimising production conditions.

The problems of converting flow cytometry into an on-line technique are discussed by Degelau and colleagues [150], however, more recently a flow injection flow cytometer for on-line monitoring of bioreactors has been developed by Zhao and colleagues [151]. In the system described a sample is removed from the fermentor under computer control. The sample is degassed prior to passing into a microchamber where it is automatically diluted if necessary prior to the addition of stains or other reagents. Following an appropriate incubation in the microchamber the sample is delivered to the flow cytometer for analysis. This instrument has been used successfully to monitor both the production of green fluorescent protein (Gfp) in *E. coli* and to determine the distribution of DNA content of a *S. cerevisiae* population without the necessity for operator input. With continuing decrease in costs and increase in automation flow cytometry is likely to play an increased role in bioprocess monitoring and control.

6
Data Analysis

Whilst modern instruments may provide much more accurate data than those of years ago, new types of instrument are being developed which provide data of somewhat lesser accuracy, but which have other advantages (e.g. speed, throughput, on-line). Advances in computing methods help in the extraction of meaningful information from such data, which in the past would have been impossible, and so bioinformatics has become an essential part of the experimental procedure.

6.1
Data Pre-processing

Before carrying out any statistical analysis on multivariate data, it is important to ensure that the data are valid, and in a suitable format. This means:

- Ensuring that there are no errors in the data
- Normalising, when necessary

Errors may be caused by data input error (where this is done by hand), or by an incorrectly analysed sample. In the former case, this is typically a wrong number, or a decimal point missed or wrongly placed. Such errors may usually be found by testing the maximum and minimum values of a variable. If one value is found to be significantly different to the others, it is suspect, and should either be corrected (e.g. by referring back to the original experimental results, where available, or moving a decimal point), or the whole object affected deleted. If the measurements for one sample are consistently found to be suspect, normalisation may solve this problem. If it is suspected that the sample was incorrectly analysed, and cannot easily be reanalysed, it should be deleted from the data set.

Many spectroscopic methods will produce results whose magnitude depends upon the amount of sample present during the analysis or prevailing experimental conditions (e.g. Pyrolysis Mass Spectrometry, Raman spectroscopy). In such cases, the samples should be *normalised*, either to an internal standard or variable of consistent value, or, where the totals are expected to be about the same for each object (PyMS), to the total.

For example, to normalise the total of all objects to 1000, each variable x_{ib} before normalisation in object x with n variables becomes after normalisation (x_{ib}):

$$x_{in} = \frac{1000}{\sum_{j=1}^{n} x_i} \times x_{ib}$$

Where the result does *not* depend on such factors, or a normalisation to an internal standard is carried out by the spectrometer or accompanying software automatically (e.g. in Nuclear Magnetic Resonance – NMR), further normalisation *should not be carried out*.

If after normalisation to the total, a variable is found to be suspect and deleted, normalisation must be carried out again. It is possible, when normalising to the total, that such re-normalisation may adversely affect the remaining data. If this is judged to be the case, the whole experiment will need to be repeated.

Most statistical packages will carry out normalisation of the *variables*, typically to $\frac{1}{StDev}$ *for each variable*. The purpose of this is to negate the effect of large variables on the model formed [152]. If the package being used does not provide this facility, or if for some other reason it is believed that a better result will be obtained by using a different normalisation of the variables, this should be carried out at this stage. Such normalisation should always be carried out *after* any normalisation of the objects has been performed.

6.2
Model Simplification

When performing multivariate statistical analysis on a set of data for classification or quantification, it is common practice to use all the variables available.

The belief is that the statistical method used (such as PLS, PCR, MLR, PCA, ANNs) will extract from the data those variables which are most important, and discard irrelevant information. Statistical theory shows that this is incorrect. In particular, the *principle of parsimony* states that a simple model (one with fewer variables or parameters), if it is just as good at predicting a particular set of data as a more complex model, will tend to be better at predicting a new, previously unseen data set [153–155]. Our work has shown that this principle holds.

There have been a number of methods of data reduction proposed, some of which are briefly described here.

One method is to use a *variable ranking* system, in which the best n variables (where n ranges from 1 to the total number of variables), are tested. The variables used for the value of n at which the best model is formed will then be taken to be the optimal. This method has proved very successful, particularly for relatively low noise NMR data from olive oils [156–158], the results clearly showing that the use of all variables in model creation does not yield an optimal result in most cases, and for Raman data [159], where the variables are peak height, width, area and position, the peaks initially chosen being representative of certain bonds within the substance being analysed. It has the advantage of being relatively quick (only n models need be formed), and simple to understand. It can also be a great aid to understanding the data being analysed. However, it does not take account of collinearity in the data, nor the possibility that two variables may be additively, but not individually, important

Taking *Fourier transforms* of spectra (e.g. [160, 161]) and selecting a suitable cut-off will eliminate most of the noise whilst retaining most of the information. The precise point of the cut-off is not easy to determine, as there is a trade-off between eliminating noise and losing data. It is also likely that many of the remaining variables will be collinear (essentially saying the same thing), and therefore make the model unnecessarily complicated.

Using the first n *principal components* (where n is determined by some metric which attempts to remove components containing only noise) also suffers from the problem of this trade-off, but does have the advantage that no variables remaining will be collinear (therefore they all contribute different information to the model).

Genetic programming, described earlier, picks only certain variables from the model. The rules, which may be in the form of a computer language such as Lisp, or easily interpretable equations, produce a formula from which a result can be calculated (e.g. if *(measurement_1 > 2.37* and *measurement_2 < 0.53)* or *measurement_3 > 4.28* then *sample is adulterated* else *sample is clean)* [162–165]. Rather than being a pre-processing step before statistical analysis, this method combines the variable selection and model formation stages into one.

6.3
Data Partitioning

It is, at this point, important to understand the difference between *unsupervised methods* and *supervised methods*. With the former, there is no indication given to the model creation program (e.g. PCA, self-organising maps) of where any of the data should lie, or its class or value. With such a technique, therefore, one set of data is sufficient. However, if variable selection is being used to produce the optimum variables for the model, it is better to use two data sets, using one for establishing the best number of variables, and the second for producing the results.

The remainder of this section deals with supervised methods.

6.3.1
Training and Testing

In order to create a prediction, the data must be divided up into a *training set* (on which the model is formed) and a *query* or *test set* (using which the model is tested, and the best number of factors or epochs established).

Since most supervised methods of forming a model will use the query set in order to establish the optimal number of factors (or epochs, in the case of an artificial neural network), a completely independent *validation* set is required, to ensure that the model is valid. This data set will not have been seen by the model in any form at any time. The only reason for not using a third data set is where there are insufficient objects to form a meaningful model if the data are divided into three. In such cases, it must be remembered that the results may appear better than they really are, and this fact should be noted in any results. Other methods of forming a model are able to establish the optimal factors or epochs from the training set alone, for example by dividing the training set into two and alternately training the model on one section and testing on the other. In such cases, two data sets are probably sufficient.

Replicates should always be kept in the same data set; not to do so would definitely classify as 'cheating'. If one of two replicates were in the training set, it would be expected that its partner in the validation set would be predicted with accuracy.

6.3.2
The Extrapolation Problem

Statistical models are not in general able to extrapolate; that is to say, if for a given variable, the training set data are in the range 3 to 4, there is no way a meaningful prediction can be made if the validation data contains a 5. This means that the training set should encompass the whole of the query and validation sets.

For quantification (e.g. prediction of concentration in a solution), the solution is easy: objects should be placed alternately in the training, query and (where there are sufficient objects) validation sets, ensuring that the objects with the lowest and highest value in the target being predicted are in the training set.

For classification (e.g. identification of country of origin or variety of a sample, or the bacterial strain), it is not quite so straightforward, as it is difficult to know within a class (country of origin, etc.) which data lie at the edge of the spectrum. This may, however, be achieved, by examining the data for each object within a group, and determining how the variables lie with respect to those of other objects in the same class. With n variables, this means looking in n dimensional space; clearly not a task that is possible for the mere human. To facilitate this, a program called MultiPlex has been written by Dr. Alun Jones of UWA (an extension of the duplex algorithm described in [166]). Using this program will ensure that the objects are divided between training, query and, if desired, validation, sets appropriately. Provided that any replicates in the samples are correctly identified, it will also ensure that replicates are placed in the same set.

7
Concluding Remarks

"Organisms are not billiard balls, struck in deterministic fashion by the cue of natural selection and rolling to optimal positions on life's table. They influence their own destiny in interesting, complex and comprehensible ways." – S.J. Gould (1993) Evolution of organisms. In: Boyd CAR, Noble D (eds) The logic of life. Oxford University Press, p 5

Biological systems are indeed complex (and this differs from 'complicated' – [167]), but many of their most important features that are of interest to us for specific purposes are in fact of low dimensionality. The key to understanding them then lies in acquiring large amounts of the right kind of data which can act as the inputs to intelligent and sophisticated data processing and machine learning algorithms. These approaches alone – especially those based on induction – will help us unravel their workings [168].

Acknowledgments. We thank the BBSRC, the EPSRC and HEFCW for financial support of our collaborative programme in Analytical Biotechnology, Spectrometry, Chemometrics and Machine Learning.

References

1. Kell D (1998) Trends in Biotechnology 16:491
2. Oliver SG, Winson MK, Kell DB, Baganz F (1998) Trends in Biotechnology 16:373
3. DeRisi JL, Iyer VR, Brown PO (1997) Science 278:680
4. de Saizieu A, Certa U, Warrington J, Gray C, Keck W, Mous J (1998) Nature Biotechnol 16:45
5. Humphery-Smith I, Cordwell SJ, Blackstock WP (1997) Electrophoresis 18:1217
6. Wilkins MR, Williams KL, Appel RD, Hochstrasser DF (1997) Proteome research: new frontiers in functional genomics. Springer, Berlin Heidelberg New York
7. Blackstock WP, Weir MP (1999) Tibtech 17:121
8. Eisen MB, Spellman PT, Brown PO, Botstein D (1998) Proc Natl Acad Sc 95:14863
9. Kell DB (1980) Process Biochemistry 15:18
10. Clarke DJ, Kell DB, Morris JG, Burns A (1982) Ion-Selective Electrode Rev 4:75
11. Lee MS, Hook DJ, Kerns EH, Volk KJ, Rosenberg IE (1993) Biological Mass Spectrometry 22:84
12. Heinzle E, Moes J, Griot M, Kramer H, Dunn IJ, Bourne JR (1984) Analytical Chimica Acta 163:219
13. Heinzle E, Oeggerli A, Dettwiler B (1990) Analytica Chimica Acta 238:101

14. Matz G, Loogk M, Lennemann F (1998) Journal of Chromatography A 819:51
15. Namdev PK, Alroy Y, Singh V (1998) Biotechnology Progress 14:75
16. Bohatka S, langer G, Szilagyi J, Berecz I (1983) International Journal of Mass Spectrometry 48:277
17. Dongre AR, Hayward MJ (1996) Analytica Chimica Acta 327:1
18. Heinzle E, Kramer H, Dunn IJ (1985) Biotechnology and Bioengineering 27
19. Lauritsen FR, Choudhury TK, Dejarme LE, Cooks RG (1992) Analytica Chimica Acta 266:1
20. Lauritsen FR, Nielsen LT, Degn H, Lloyd D, Bohatka S (1991) Biological Mass Spectrometry 20:253
21. Lloyd D, Ellis JE, Hillman K, Williams AG (1992) Journal of Applied Bacteriology 73
22. Weaver JC (1982) Continuous monitoring of volatile metabolites by a mass spectrometer. In: Cohen JS (ed) Noninvasive Probes of Tissue Metabolism. J Wiley, New York
23. Irwin WJ (1982) Analytical Pyrolysis: A Comprehensive Guide. Marcel Dekker, New York
24. Meuzelaar HLC, Haverkamp J, Hileman FD (1982) Pyrolysis Mass Spectrometry of Recent and Fossil Biomaterials. Elsevier, Amsterdam
25. Goodacre R, Kell DB (1996) Current Opinion in Biotechnology 7:20
26. Magee JT (1993) Whole-organism fingerprinting. In: Goodfellow M, O'Donnell AG (eds). Handbook of New Bacterial Systematics. Academic Press, London, p 383
27. Tas AC, Vandergreef J (1994) Mass Spectrometry Reviews 13:155
28. Goodacre R, Berkeley RCW (1990) FEMS Microbiology Letters 71:133
29. Goodacre R, Berkeley RCW, Beringer JE (1991) Journal of Analytical and Applied Pyrolysis 22:19
30. Goodacre R, Rooney PJ, Kell DB (1998) Journal of Antimicrobial Chemotherapy 41:27
31. Goodacre R, Trew S, Wrigley-Jones C, Neal MJ, Maddock J, Ottley TW, Porter N, Kell DB (1994) Biotechnology and Bioengineering 44:1205
32. Schulten H-R, Lattimer RP (1984) Mass Spectrometry Reviews 3:231
33. Van de Meent D, de Leeuw JW, Schenck PA, Windig W, Haverkamp J (1982) Journal of Analytical and Applied Pyrolysis 4:133
34. Heinzle E, Kramer H, Dunn IJ (1985) Analysis of biomass and metabolites using pyrolysis mass spectrometry. In: Johnson A (ed) Modelling and Control of Biotechnological Processes. Pergamon, Oxford
35. Sandmeier EP, Keller J, Heinzle E, Dunn IJ, Bourne JR (1988) Development of an on-line pyrolysis mass spectrometry system for the on-line analysis of fermentations. In: Heinzle E, Reuss M (eds). Mass Spectrometry in Biotechnological Process Analysis and Control. Plenum, New York, p 209
36. Heinzle E (1992) Journal of Biotechnology 25:81
37. Goodacre R, Edmonds AN, Kell DB (1993) Journal of Analytical and Applied Pyrolysis 26:93
38. Goodacre R, Kell DB (1993) Analytica Chimica Acta 279:17
39. Goodacre R, Neal MJ, Kell DB (1994) Analytical Chemistry 66:1070
40. Goodacre R, Karim A, Kaderbhai MA, Kell DB (1994) Journal of Biotechnology 34:185
41. McGovern AC, Ernill R, Kara BV, Kell DB, Goodacre R (1999) Journal of Biotechnology 72:157–167
42. Broadhurst D, Goodacre R, Jones A, Rowland JJ, Kell DB (1997) Anal. Chim. Acta 348:71
43. Goodacre R, Trew S, Wrigley-Jones C, Saunders G, Neal MJ, Porter N, Kell DB (1995) Analytica Chimica Acta 313:25
44. McGovern AC, Broadhurst D, Taylor J, Gilbert RJ, Kaderbhai N, Small DAP, Kell DB, Goodacre R (1999) (in preparation)
45. Kang SG, Lee DH, Ward AC, Lee KJ (1998) Journal of Microbiology and Biotechnology 8:523
46. Kang SG, Kenyon RGW, Ward AC, Lee KJ (1998) Journal of Biotechnology 62:1
47. Schrader B (1995) Infrared and Raman spectroscopy: methods and applications. Verlag Chemie, Weinheim.
48. Ingle Jr JD, Crouch SR (1988) Spectrochemical Analysis, Prentice-Hall, London

49. Martin KA (1992) Applied spectroscopy reviews 27:325
50. Howard WW, Sekulic S, Wheeler MJ, Taber G, Urbanski FJ, Sistare FE, Norris T, Aldridge PK (1998) Applied Spectroscopy 52:17
51. Macaloney G, Draper I, Preston J, Anderson KB, Rollins MJ, Thompson BG, Hall JW, McNeil B (1996) Food and Bioproducts Processing 74:212
52. Yano T, Harata M (1994) Journal of Fermentation and Bioengineering 77:659
53. Marquardt LAA, M.A. Small,G.W. (1993) Anal chemistry 65:3271
54. Macaloney G, Hall JW, Rollins MJ, Draper I, Thompson BG, McNeil B (1994) Biotechnology Techniques 8:281
55. Brimmer PJ, Hall JW (1993) Canadian Journal of Applied Spectroscopy 38:155
56. Yano T, Aimi T, Nakano Y, Tamai M (1997) Journal of fermentation and bioengineering 84:461
57. Norris T, Aldridge PK (1996) Analyst 121:1003
58. Hall JW, McNeill B, Rollins MJ, Draper I, Thompson BG, Macaloney G (1996) Applied Spectroscopy 50:102
59. Riley MR, Rhiel M, Zhou X, Arnold MA (1997) Biotechnology and Bioengineering 55:11
60. McShane MJ, Cote GL (1998) Applied Spectroscopy 52:1073
61. Swierenga H, Haanstra WG, deWeijer AP, Buydens LMC (1998) Applied Spectroscopy 52:7
62. Cavinato AG, Mayes DM, Ge ZH, Callis JB (1990) Analytical Chemistry 62:1977
63. Ge ZC, AG Callis, JB (1994) Analytical Chemistry 66:1354
64. Vaccari G, Dosi E, Campi AL, Gonzalezvara A, Matteuzzi D, Mantovani G (1994) Biotechnology and Bioengineering 43:913
65. Vaccari G, Dosi E, Campi AL, Mantovani G (1993) Zuckerindustrie 118:266
66. Varadi M, Toth A, Rezessy J (1992) Application of NIR in a fermentation process. VCH Publishers, New York
67. Hammond SV (1992) NIR Analysis of Antibiotic Fermentations. In: Murray I, Cowe IA (eds) Making Light Work: Advances in Near-Infrared Spectroscopy. VCH Publishers, New York, p 584
68. Hammond SV (1992) Near-Infrared Spectroscopy – A Powerful Technique for At-Line and Online Analysis of Fermentations. In: Bose A (ed) Harnessing Biotechnology for the 21st Century: Proceedings of the Ninth International Symposium and Exhibition. American Chemical Society, Washington D.C., p 325
69. Validyanathan S, Macaloney G, McNeill B (1999) Analyst 124:157
70. Koza JR (1995) Proceedings of Wescon 95:E2. Neural-Fuzzy Technologies and Its Applications
71. McShane MJ, Cote GL, Spiegelman C (1997) Applied Spectroscopy 51:1559
72. Bangalore AS, Shaffer RE, Small GW, Arnold MA (1996) Analytical Chemistry 68:4200
73. Shaffer RE, Small GW, Arnold MA (1996) Analytical Chemistry 68:2663
74. Hassell DC, Bowman EM (1998) Applied Spectroscopy 52:A18
75. Guzman M, deBang M, Ruzicka J, Christian GD (1992) Process Control and Quality 2:113
76. Alberti JC, Phillips JA, Fink DJ, Wacasz FM (1985) Biotechnology and Bioengineering Symp. 15:689
77. Picque D, Lefier D, Grappin R, Corrieu G (1993) Analytica Chimica Acta 279:67
78. Fayolle P, Picque D, Corrieu G (1997) Vibrational Spectroscopy 14:247
79. Hayakawa K, Harada K, Sansawa H (1997) Abstracts of the 8th European Congress on Biotechnology 275
80. Wilson RH, Holland JK, Potter J (1994) Chemistry in Britain 30:993
81. Winson MK, Goodacre R, Timmins ÉM, Jones A, Alsberg BK, Woodward AM, Rowland JJ, Kell DB (1997) Analytica Chimica Acta 348:273
82. Kell DB, Winson MK, Goodacre R, Woodward AM, Alsberg BK, Jones A, Timmins ÉM, Rowland JJ (1998) DRASTIC (Diffuse Reflectance Absorbance Spectroscopy Taking In Chemometrics). A novel, rapid, hyperspectral, FT-IR-based approach to screening for biocatalytic activity and metabolite overproduction. In: Kieslich K (ed) New Frontiers in Screening for Microbial Biocatalysts. Elsevier Science B.V., The Netherlands, p 61

83. Winson MK, Todd M, Rudd BAM, Jones A, Alsberg BK, Woodward AM, Goodacre R, Rowland JJ, Kell DB (1998) A DRASTIC (Diffuse Reflectance Absorbance Spectroscopy Taking in Chemometrics) approach for the rapid analysis of microbial fermentation products: quantification of aristeromycin and neplanocin A in *Streptomyces citricolor* broths. In: Kieslich, K (ed) New Frontiers in Screening for Microbial Biocatalysts. Elsevier Science B.V., The Netherlands, p 185
84. Kell DB, Sonnleitner B (1995) Trends Biotechnol. 13:481
85. Montague GA (1997) Monitoring and control of fermenters, Institute of Chemical Engineers London
86. Pons M-N (1991) Bioprocess monitoring and control. Hanser, Munich
87. Kell DB, Markx GH, Davey CL, Todd RW (1990) Trends in Analytical Chemistry 9:190
88. Adar F, Geiger R, Noonan J (1997) Applied Spectroscopy Reviews 32:45
89. Chase B (1994) Appl Spectrosc 48:14 A
90. Gerrard DL (1994) Analytical Chemistry 66:R 547
91. Góral J, Zichy V (1990) Spectrochimica Acta 46 A:253
92. Graselli JG, Bulkin BJ (1991) Analytical Raman spectroscopy, John Wiley, New York
93. Hendra P, Jones C, Warnes G (1991) Fourier Transform Raman Spectroscopy. Ellis Horwood, Chichester
94. Hendra PJ, Wilson HMM, Wallen PJ, Wesley IJ, Bentley PA, Arruebarrena Baez M, Haigh JA, Evans PA, Dyer CD, Lehnert R, Pellow-Jarman MV (1995) Analyst 120:985
95. Hirschfeld T, Chase B (1986) Appl Spectrosc 40:133
96. Keller S, Schrader B, Hoffmann A, Schrader W, Metz K, Rehlaender A, Pahnke J, Ruwe M, Budach W (1994) J Raman Spectrosc 25:663
97. Naumann D, Keller S, Helm D, Schultz C, Schrader B (1995) Journal of Molecular Structure 347:399
98. Parker SF (1994) Specrochim. Acta 50 A:1841
99. Puppels GJ, Colier W, Olminkhof JHF, Otto C, Demul FFM, Greve J (1991) Journal of Raman Spectroscopy 22:217
100. Puppels GJ, Greve J (1993) Adv Spectrosc 20 A:231
101. Puppels GJ, Schut TCB, Sijtsema NM, Grond M, Maraboeuf F, Degrauw CG, Figdor CG, Greve J (1995) Journal of Molecular Structure 347:477
102. Schrader B, Baranovic G, Keller S, Sawatzki J (1994) Fresenius Journal of Analytical Chemistry 349:4
103. Treado PJ, Morris MD (1994) Applied spectroscopy reviews 29:1
104. Twardowski J, Anzenbacher P (1994) Raman and infrared spectroscopy in biology and biochemistry. Ellis Horwood, Chichester
105. Carrabba MM, Spencer KM, Rich C, Rauh D (1990) Appl Spectrosc 44:1558
106. Kim M, Owen H, Carey PR (1993) Applied Spectroscopy 47:1780
107. Puppels GJ, Huizinga A, Krabbe HW, Deboer HA, Gijsbers G, Demul FFM (1990) Review of Scientific Instruments 61:3709
108. Tedesco JM, Owen H, Pallister DM, Morris MD (1993) Analytical Chemistry 65:A 441
109. Treado PJ, Morris MD (1990) Spectrochimica Acta Reviews 13:355
110. Turner JF, Treado PJ (1996) Applied Spectroscopy 50:277
111. Griffiths PR, de Haseth JA (1986) Fourier transform infrared spectrometry. John Wiley, New York
112. Williams KPJ, Pitt GD, Batchelder DN, Kip BJ (1994) Applied spectroscopy 48:232
113. Williams KPJ, Pitt GD, Smith BJE, Whitley A, Batchelder DN, Hayward IP (1994) Journal of Raman Spectroscopy 25:131
114. Erckens RJ, Motamedi M, March WF, Wicksted JP (1997) Journal Of Raman Spectroscopy 28:293
115. Wicksted JP, Erckens RJ, Motamedi M, March WF (1995) Applied Spectroscopy 49:987
116. Shope TB, Vickers TJ, Mann CK (1987) Applied Spectroscopy 41:908
117. Gomy C, Jouan M, Dao NQ (1988) Analytica Chimica Acta 215:211
118. Gomy C, Jouan M, Dao NQ (1988) Comptes Rendus De L Academie Des Sciences Serie Ii-Mecanique Physique Chimie Sciences De L Univers Sciences De La Terre 306:417

119. Spiegelman CH, McShane MJ, Goetz MJ, Motamedi M, Yue QL, Cote GL (1998) Anal Chem 70:35
120. Harris CM, Todd RW, Bungard SJ, Lovitt RW, Morris JG, Kell DB (1987) Enzyme Microbial Technol 9:181
121. Kell DB (1987) The principles and potential of electrical admittance spectroscopy: an introduction. In: Turner APF, Karube I, Wilson GS (eds). Biosensors; fundamentals and applications. Oxford University Press, Oxford, p 427
122. Pethig R, Kell DB (1987) Phys Med Biol 32:933
123. Kell DB, Kaprelyants AS, Weichart DH, Harwood CL, Barer MR (1998) Antonie van Leeuwenhoek 73:169
124. Kell DB, Davey CL (1992) Bioelectrochemistry and Bioenergetics 28:425
125. Nicholson DJ, Kell DB, Davey CL (1996) Bioelectrochemistry and Bioenergetics 39:185
126. Davey CLK, D. B. (1998) Bioelectrochemistry and Bioenergetics 46:91
127. Davey CL, Kell DB (1998) Bioelectrochemistry and Bioenergetics 46:105
128. Debye P (1929) Polar Molecules. Dover Press, New York
129. Woodward AM, Kell DB (1990) Bioelectrochemistry and Bioenergetics 24:83
130. Davey CL, Kell D B (1990) The dielectric properties of cells and tissues what can they tell us about the mechanisms of field/cell interactions. In: O'Connor ME, Bentall RHC, Monahan JC (eds). Emerging Electromagnetic Medicine. Springer, Berlin Heidelberg New York, p 19
131. Kell DB, Astumian RD, Westerhoff HV (1988) Ferroelectrics 86:59
132. Martens H, Næs T (1989) Multivariate calibration. John Wiley, Chichester
133. Woodward AM, Gilbert RJ, Kel DB (1999) Bioelectrochemistry and Bioenergetics (in press)
134. Woodward AM, Davies EA, Denyer S, Olliff C, Kell DB (1999) Submitted for publication in Journal of Electroanalytical Chemistry
135. Woodward AM, Jones A, Zhang X-Z, Rowland JJ, Kell DB (1996) Bioelectrochemistry and Bioenergetics 40:99
136. Jeon SI, Lee JH, Andrade JD, de Gennes PG (1991) Journal of Colloidal and Interface Science 142:149
137. Koza JR (1992) Gentic programming: on the programming of computers by means of natural selection, MIT press Cambridge, MA
138. McShea A, Woodward AM, Kell DB (1992) Bioelectrochemistry and Bioenergetics 29:205
139. Davies EA, Olliff C, Wright I, Woodward AM, Kell DB (1999) Bioelectrochemistry and Bioenergetics (in press)
140. Davies EA, Woodward AM, Kell DB (1999) Bioelectromagnetics (in press)
141. Davey HM, Kell DB (1996) Microbiol Rev 60:641
142. Shapiro HM (1995) Practical flow cytometry, 3rd edn. Alan R. Liss, New York
143. Münch T, Sonnleitner B, Fiechter A (1992) Biotechnology 22:329
144. Münch T, Sonnleitner B, Fiechter A (1992) Journal of Biotechnology 24:299
145. Srienc F, Arnold B, Bailey JE (1984) Biotechnology and Bioengineering 26:982
146. Müller S, Lösche A, Bley T, Scheper T (1995) Applied Microbiology and Biotechnology 43:93
147. Müller S, Lösche A, Bley T (1993) Acta Biotechnol 13:289
148. Gjelsnes O, Tangen R (1994) Norway patent WO 94/29695
149. Rønning Ø (1999) Genetic Engineering News 19:18
150. Degelau A, Freitag R, Linz F, Middendorf C, Scheper T, Bley T, Müller S, Stoll P, Reardon KF (1992) Journal of Biotechnology 25:115
151. Zhao R, Natarjan A, Srienc F (1999) Biotechnology and Bioengineering 62:609
152. Neal MJ, Goodacre R, Kell DB (1994) On the analysis of pyrolysis mass spectra using artificial neural networks. Individual input scaling leads to rapid learning. in Proceedings of the World Congress on Neural Networks. International Neural Network Society San Diego
153. de Noord OE (1994) Chemometrics and Intelligent Laboratory Systems 23:65

154. Flury B, Riedwyl H (1988) Multivariate Statistics: A Practical Approach. Chapman and Hall, London
155. Seasholtz MB, Kowalski B (1993) Analytica Chimica Acta 277:165
156. Shaw AD, di Camillo A, Vlahov G, Jones A, Bianchi G, Rowland J, Kell DB (1996) Discrimination of Different Olive Oils using ^{13}C NMR and Variable Reduction. in Food Authenticity '96. Norwich, UK
157. Shaw AD, di Camillo A, Vlahov G, Jones A, Bianchi G, Rowland J, Kell DB (1997) Analytica Chimica Acta 348:357
158. Vlahov G, Shaw AD, Kell DB (1999) Accepted for publication in Journal of the American Oil Chemists Society
159. Shaw AD, Kaderbhai N, Jones A, Woodward A, Goodacre R, Rowland J, Kell DB (1999) Accepted for publication in Applied Spectrometry
160. Boschelle O, Giomo A, Conte L, Lercker G (1994) La Rivista Italiana delle Sostanze Grasse 71:57
161. Hazen KHA, MA Small, GW (1994) Applied spectroscopy 48:477
162. Gilbert RJ, Goodacre R, Woodward AM, Kell DB (1997) Analytical Chemistry 69:4381
163. Gilbert RJ, Goodacre R, Shann B, Taylor J, Rowland JJ, Kell DB (1998) Genetic Programming based Variable Selection for High Dimensional Data in Proceedings of Genetic Programming 1998. Morgan Kaufmann, Madison, Wisconsin, USA
164. Taylor J, Winson MK, Goodacre R, Gilbert RJ, Rowland JJ, Kell DB (1998) Genetic Programming in the Interpretation of Fourier Transform Infrared Spectra: Quantification of Metabolites of Pharmaceutical Importance in Genetic Programming 1998. Morgan Kaufmann, Madison, Wisconsin, USA
165. Taylor J, Goodacre R, Wade W, Rowland JJ, Kell DB (1998) FEMS Microbiology Letters 160:237
166. Snee RD (1977) Technometrics 19:415
167. Bialy H (1999) Nature Biotechnology, in the press
168. Kell DB, Mendes P (1999) Snapshots of systems: metabolic control analysis and biotechnology in the post-genomic era. In: Cornish-Bowden A, Cardenás ML (eds). Technological and Medical Implications of Metabolic Control Analysis (in press) (and see http://gepasi.dbs.aber.ac.uk/dbk/mca99.htm). Plenum Press, New York

On-line and Off-line Monitoring of the Production of Cephalosporin C by *Acremonium chrysogenum*

G. Seidel, C. Tollnick, M. Beyer and K. Schügerl*

Institut für Technische Chemie der Universität Hannover, Callinstr. 3, D-30167 Hannover, Germany
* E-mail: Schuegerl@mbox.iftc.uni-hannover.de

Process monitoring of cephalosporin C formation by *Acremonium chrysogenum* in laboratory investigations is considered. The goal of these investigations is the identification of bottlenecks in the biosynthesis and the improvement of the process performance. Based on reports of other research groups and own experience the key parameters were selected, which influence the process performance. They are: dissolved oxygen and pH values. In addition the concentrations of biomass, DNA, glucose and reducing sugars (glucose, maltose, maltotriose and oligosaccharides), methionine, other nitrogen sources (ammonium ion, other amino acids), organic acids, phosphate, sulfate, dissolved organic carbon, proteins, product and precursors in the cell free cultivation medium are monitored. In addition the intracellular concentrations of RNA, DNA, proteins, amino acids as well as the activities of the enzymes of the biosynthesis of cephalosporin C are determined. The influence of these parameters on the biosynthesis is discussed.

Keywords. Cephalosporin C, *Acremonium chrysogenum* Extracellular components, Intracellular components, Enzyme activities of biosynthesis, Bottleneck of biosynthesis

1	Introduction	115
2	Analysis of Extracellular Components	118
2.1	Off-Line Monitoring of Medium Components	118
2.2	On-Line Monitoring of Medium Components	118
2.3	On-Line and Off-Line HPLC Monitoring	119
2.3.1	On-Line Product and Precursor Analysis	120
2.3.2	Off-Line Amino Acid and ACV Analysis	122
2.3.3	On-Line Carboxylic Acid Analysis	123
3	Analysis of Intracellular Components	124
3.1	Cell Disintegration	124
3.2	Measurement of RNA	125
3.3	Preconditioning, Amino Acid Concentrations and Enzyme Activity Measurement	126
4	2D-Fluorescence Spectroscopy	127
5	Morphology	127

6   Discussion 128

7   Conclusion 130

References 130

List of Symbols and Abbreviations

AA	amino acid
AAA	α-amino adipinic acid
ACV	δ-(α-aminoadipyl-L-cysteinyl-D-valine
ACVS	ACV synthase
Arg	arginine
Asp	asparagine
CPR	carbon dioxyde formation rate
CPC	cephalosporin C
Cys	cystein
DAC	deacetylcephalosporin C
DAOC	deacetoxycephalosporin C
DAOCS	DAOC synthase
DNA	desoxyribonucleic acid
DO	dissolved oxygen concentration
FIA	flow injection analysis
Glu	glutamate
Gln	glutamine
Gly	glycine
GOD	glucose oxidase
HPLC	high performance liquid chromatography
His	histidin
Ile	isoleucine Ile
IPN	isopenicillin N
IPNS	IPN synthase
Leu	leucine
Lys	lysine
NAD(P)H	nicotinamid-adenin-dinucleide(phosphate)
OPA	o-phtalaldehyde
OTR	oxygen transfer rate
PEN	penicillin N
Phe	phenylalanine
Pro	proline
RNA	ribonucleic acid
Ser	serine
Trp	tryptophan
Tyr	tyrosine
Val	valine

1
Introduction

Process monitoring is a prerequisite for the improvement of process performance, that means the identification of bottlenecks in biosynthesis, to improve the process strategy and to use suitable process control. In situ and fast on-line monitoring is needed for process control, but no fast analysis is necessary for identification of bottlenecks and for evaluation of the optimal process strategy.

For the production process only the monitoring of the main key parameters are needed when the optimal process has been evaluated in laboratory or in pilot plant.

In the present paper the process monitoring of cephalosporin C (CPC) formation by *Acremonium chrysogenum* in laboratory investigations are considered. The goal of these investigations was the identification of bottlenecks in the biosynthesis and the improvement of process performance by maximizing the product concentration and yield coefficient and by minimizing the by-products, which impair the product purification.

Based on reports of other groups and own experience key parameters were selected, which influence the process performance. Easily-utilized carbon sources such as glucose have been known for long time ago to exert negative effects on β-lactam biosynthesis [1–4]. Methionine markedly stimulates CPC and penicillin N (PEN) formation by *A. chrysogenum* [5, 6] and influences the morphology [7].

The biosynthesis of β-lactam antibiotics is subject to nitrogen source regulation [8, 9], so ammonium ions repress β-lactam synthetases, especially expandase in *A. chrysogenum* [10] and ammonium sulfate represses both ACVS and expandase [11]. Excess of phosphate exerts its negative effect indirectly by increasing the rate of glucose consumption, thus enhancing carbon source regulation [12]. However, Martin et al. [13] found that phosphate interferes with CPC and PEN production in absence of glucose in the same fungus. ACVS, cyclase and expandase are repressed by higher concentration of phosphate [14]. The dissolved oxygen concentration (DO) level strongly influences the biosynthesis of CPC. At low DO PEN concentration increases and CPC production is repressed [4, 15–17]. In contrast to this the growth rate is only moderately influenced because of the alternative (cyanide resistant) respiration of this fungus, which becomes active at low DO values [18–20]. The cephalosporin concentration is considerably influenced by the amount of peanut flour [21] as well as by the amount of corn steep liquor [22]. At the end of the cultivation a significant amount of deacetycephalosporin C (DAC) is formed. The deviation of the pH value of the cultivation medium from its optimum enhances the decomposition of CPC to DAC (unpublished results). Bottlenecks of the biosynthesis of CPC were investigated using low producing *A. chrysogenum* strains [23, 24]. Morphological differentiation is important for the product formation: arthrospore and product formation are closely connected [25–27]. The effects of various ions on the activity of the enzymes of biosynthesis were investigated by several research groups: Mg^{2+} on ACVS [28] Fe^{2+} on cyclase [29] Fe^{2+} on expandase/hydrolase [10, 30], Mg^{2+} on acyltransferase [31, 32].

The key components for process monitoring were selected according to these data and some additional information. The in situ monitoring of the DO as well as the oxygen and carbon dioxide concentrations in the off gas allowed the evaluation of the oxygen transfer rate (OTR), the CO_2 production rate (CPR) and the respiratory quotient (RQ). The control of the pH-value and the DO was the prerequisite for the maintenance of the optimal growth and product formation conditions.

2
Analysis of Extracellular Components

2.1
Off-Line Monitoring of Medium Components

Before and after the sampling, the sampling valve of the bioreactor was sterilized with wet steam. It was necessary to determine the pH of the samples to calibrate the pH values monitored in situ with the installed electrodes, because of the their drift during the wet steam sterilization of the whole reactor.

The biomass determination by weight of the sediment after centrifugation at 3500 rpm at 10 min and drying at 110 °C for 48 h was not possible because of the solid content of the cultivation medium. However, the solid medium components were consumed gradually, thus after 100 h the sediment amount was identical to the biomass one. The variation of the sediment amount with the time in the first 100 h gave only a qualitative picture on the increase of the biomass amount.

Protein was analyzed according to Bradford [33] from the supernatant of 3×2 ml samples after storage at -18 °C. DNA concentration was determined according to the method of Dische modified by Burton [34]. One of the C-source was dry glucose syrup consisting of 1.5 % glucose, 35 % maltose, 21 % maltotriose and 42.5 % oligosaccharides. The sample was treated with amyloglucosidase and the glucose formed was determined by p-hydroxybenzoic acid hydrazide (pHBAH) as described by Lever [35] and Schmidt [36]. The test set of Merck (Spectroquant P14848) was used for the determination of phosphate concentration. The method of the German Industrial Standard (DIN 1983) was applied for the measurement of the ammonium ion concentration. To render the carbon balance, the dissolved organic carbon (DOC) and dissolved inorganic carbon (DIC) were measured with a TOCOR 3-Analyzer (Maihak Co).

2.2
On-Line Monitoring of Medium Components

The samples for on-line analysis were taken through a polysulfone membrane [37]. The seven channel automatic analyzer system consisted of six air-segmented continuous flow-wet chemical analyzers (Skalar analytics) to measure the concentration of ammonia with ion selective electrode (Philips IS-570), phosphate with ammonium molybdate at 880 nm, reducing sugar with pHBAH at 410 nm, methionine with Na-nitroprusside at 505 nm, cephalosporin C with cephalosporinase at 254 nm, sulfate with methylthymolblue complex at 450 nm [38, 39].

The glucose was determined enzymatically (Yellow Springs Instruments). The response time of the complete analyzer system was ca. 37 min, of which 5 min are due to the transfer distance of 3 m between the reactor, filtration module and analyzer system. The continuous air segmented flow analyzers have the disadvantage that after 3-4 days operation, microorganisms growth was detected in the non aseptic analyzer system.

Better experience was gained with flow injection analyzers (FIAs), in which a small amount of sample is injected into the carrier stream of reagent and the reaction product concentration is monitored in a flow through detector [40]. If detectors with short response time are used and depending on the type of analyte and matrix, there is a capacity for 15-20 measurements per hour. Biosensors are especially suitable as detectors. On-line determination of cephalosporin C was determined with a β-lactamase field effect transistor (FET)-FIA [41] and urea with a urease FET-FIA [42]. Glucose was determined by a glucose oxydase (GOD) FET-FIA. During the reaction gluconic acid is formed and the concentration is determined by the pH sensitive FET [43]. Because this detector is sensitive to the buffer capacity of the sample as well, different fluoride sensitive FETs were applied [44]. The hydrogen peroxyde formed during the reaction of glucose with GOD reacts with 4-fluoroaniline in presence of peroxidase (POD) and forms a fluoride ion which is detected by the fluoride sensitive FET [44].

Another FIA system uses a cartridge of 1 ml volume with immobilized enzyme, e.g. glucose oxidase (GOD) which converts glucose to gluconic acid. This reaction produces hydrogen peroxide and consumes oxygen, the concentration of which is measured by a mini oxygen electrode. The concentration of several other sugars can be measured with this method after they were converted enzymatically to glucose, glucopyranose, xylopyranose, galactopyranose, etc. [45] which are converted by the corresponding oxidases to gluconic acid or keto-sugars and hydrogen peroxide. These reactions consume oxygen as well. Also alcohols, amino acids, lactate were measured by means of the corresponding oxidases [46].

2.3
On-Line and Off-Line HPLC Analyses

The biosynthesis of CPC is well known. The first step of CPC synthesis is the formation of the tripeptide δ-(L-α-aminoadipyl)-L-cysteinlyl-D-valine (LLD-ACV) from L-α-aminoadipinic acid (AAA), L-cystein (CYS) and D-valine (VAL) by the LLD-ACV synthetase (ACVS). LLD-ACV is converted with the enzyme isopenicillin N-synthetase (cyclase) to isopenicillin N (IPN) and then with isopenicillin N-epimerase to penicillin N (PEN). The 6-ring is formed by deacetoxycephalosporin C-synthase (expandase). Deacetoxycephalosporin C (DAOC) is converted with deacetoxycephalosporin C-hydroxylase to decacetylcephalosporin C (DAC) and the latter with deacetylcephalosporin C-acyltransferase to CPC (Fig. 1). Except the amino acid and ACV, all others were monitored by on-line HPLC.

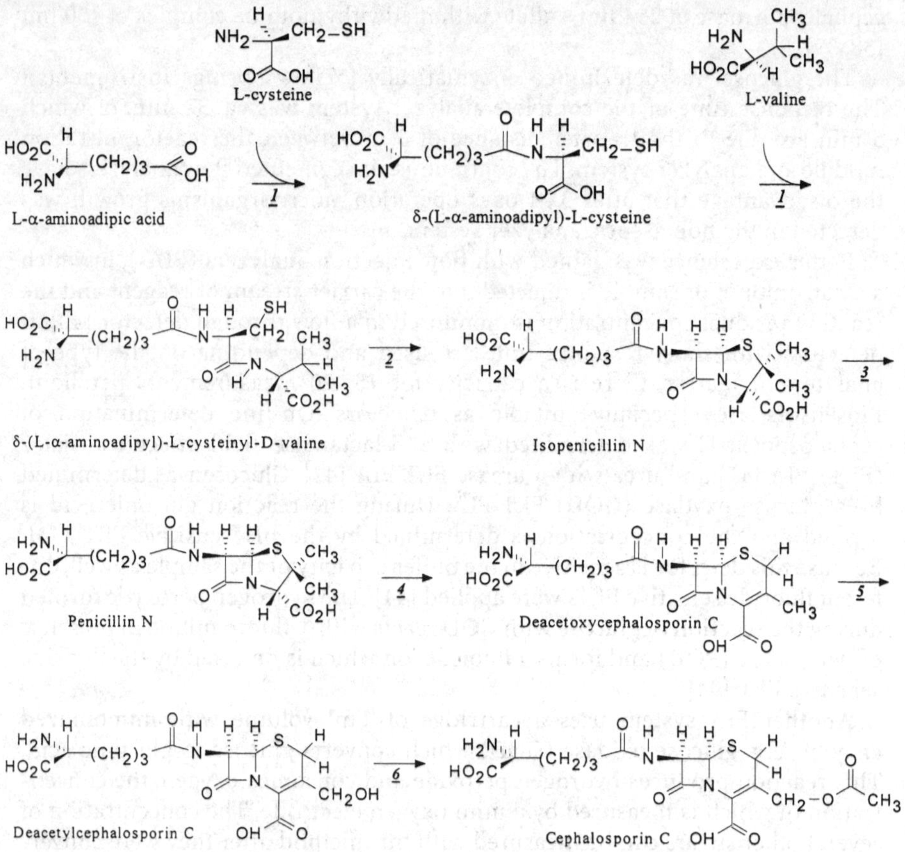

Fig. 1. Biosynthesis of Cephalosporin C in *Acremonium chrysogenum*

<u>1</u> ACV-Synthetase

<u>2</u> Isopenicillin N-Synthetase

<u>3</u> Isopenicillin N-Epimerase

<u>4</u> Deacetoxycephalosporin C-Synthase (Expandase)

<u>5</u> Deacetoxycephalosporin C-Hydroxylase

<u>6</u> Deacetylcephalosporin C-Acetyltransferase

2.3.1
On-Line Product and Precursor Analyses

The samples were taken directly from the reactor through two modules shown in Fig. 2. The modules were inserted in the reactor and connected via two peristaltic pumps with an autosampler including a rack, which was refrigerated to −18 °C. The samples could be stored at −20 °C for some weeks without a

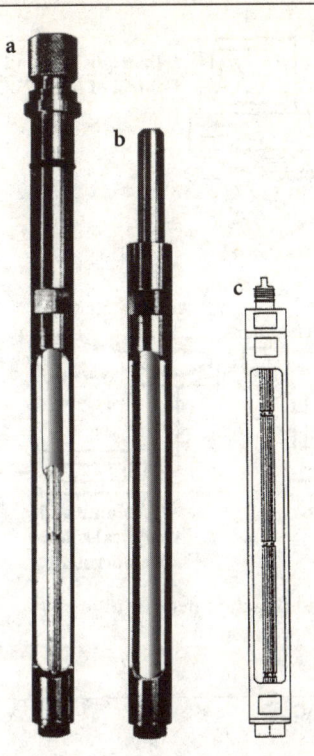

Fig. 2. Modules for on-line sampling of cell-free cultivation medium

detectable change in composition. Before the analyses the samples were deproteinated with methanol, stored on ice overnight and then centrifuged. The product and precursor analysis was performed with an automated HPLC-system (Fig. 3), even the later described amino acid detection was carried out with an equal scheme.

The system consisted of an autosampler (Gilson XL 222, Abimed), pump (Irica), injection valve (Valco), precolumn, column (Macherey and Nagel), thermostate (Julabo), two channel UV-detector (Gynkotek), interface (CSI-box) for control of sampling and data monitoring and a PC (486 DX66) containing the APEX-chromatography Workstation (Autochrom inc. Version 2.11). A reversed-phase Nucleosil $5^R C_{18}$ packed column with 12.5 or 25 cm length was used for the separation of the substances. Methanol containing aqueous phosphate buffer supplemented with tetrabutylammoniumhydrogen sulfate was applied as eluent. The detection was performed at 260 and 220 nm and chromatograms were monitored and evaluated by APEX workstation. The further evaluation of the analysis results were carried out with Program Origin 5.0 (MicroCal Software). Figure 4 shows a typical chromatogram of the product and precursors.

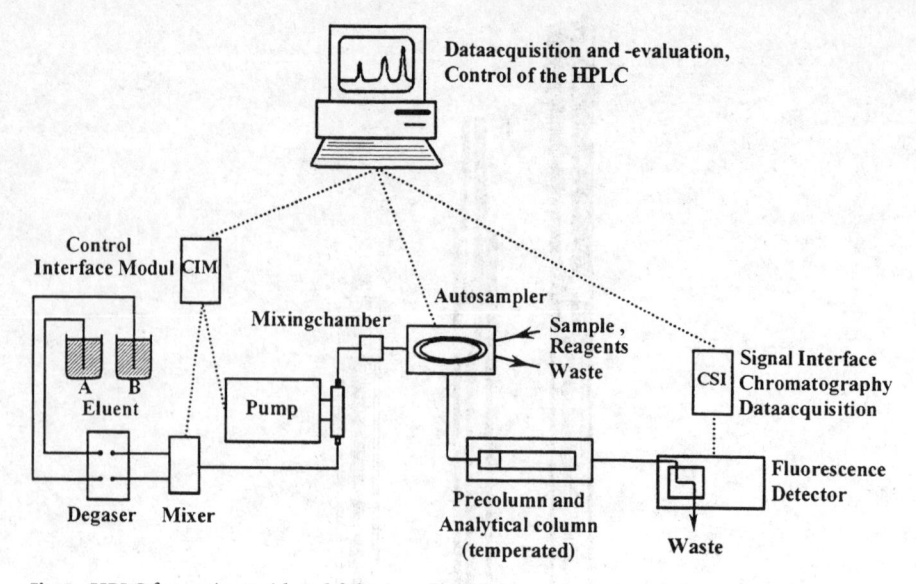

Fig. 3. HPLC for amino acid and β-lactam detection

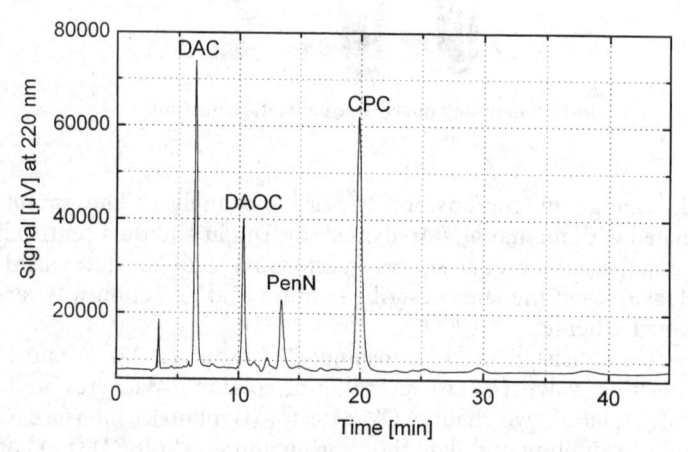

Fig. 4. Separation of the four main products appearing during the cultivation of *Acremonium chrysogenum* at 220 nm

2.3.2
Off-Line Amino Acid and ACV Analyses

The method recommended by Cooper et al [47] was applied for the analysis of the amino acids: Asp, Glu, AAA, Asn, Ser, Tyr, Met, Val, Trp, Phe, Ile, Leu and Lys. The HPLC-system consisted of an autosampler with injection valve (Pharmacia Biosystems), pump (Irica), degaser (Gastorr), thermostat (Julabo), precolumn and column (Waters), fluorescence detector (Shimadzu) and a PC (486 DX66)

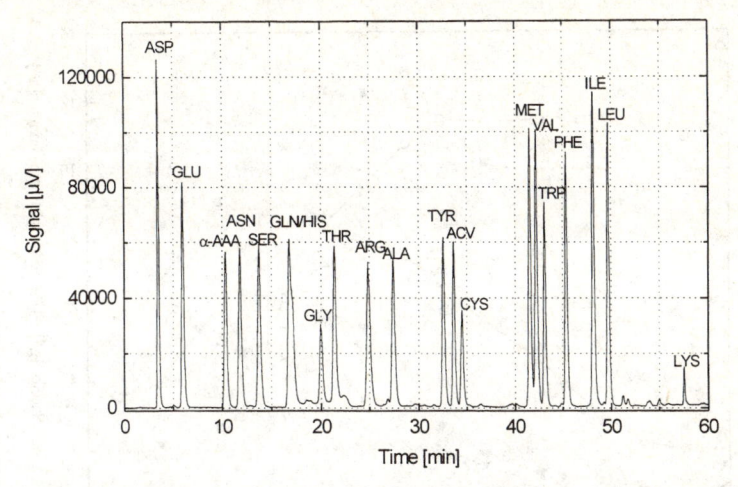

Fig. 5. A typical chromatogram of the analysis of amino acids after treating the samples with methyliodide (necessary for the detection of ACV and CYS)

containing the APEX-chromatography Workstation (Autochrom Inc. Version 2.11). The separation was performed on a Resolve $^{TM}C_{18}$ 90 A, 5 µm column (Waters) at 30 °C. Amino acid standard (Sigma) was applied for the evaluation of the chromatograms. Samples of cell free extract were deproteinated with methanol diluted with borate buffer. The precolumn derivatization with OPA was performed in the autosampler (Pharmacia Biosystems). Linear gradient of eluent A (sodium acetate, sodium dihydrophosphate and tetrahydrofuran) and eluent B (54% methanol) were applied. The concentration measurement was carried out with fluorescence detector and the evaluation with APEX software.

The Gln, His, Gly and Thr separation could be impaired by increasing amount of antibiotics. The separation of Cys and ACV was carried out according to Holzhauer-Rieger [48]. The chromatographic separation was performed after reduction and methylation of the sample. After automatic precolumn OPA derivatization the same gradient/eluents were used as for amino acid analysis. The gradient control was performed by Autochrom CIM Interface. The concentrations were measured with fluorescence detector and the data monitoring carried out with Autochrom CSI2 Interface. Figure 5 shows a typical chromatogram after reduction and methylation.

2.3.3
On-Line Carboxylic Acid Analyses

The organic acids were analyzed with an HPLC system consisting of degassing unit (Gastorr GT-103), pump (Pharmacia LKB), autosampler (Pharmacia, LKB), precolumn (Shodex R Spak KC-LG), separation column (Shodex R Spak KC-811), water bath (Julabo U3) and a UV detector (Soma Optics LTD S-3702). The data monitoring was performed using a PC-chromatography data system Andromeda 1.6 (Techlab).

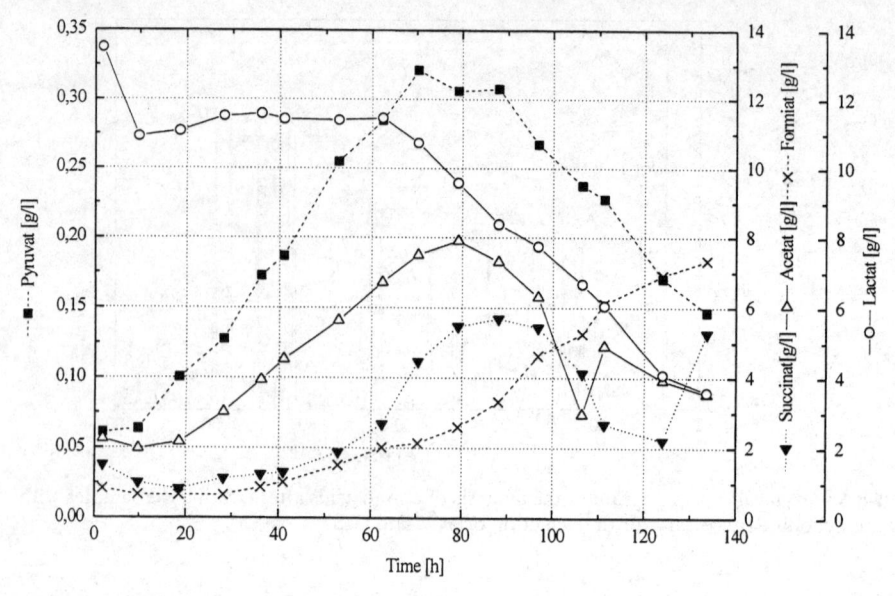

Fig. 6. A typical chromatogram of the analysis of organic acids detected in the cultivation medium

The cell free samples were mixed with perchloric acid/perchlorate and stored at low temperature for 48 h, at 6000 rpm centrifuged, decanted, stored at −22 °C for a week and centrifuged again for deproteination. The separation was performed on an ion exclusion polystyrene/divinylbenzene column at 40 °C with 15 mM sulfuric acid as eluent and detection at 205 nm. For the quantification a standard was applied. A typical chromatogram is shown in Figure 6. Pyruvate, succinate, lactate, formate and acetate were detected in the cultivation medium.

3
Analysis of Intracellular Components

3.1
Cell Disintegration

For measurement of intracellular components it is necessary to stop the metabolism very quickly and to disintegrate the cells quantitatively. This was a problem in the past when using ultrasonic modules – because of the different morphological structures is was nearly impossible to destroy all the structures in one step. Our recent investigations showed that there are really great differences according to which method is used for cell disruption. The best results were obtained by treatment of the samples with a French Press [49] or with a mortar under liquid nitrogen [22]. The mortar method was applied for routine cell disintegration. The frozen sediments were placed in a cooled mortar with liquid nitrogen and then treated for 5–10 min. It is very important that there is no warm up of the samples because of handling after cell disintegration. After disintegration, the cell powder is placed in a reaction tube and slowly warmed up. After centrifugation (3500 rpm, 3 min, 2 °C) the assays could be carried out.

3.2
Measurement of RNA

The determination of RNA was carried out with a method according to Kuenzi [50]. The method depends on the reaction of DNA and RNA with alkalihydroxide. According to the pH-value only the RNA-monomers could be detected quantitatively in solution at 260 nm. By using a standard solution of RNA from *Torula utilis* it is possible to calibrate the system for a reproducible process and an optimal dilution. To eliminate the dimming effects of the solution another maximum at 320 nm is detected. The evaluation is performed on the spectrometer by using a simple equation, namely $OD_{sample} = E_{260} - E_{320}$, the conversion factor depends on the standard solution. Fig. 7 shows the typical correlation between RNA and biomass.

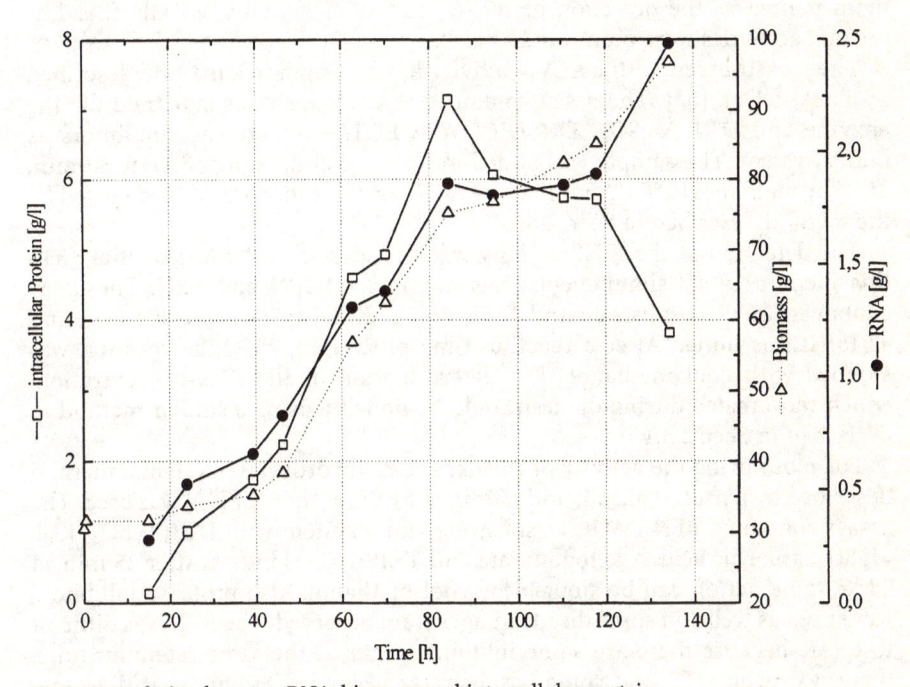

Fig. 7. Correlation between RNA, biomass and intracellular protein

3.3
Preconditioning, Amino Acid Concentrations and Enzyme Activity Measurement

For the separation on the HPLC System it was necessary to prepare the samples, because they contained glycerol from the disintegration buffer, which causes some trouble with measuring the enzyme activities, especially for determination of AVCS-activity. The samples were centrifuged (14000 pm, 2 °C and then filtered over a Sephadex Gel column (NAP™-10/Pharmacia) with an eluent containing TRIS/HCl, KCl, $MgSO_4$ and DTT. After that the samples could be stored at $-80\,°C$ for weeks.

For measurement of the intracellular amino acid concentrations the same HPLC-system and method were used as described in Sect. 2.3.2. The only differences were the very low concentrations. It was possible to detect all amino acids but not over the whole cultivation time, because of the detection level. In this case a extrapolation program (Origin 5.0, MicroCal) was used.

The main aspect for monitoring intracellular components was to find the bottlenecks in cephalosporin synthesis. For this it was important to measure the activities of the relevant enzymes in the biosynthesis. After tests in the past [49, 51] and new investigations [22] it was known that the use of protease inhibitors in the disintegration buffer was necessary for routine analysis of activities in crude material. The expense for such measurements was very high and so the main point was the detection of ACVS- and IPNS-activity. But all other important activities were monitored as well.

The measurement of the ACVS-activity has its origin in a method described by Banko et al. [52]. The assay contained AAA, Val, Cys as substrate for the enzyme and DTT, $MgSO_4$, TRIS/HCl, ATP, EDTA and protease inhibitors as buffer system. The sample was added and the reaction stopped after 60 min, 25 °C by methanol (0 °C). After that the ACV-concentration could be detected by the method described in Sect. 2.3.2.

The detection of the IPNS-activity was first described by Kupka et al. [53], this method is for simultaneous measurement of IPN and PEN. The assay contained ACV as substrate and DTT, $FeSO_4$, KCl, $MgSO_4$, ascorbic acid and TRIS/HCl as buffer. After a reaction time of 20 min, 25 °C the reaction was stopped with cool methanol. The determination of the PEN-concentration, which was created during the test, could be done by using a similar method as described in Sect. 2.3.1.

For monitoring the activity of the expandase/hydroxylase-system a method described by Jensen et al. [54] and modified by the authors of [49] was used. The assays contained PEN DAOC as substrate and a system with DTT, $FeSO_4$, KCl, $MgSO_4$, ascorbic acid, α-ketoglutarate and TRIS/HCl as buffer. After 25 min, at 25 °C, the reaction can be stopped by cool methanol. Also protease inhibitors were used as well and some disadvantages were observed. The PEN was difficult to locate because there are some inhibitor peaks at the same retention time, there was even DTT and some cosubstrates leaving the column at the same time. However, it is possible to measure the activity over the whole cultivation time if many tests are done. The detection of DAOC and DAC was carried out with the same method as the cultivation byproducts.

Matsuyama et al. [55] described a method for measuring the acetyltransferase-activity which was used. The essay contained DAC as substrate and $MgSO_4$, acetyl-CoA and TRIS/HCl as buffer. After 20 min, at 25 °C, the reaction was stopped with methanol (0 °C). The detection of CPC formed was carried out with the same method as the cultivation main product.

The evaluations of the activities were done by extracting the necessary product concentration out of the chromatogram, calculating conversion of the enzymatic reaction, and using the intracellular protein concentration, measured by the method of Bradford [33].

4
2D-Fluorescence Spectroscopy

Fluorescence sensors have been used since 1957 to measure cell internal NAD(P)H at 450 nm. Later on they were applied for in situ determination of the cell concentration. However, the culture fluorescence intensity is not only influenced by the cell concentration, but also by the physiological state of the cells [56] and, in addition to that, there are several other compounds that participate in the fluorescence emission besides NAD(P)H. To identify the fluorophores in the cells and cultivation medium, the excitation and the emission wave lengths are varied in a broad range [57, 58]. Two instruments were applied for the 2D-fluorescence spectroscopy: Model F-4500 (Hitachi) and the BioView Sensor (Delta light & Optics). Each of them uses an excitation range of 250–560 nm, an emission range of 260/300–600 nm and the measuring time of 1 min [59, 60]. The application of this technique for CPC production was performed by Lindemann [61].

The evaluation of the measurements, the correlation between the medium components and the various ranges of the 2D-fluorescence spectrum was performed by Principal Component Analysis (PCA), Self Organized Map (SOM) and Discrete Wavelet Transformation (DWT), respectively. Back Propagation Network (BPN) was used for the estimation of the process variables [62]. By means of the SOM the courses of several process variables and the CPC concentration were determined.

5
Morphology and Rheology

Nash and Huber [63] reported that submerged cultures of *A. chrysogenum* in synthetic media showed four distinct morphological states: hyphae, swollen hyphal fragments, conidia and germlings. Queener and Ellis [64] reported that differentiation of swollen hyphal fragments into spherical unicellular arthrospores accompanied the synthesis of CPC in complex medium. Masazumi et al. [65] and Natsume and Marumo[66] investigated the relationship between morphological differentiation and CPC synthesis in synthetic media. According to them, the depletion of glucose in the media caused hyphal fragmentation and production of CPC. A one-to-one correspondence was observed between the final percentage of spherical arthrospores and the final titer of CPC. Moreover the differentiation was enhanced by methionine [27]. In synthetic and semisynthetic media the morphology can be determined by image analysis [65, 66]. Image analysis allows the characterization of the spore germination from the inoculum stage and of the subsequent dispersed pellet forms. Further methods include characterizing vacuolation and simple structural differentiation of mycelia.

The basic hardware components of image processing are: input devices (microscope, macroviewer with video camera, scanner) and a high-performance PC or work station with optional image processing. The software consists of a program that controls the hardware to carry out a variety of enhancement, manipulation, filtering and smoothing operations, and measurements. Parameters for dispersed forms are minimum branch length, mean hyphal width, minimum length, number of loops, maximum fullness ratio, circulatory factor. In addition one distinguishes between loose cell aggregates (clumps) and dense aggregates (pellets). However, in complex media the pellets include the solids of corn steep liquor (CSL), the mycelium partly grow on the surface of peanut flour (PF) particles or form clumps enhanced by soy oil. Therefore, a quantitative image analysis in the first 100 h is not possible.

After 100 h only cell fragments prevail. The morphology influences not only the product formation, but the rheology of the culture medium. The rheology of the cultivation medium was determined by Searle Rheometer (Rheomat 115, Contraves) [67]. The rheology of a *A. chrysogenum* cultivation medium with $30-100$ g l^{-1} peanut flour can be described by the Bingham model. The yield stress (flow limit) increased strongly with the biomass concentration up to 90 h, and after that it quickly decreased because the peanut flour was consumed. The Bingham viscosity passed a maximum at the same time, but its maximum is much less than that of the yield stress. The oxygen supply of the cells was strongly impaired by the high yield stress and viscosity. The lack of oxygen caused a interruption of the biosynthesis at PEN.

6
Results and Discussion

The determination of the biomass concentration in complex media by weight is not possible. Therefore, indirect methods were tested. The OTR and CPR are not suitable for the biomass evaluation, because RQ considerably varies during the cultivation. In the first 60 h, it increases, between 60 and 80 h it drops sharply and only after 100 h does it attain a constant value. However, after 100 h, the biomass concentration can be measured by weight in any case. The best agreements were obtained between RNA and biomass concentrations in semisynthetic media. The biomass concentration increases with cultivation time up to 140 h.

By supplementing the glucose, the growth rate increased and the cultivation time was reduced, but the CPC concentration diminished considerably, because the CPC formation was repressed. Short cultivation time is desirable, because after 100 h the CPC concentration nearly stagnates and the concentrations of byproducts (DAC, PEN and DAOC) increase. By moderate dilution of the cultivation medium and amino acid supplement the cultivation time was reduced and the productivity was increased. The best yield coefficient was obtained by considerable dilution of the medium and supplementing the amino acid. This procedure increased the cultivation time and reduced the productivity. By means of amino acid supplements the CPC concentration was increased by 50% and the byproducts reduced. However, it was not possible to find out the basis for this positive effect on the biosynthesis.

The determination of the intracellular enzyme activities is impaired by the proteases and hydrolases, which are released by the disintegration of the cells. The suppression of these enzymes by protease/hydrolase inhibitors influences the activity of the investigated enzyme as well.

The determination of the activity of the bottleneck enzyme ACVS is especially difficult, because the activity of the successive enzyme IPNS is higher by two orders of magnitude. Again, the suppression of the IPNS impairs the activity of ACVS as well. Therefore, it is not possible to evaluate the true enzyme activities. However, these investigations show qualitatively how the enzyme activities vary during the cultivation (Fig. 8). ACVS activity is low and it attains a maximum at 60 h and then gradually diminishes. IPNS activity is intermediate and nearly constant up to 95 h and then decreases. EXP activity is high and that of HYD intermediate. EXP activity is nearly constant in the range 40–100 h. HYD activity is nearly constant and increases after 80 h. ATC activity is very high and obtains its maximum during 60–90 h. A comparison of CPC and precursor formation rates with the activities of these enzymes showed that CPC formation rate obtains a maximum at 70–80 h, somewhat later than the

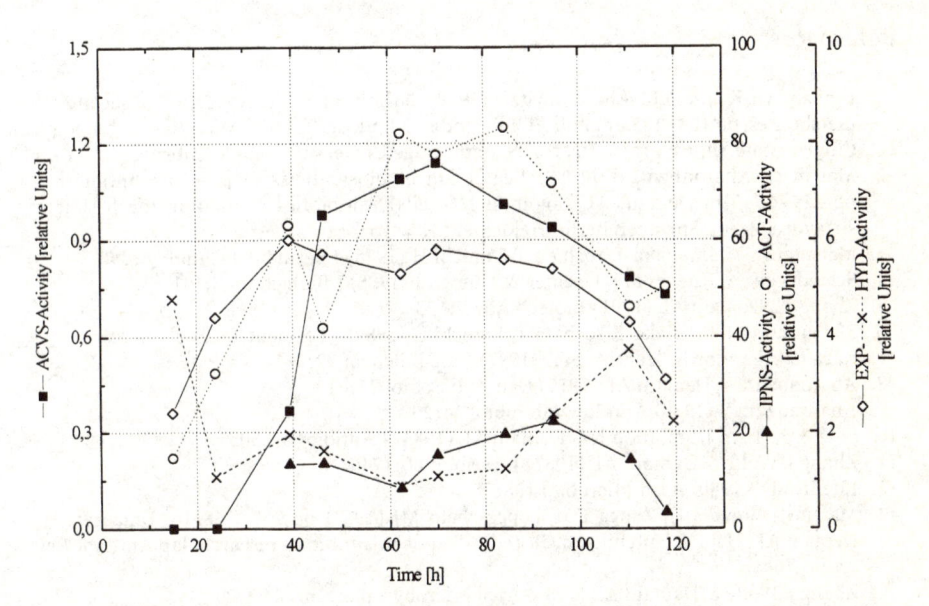

Fig. 8. Comparison of the detected enzyme activities during the cultivation of *Acremonium chrysogenum*

ACVS. PEN formation rate has a maximum at 110 h, and at 115 h it increases again. The formation rates of all other precursors increase with the cultivation time. This clearly indicates, that the low ACVS activity and its dramatic reduction after 90 h caused the bottleneck in the biosynthesis. By maintaining a high CPC formation rate after 80 h due to keeping the ACVS activity at high level after 90 h, the CPC concentration and productivity could be increased considerably.

7
Conclusions

The investigations indicated that the productivity in complex media is much higher than in synthetic and semi synthetic media. With 100 g l^{-1} PF, the product concentration in the stirred tank was twice as high as with 30 g l^{-1} PF and three times as high as one without PF. With another medium, the CPC concentration was twice as high with CSL than without it. The maximum CPC concentration was obtained with CSL and amino acid supplement. The medium compositions for optimal CPC concentration, CPC yield coefficient and CPC space time yield are different. ACVS is the bottleneck in the biosynthesis in the first phase of the cultivation. After 100 h, the expandase/hydroxylase becomes the bottleneck. In this phase, DAC and PEN concentrations considerably increase, but the CPC concentration varies only slightly. By the suppression of the PEN formation the production of DAOC, which impairs the purification of CPC, is reduced as well.

References

1. Demain AL, Kennel YM, Aharonowitz Y (1979) Carbon catabolite regulation of secondary metabolites. pp 163–185. In: Bull AT, Ellwood DC, Ratledge C (eds) Microbial technology: Current state, future prospects, vol 29. Cambridge University Press, Cambridge.
2. Martin JF, Aharonowitz Y (1983) Regulation of biosynthesis of β-lactam antibiotics. pp 229–254. In: Demain AL, Solomon NA (eds) Antibiotics containing the β-lactam structure, Part I. Springer, Berlin Heidelberg New York
3. Matsumura M, Imanaka T, Yoshida T, Taguchi H (1978) J Ferment Technol 56:345
4. Scheidegger A, Küenzi MT, Fiechter A, Nüesch J (1988) J Biotechnol 7:131
5. Zhang JS, Demain AL (1991) Biotech Adv 9:623
6. Matsumura M, Imanaka T, Yoshida T, Taguchi H (1989) J Ferment Technol 58:205
7. Karaffa L, Sandor E, Szentirmai A (1997) Proc Biochem 32:495
8. Aharonowitz Y, Demain AL (1979) Can J Microbiol 25:61
9. Aharonowitz Y (1980) Ann Rev Microbiol 34:209
10. Shen YQ, Heim J, Solomon NA, Demain AL (1984) J Antibiot 37:503
11. Zhang J, Wolfe S, Demain AL (1987) J Antibiot 40:1746
12. Kuenzi MT (1980) Arch Microbiol 128:78
13. Martin JF, Revilla G, Zanca DM, Lopez-Nieto MJ (1982) pp 258–268. In: Umezewa H, Demain AL, Hata T, Hutchinson CR (eds) Trends in antibiotic research. Jap Antibiot Res Assoc, Tokyo
14. Zhang J, Wolfe S, Demain AL (1988) Appl Microbiol Biotechnol 29:242
15. Rollins MJ, Jensen SE, Wolfe S, Westlake DWS (1990) Enzyme Microb Technol 12:40
16. Bainbridge ZA, Scott RI, Perry D (1992) J Chem Technol Biotechnol 55:233

17. Zhou W, Holzhauer-Rieger K, Dors M, Schügerl K (1992) Enzyme Microb Technol 14:848
18. Kozma J, Lucas L, Schügerl K (1991) Biotechnol Lett 13:899
19. Kozma J, Lucas L, Schügerl K (1993) Appl Microbiol Biotechnol 40:463
20. Kozma J, Karaffa L (1996) J Biotechnol 48:59
21. Zhou W, Holzhauer-Rieger K, Dors M, Schügerl K (1992) J Biotechnol 23:315
22. Seidel G (1999) PhD thesis, University Hannover
23. Malmberg LH, Hu WS (1992) Appl Microbiol Biotechnol 38:122
24. Malmberg LH, Sherman DH, Hu WS (1992) Ann New York Acad Sci 665:16
25. Matsumura M, Imanaka T, Yoshida T, Taguchi H (1980) J Ferment Technol 58:197
26. Natsume M, Marumo S (1984) Agric Biol Chem 48:567
27. Batoshevich YuE, Zaslavskaya PI, Novak MJ, Yudina OD (1990) J Basic Microbiol 30:313
28. Lopez-Nieto MJ, Ramos FR, Luengo FR, Martin JF (1985) J Appl Microbiol Biotechnol 22:343
29. Baldwin JE, Abraham E (1988) The biosynthesis of penicillins and cephalosporins. Nat Prod Rep pp 129–145
30. Lübbe C, Wolfe S, Demain AL (1985) Enzyme Microb Technol 7:353
31. Fujisawa Y, Kanazaki T (1975) Agric Biol Chem 39:2043
32. Felix HR, Nüesch J, Wehrli W (1980) FEMS Microbiol Lett 8:55
33. Bradford M (1976) Anal Biochem 72:248
34. Burton K (1956) J Biochem 62:315
35. Lever M (1972) Anal Biochem 47:273
36. Schmidt WJ, Meyer HD, Schügerl K (1984) Anal Chim Acta 163:101
37. Lorenz T, Schmidt W, Schügerl K (1987) Chem Eng J 35:B15
38. Bayer T, Herold T, Hiddessen R, Schügerl K (1986) Anal Chim Acta 190:213
39. Bayer T, Zhou W., Holzhauer K, Schügerl K (1989) Appl Microbiol Biotechnol 30:26
40. Ruzicka J, Hansen EH (1988) Flow Injection Analysis. 2nd edn. Wiley, New York
41. Ulber R (1996) PhD Thesis University Hannover
42. Wieland M (1995) Master's Thesis, University Hannover
43. Brand U, Reinhardt B, Rüther F, Scheper T, Schügerl K (1990) Anal Chim Acta 238:201
44. Menzel C, Lerch T, Scheper T, Schügerl K (1995) Anal Chim Acta 317:259
45. Weigel B, Hitzmann B, Kretzmer G, Schügerl K, Huwig A, Giffhorn F (1996) J Biotechnol 50:93
46. Jürgens H, Kabus R, Plumbaum T, Weigel B, Kretzmer G, Schügerl K, Andres K, Ignatzek E, Giffhorn F (1994) Anal Chim Acta 298:141
47. Cooper JHD, Odgen G, Mcintosh J, Turnell DC (1984) Anal Biochem 142:98
48. Holzhauer-Rieger K, Zhou W, Schügerl K (1990) J Chromatogr 499:609
49. Beyer M, PhD thesis (1996), University Hannover
50. Kuenzi MT (1979) Biotechnol Lett 1:127
51. Tollnick C, PhD thesis (1996), University Hannover
52. Banko G, Wolfe S., Demain A.L (1986) Biochem Biophys Res Commun 137:528
53. Kupka J, Shen Y (1983) Can J Microbiol 29:488
54. Jensen SE, Westlake, DWS., Wolfe, S (1982) J Antibiot 35, 483
55. Matsuyama K, Matsumoto H., Matsuda A, Sugiura H, Komatsu K Ichikawa, S (1992) Biosc Biotech Biochem 56:1410
56. Reardon KF, Scheper T (1991) Determination of cell concentration and characterization of cells. In: Schügerl K (ed) Biotechnology 2nd edn, vol 4. Measuring, Modelling and Control. VCH, Weinheim, pp 179
57. Tartakovsky B, Sheintuch M, Hilmer JM, Scheper T (1996) Biotechnol Progr 12:126
58. Tartakovsky B, Sheintuch M, Hilmer JM, Scheper T (1997) Bioproc Eng 16:323
59. Marose S, Lindemann C, Scheper T (1998) Biotechnol Progr 14:63
60. Schügerl K, Lindemann C, Marose S, Scheper T (1998) Two-dimensional fluorescence spectroscopy for on-line bioprocess monitoring. In: Berovic (ed) Bioprocess Engineering Course. Supetar, Croatia, Natl Inst of Chem, pp 400–415
61. Lindemann C (1998) PhD Thesis, University Hannover
62. Wei J (1998) PhD Thesis, University Hannover

63. Nash CH, Huber FM (1971) Appl Microbiol 22:6
64. Queener SW, Ellis LF (1975) Can J Microbiol 21:1981
65. Paul GC, Thomas CR (1998) Adv Biochem Eng Biotechnol 60:1
66. Krabben P, Nielsen J (1998) Adv Biochem Eng Biotechnol 60:125
67. Schügerl K, Bayer T, Niehoff J, Möller J, Zhou W (1988) Influence of cell environment on the morphology of molds and the biosynthesis of antibiotics in bioreactors, pp 229–243. In: King R (ed) 2nd Conference on Bioreactor Fluid Dynamics. Elsevier, Amsterdam

Biomass Quantification by Image Analysis

Marie-Noëlle Pons*, Hervé Vivier

Laboratoire des Sciences du Génie Chimique, CNRS-ENSIC-INPL, rue Grandville,
BP 451, F-54001 Nancy cedex, France
* E-mail: pons@ensic.u-nancy.fr

Microbiologists have always rely on microscopy to examine microorganisms. When microscopy, either optical or electron-based, is coupled to quantitative image analysis, the spectrum of potential applications is widened: counting, sizing, shape characterization, physiology assessment, analysis of visual texture, motility studies are now easily available for obtaining information on biomass. In this chapter the main tools used for cell visualization as well as the basic steps of image treatment are presented. General shape descriptors can be used to characterize the cell morphology, but special descriptors have been defined for filamentous microorganisms. Physiology assessment is often based on the use of fluorescent dyes. The quantitative analysis of visual texture is still limited in bioengineering but the characterization of the surface of microbial colonies may open new prospects, especially for cultures on solid substrates. In many occasions, the number of parameters extracted from images is so large that data-mining tools, such as Principal Components Analysis, are useful for summarizing the key pieces of information.

Keywords. Image analysis, Microscopy, Visualization, Morphology, Physiology, Motility, Visual texture, Data mining

1	Introduction	135
2	Visualization	137
2.1	Microscopes	137
2.2	Mode of Operation	140
3	Image Capture	142
4	Basic Treatment	144
5	Size and Shape Description	150
6	The Special Case of Filamentous Species	158
7	Shape Classification	166
8	Densitometry and Visual Texture	168
8.1	Physiology Assessment	168
8.2	Visual Texture	174

9	Motility	177
10	Conclusions and Perspectives	178
References		179

List of Symbols and Abbreviations

BCEF-AM	2′,7′-bis(carboxyethyl)-5-carboxyfluorescein-acetoxyl methyl
CAM	Calcein acetoxy methyl ester
CCD	Charge-Coupled Device
CF	5-carboxyfluorescein and 6-carboxyfluorescein
CID	Charge-Injection Device
CMD	Charge Modulated Device
CMOS	Complementary Metal Oxide Silicon
CSLM	Confocal Laser Scanning Microscopy
DAPI	4′6-diamino-2-phenylindole
EDM	Euclidian Distance Map
ESEM	Environmental Scanning Electron Microscopy
FDA	Fluorescein Diacetate
FISH	Fluorescence In-Situ Hybridization
GFP	Green Fluorescent Protein
HSI	Hue-Saturation-Intensity color model
INT	2-(p-iodophenyl-)3)(p-nitrophenyl)-5-phenyl tetrazolium chloride
MPN	Most Probable Number
OMA	Oval Major Axis
RGB	Red – Green – Blue color model
SEM	Scanning Electron Microscopy
SGLDM	Spatial Gray Level Dependence Matrix
A	projected area
A_C	projected area of the convex bounding polygon
b	brightness histogram of EDM
c_{ij}	frequency of occurrence
C_i	Fourier shape descriptor
C	co-occurrence matrix
D_{eq}	equivalent diameter
D_h	hyphal diameter
F	length
$F_{\text{free}}^{\lambda_{ex}}$	fluorescence signal at zero calcium concentration
$F_{\text{sat}}^{\lambda_{ex}}$	fluorescence signal at maximal calcium concentration
h	threshold
IC_A	index of concavity in surface
IC_P	index of concavity in length
M_{2X}, M_{2Y}, M_{XY}	second order moments
N	total number of pixels in the image

N_i	number of pixels of gray-level i
P	perimeter
P_C	perimeter of the convex bounding polygon
p_i	probability for a pixel to have a gray-level equal to i
R	roughness
$R(s)$	contour signature
R_c	core radius
R_g	gyration radius
R_m	fluorescence ratio
R_{min}	fluorescence ratio at zero calcium concentration
R_{max}	fluorescence ratio at maximal calcium concentration
s	normalized curvilinear abscissa
s'	curvilinear abscissa
V	volume
W	width
w_0	probability for a pixel to have a gray-level less than h
w_1	probability for a pixel to have a gray-level greater than h
x_g	x-coordinate of object centroid
y_g	y-coordinate of object centroid
α	direction
λ	step length for fractal dimension calculation
$\lambda_{em}, \lambda_{ex}$	emission and excitation wavelengths
$\lambda_{min}, \lambda_{max}$	minimal and maximal eigenvalues of the inertia matrix
ω_i	number of erosions
Ω_2	reduced size of the largest concavity
μ	gray-level probability function
ϕ	histogram entropy
Σ_{XY}	inertia matrix

1
Introduction

The basic aim of biotechnological processes is to make use of microbial biomass to produce various substances of pharmaceutical, nutritional, or more generally, economical interest. It is of uttermost importance to be able to quantify its content, and in some cases its nature. For control purposes, it has to be done if possible in-situ or on-line, and if not, off-line with a lag compatible with the characteristic time constant of the process. Since the 17th century, when they were first seen through a "flea-glass" [1] microorganisms have fascinated microbiologists. Their early description relied heavily on drawings but the advent of photography in the 19th century, then of video-cameras and digital imaging has increased the tracking of the "animalcules" as they were called by A. van Leeuwenhoek, the first one to describe spermatozoids, yeast and blood cells. If spermatozoids are still of interest [2], the automated analysis of biomass now covers the whole spectrum of micro-organisms: bacteria, yeast, fungi, plant and mammalian (human and animal) cells, protozoa as well as sub-parts of cells such as RNA fragments. It has now gone beyond the stage of pure technical and mathematical development and can be used in bioresearch and in production and product-quality control.

Counting of individuals (or assemblies of individuals such as microcolonies or aggregates) is the basic application of automated image analysis. The task may seem to be easy but it involves the recognition of the individuals, by size, shape or color with respect to debris and/or other individuals of different species. Size determination itself cannot be separated from shape characterization: the sphere is the only object which size is given by a single parameter, its diameter. In the bioworld non-budding *Saccharomyces cerevisiae* cells, some mammalian cells, spores are approximated by spheres. However, most microorganisms are not spherical, which means that more than one parameter is required to characterize their size. Similarly, shape will be described by a set of descriptors, especially to discriminate between shapes bearing some similarities. There will be an inherent difficulty for the operator to apprehend this multi-dimensional space of descriptors and data analysis tools are welcomed for the interpretation of the raw results.

Many visualization techniques are based on smears on glass slides when the microorganisms are living in a 3D space. Even when the cells are just imprisoned in a liquid phase between two glass surfaces, images will render a 2D view of a 3D reality: shape descriptors describe the morphology of the silhouette of the laying microorganisms. Of course changes of morphology can occur during a bioprocess: the budding of *Saccharomyces cerevisiae* and the progressive development of the daughter cell induce a distribution of morphologies. Similarly the germination of spores in filamentous species will result in a wide range of morphologies: almost spherical ungerminated spores, germinating spores with one or more germ tubes of different lengths, filaments with various degrees of branching, entanglements of one or more filaments, pellets.

The physiology of cells is equally important: in many applications the living biomass should be quantified and distinguished from the necromass. As for higher organisms, it is often difficult to cut a clear edge between life and death, as death results from a slow modification of various cellular functions. Vacuolization in filamentous species is related to an important decay of the microorganisms. In contrast, vacuoles in *Saccharomyces cerevisiae* are associated with the storage of materials [3] and not necessarily to moribund cells. The physiology is generally assessed through staining: classical stains such as Methylene Blue, Neutral Red, Ziehl Fuchsin as well as new fluorescent dyes such as Acridine Orange or Propidium Iodide are used. Recently the Green Fluorescent monomeric Protein (GFP) of the jellyfish *Aequorca victoria* has been cloned into *E. coli* strains [4–6]. Various applications combining its fluorescence properties could be foreseen such as the identification of transformed cells, the localization and the traffic monitoring of intracellular proteins. The quantification of physiology relies generally on densitometric measurements performed on a single image or on combinations of images.

Most of the reported applications concern the characterization of a given cell at a single instant. Dynamics can also be considered. The division of *Sac-*

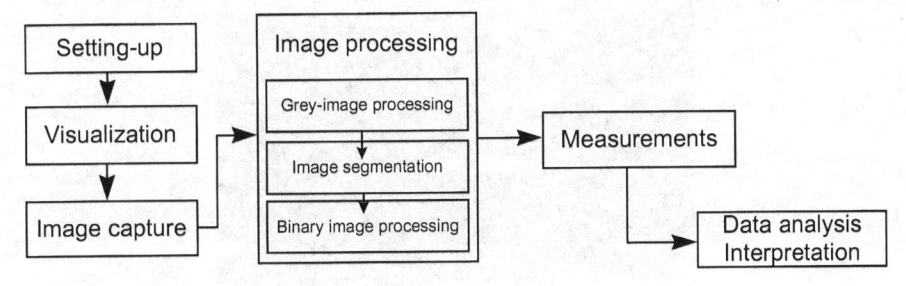

Fig. 1. The main steps of image analysis of microbes

charomyces cerevisiae cells was investigated early on by Lord and Wheals [7-9]. Motility experiments are now facilitated by the use of CCD cameras and digital video-recorders which can register sequences of images for several minutes which are then analyzed later.

Image analysis, as applied to microorganisms or to inert particles such as crystals [10] consists in a series of steps which are relatively independent of the kind of objects considered (Fig. 1). In this chapter the basics of visualization as applied to microorganisms will be presented first. The general procedure of image treatment illustrated by bioapplications will follow. Finally, quantification of size and physiology, motility and visual texture will be discussed.

2
Visualization

The first step of automated image analysis of microorganisms is the visualization. Considering their size range, microscopes (or for the largest objects, microviewers) have to be used. Two aspects should be discussed:

- the type of microscope: optical/electron
- the device for image capture/treatment: off-line/on-line/in-situ

2.1
Microscopes

Since A. van Leeuwenhoek, optical microscopes have been routinely used for microbial observations [11]. Due to the transparency of the cells, phase-contrast lenses allow diascopic visualization at the highest magnifications (×400 and above) (Fig. 2a). Cells are surrounded by a halo which can give some trouble for the subsequent treatment of images. Episcopy is largely used in combination with fluorescent dyes (Fig. 2b). Halos around the cells, photobleaching, which can make the observation difficult, and fluorescence of the background due to some remaining broth elements or to an excess of dye are typical problems met in epifluorescence microscopy [12]. A compromise should be found between the excitation intensity and the emitted light. Fresh or dried slides can be examined. Inverted microscopes avoid the use of a cover glass.

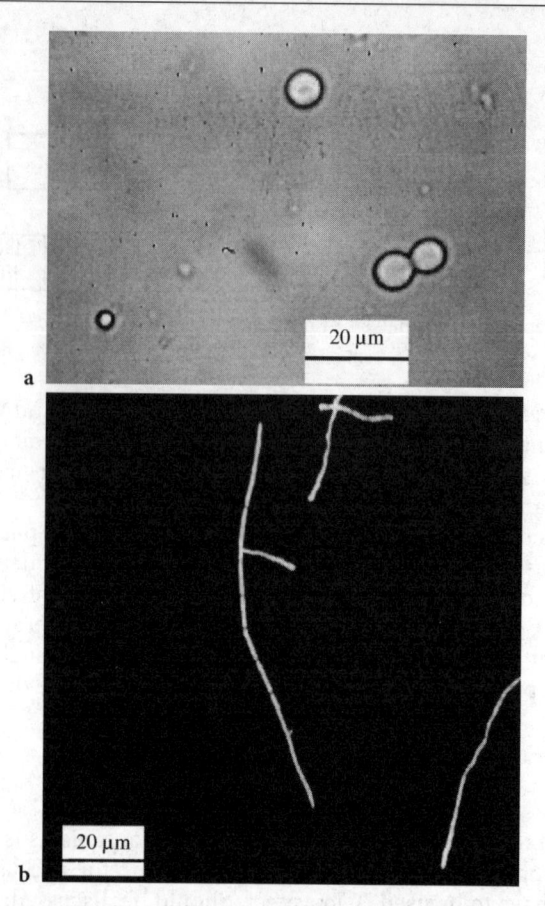

Fig. 2. *Saccharomyces cerevisiae* cells observed by phase-contrast microscopy (**a**) and *Streptomyces ambofaciens* filament observed by epifluorescence microscopy (after Propidium Iodide staining) (**b**)

This is interesting for fragile objects such as mammalian cells [13], which should not be squeezed between glass surfaces, and for motility experiments as the displacement of the cells is not limited vertically. The largest available magnification, depending upon the ocular-lens combination is about ×1500. This value limits the observation of the smallest cells such as non filamentous bacteria (*E. coli*) or plankton.

Confocal Laser Scanning Microscopy (CLSM) is particularly adapted to the study of 3D structures such as aggregates of cells, cells entrapped in or colonizing a support or biofilms, etc. [14]. It combines a traditional epifluorescence microscope with a laser light source and a special scanning equipment, which permits one to obtain optical sections (2D-images), without risk of damaging the structure by dehydration and/or slicing (Fig. 3) [15]. The key features of the system are the high-capacity storage systems and the powerful software necessary to reconstruct the 3-dimensionnal shape from the array of 2D-images.

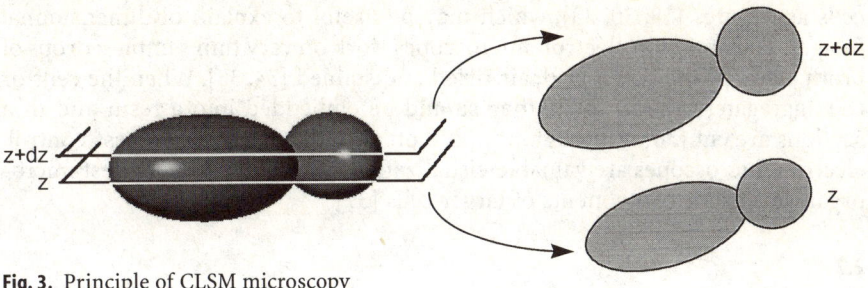

Fig. 3. Principle of CLSM microscopy

Proper alignment of the system is of course necessary. Differences in systems from various manufacturers are not so much in the technique itself but more in the user-friendliness of the system. The main fields of CLSM applications in bioengineering concern bacteria growing in soils [16, 17], on non-transparent supports such as leaves [18], in biofilms [19–21], in particular for wastewater treatment or biofouling studies [22], and mammalian cells growing on microcarriers [23, 24] or in aggregates [25, 26].

Electron microscopes give higher magnifications than optical systems: very small bacteria such as *E. faecalis* (≈ 1 µm) [27] or even phages (≈ 200 nm) [28] can be observed. However the preparation of the samples is difficult: dehydration (because of the high-vacuum in the chamber) and metal-plating (to make the sample conductive) are required by most Scanning Electron Microscopes (SEM). The water content of cells being very high, the dehydration and fixation (usually with osmium tetraoxide) should be conducted very carefully to limit the morphological changes [29]. In the best cases, if the general shape is maintained, shrinkage is observed, making size assessment difficult [30–32] (Fig. 4). A size decrease of 35% was observed by Huls et al. [29] on *Saccharomyces cerevisiae* cells. Environmental Scanning Electron Microscopes (ESEM) operate under less stringent conditions: the gas ionization detector tolerates a pressure of about 100 Pa at 277 K, when standard SEM works under about 1 Pa. The main advantages of SEM is to provide three-dimensional visualization of the cells or

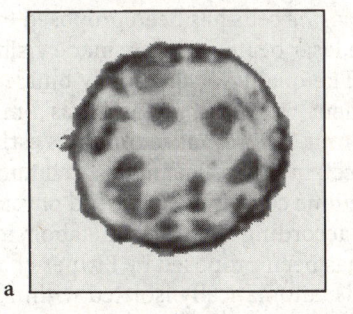

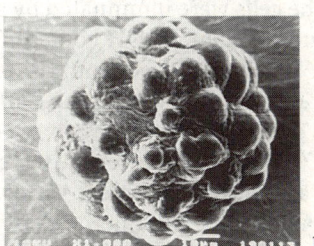

Fig. 4. CHO cells on a microcarrier: **a** optical microscopy (contrast enhancement by Crystal Violet staining) and **b** scanning electron microscopy

cells aggregates [18, 30, 33], which may be useful to explain bi-dimensionnal images. Transmission electron microscopes work on very thin samples: drops of broth are deposited on a grid, air-dried and stained [34, 35]. When the cells or cell aggregates are too thick, they should be embedded into a resin and thin sections are cut [36]. Although not used on a routine basis for process control, electron microscopes are valuable visualization systems for the smallest microorganisms or for components of larger cells [37].

2.2
Mode of Operation

Most of the actual routine image analysis applications on microorganisms are run off-line and are operator-assisted for manual sampling in the bioreactor, slide preparation (including staining) and image capture. The last step is easily automated: stage motion, focus, selection of lens, brightness can be computer-controlled. The off-line mode implies a delay between the time of sampling in the fermenter and the time the data are effectively available for monitoring and controlling the bioprocess. The acceptable delay is a function of the characteristic time constant of the bioreaction. A delay of 1 to 2 h can be considered of little importance for a 200-h process (fungi, filamentous bacteria) but will be unacceptable for fast-growing microorganisms (*E. coli*, *S. cerevisiae*). In such cases on-line and even in-situ systems are necessary.

On-line systems involve circulation loops to bring the sample to the microscope. The depth-of-focus is limited and the cells should be maintained close to the focus plane by forcing them through a capillary [38] or a flow cell [39]. For *Bacillus thuriengiensis*, the best results were obtained by Lichtfield et al. [38] with 20 µm thick glass channels. Stop flow cells have been tested by Zalewski and Buchholz [40] for yeast cells and by Treskatis et al. [41] for *S. tendae* pellets, combined with an inverted microscope. Specific cells have been designed to investigate the effect of osmotic shifts [42] and high pressure [43] on yeast cells. As in any automated sampling system for bioprocesses, there are risks of tubing clogging and fouling of the transparent optical surfaces.

If the machinery for on-line visualization seems sophisticated, the development of in-situ microscopes is even more complex. It is, however, the only way to see cells without the stress of the sampling and slide preparation. A device for the observation of *Saccharomyces cerevisiae* cells has been proposed by Suhr et al. [44]. The cells were illuminated by a laser beam and were made visible by the fluorescence of their NADH content. The quality of the highly blurred images was such that the subsequent treatment was complex and was limiting the analysis rate to an average of one cell/min. The response time was estimated to 20 min to get reliable statistics on the cell number. The highest cell number was around 10^9 cells/ml. However the technique combines sizing and physiology assessment as the NADH content varies according to the cell metabolism.

Lately another in-situ microscope has been proposed by Bittner et al. [45]. A small volume of broth ($\approx 2 \times 10^{-8}$ ml) is automatically isolated within the reactor. The quality of images is greatly improved and similar to what can be obtained off-line (Fig. 5). The treatment is faster because the images are "simpler" to

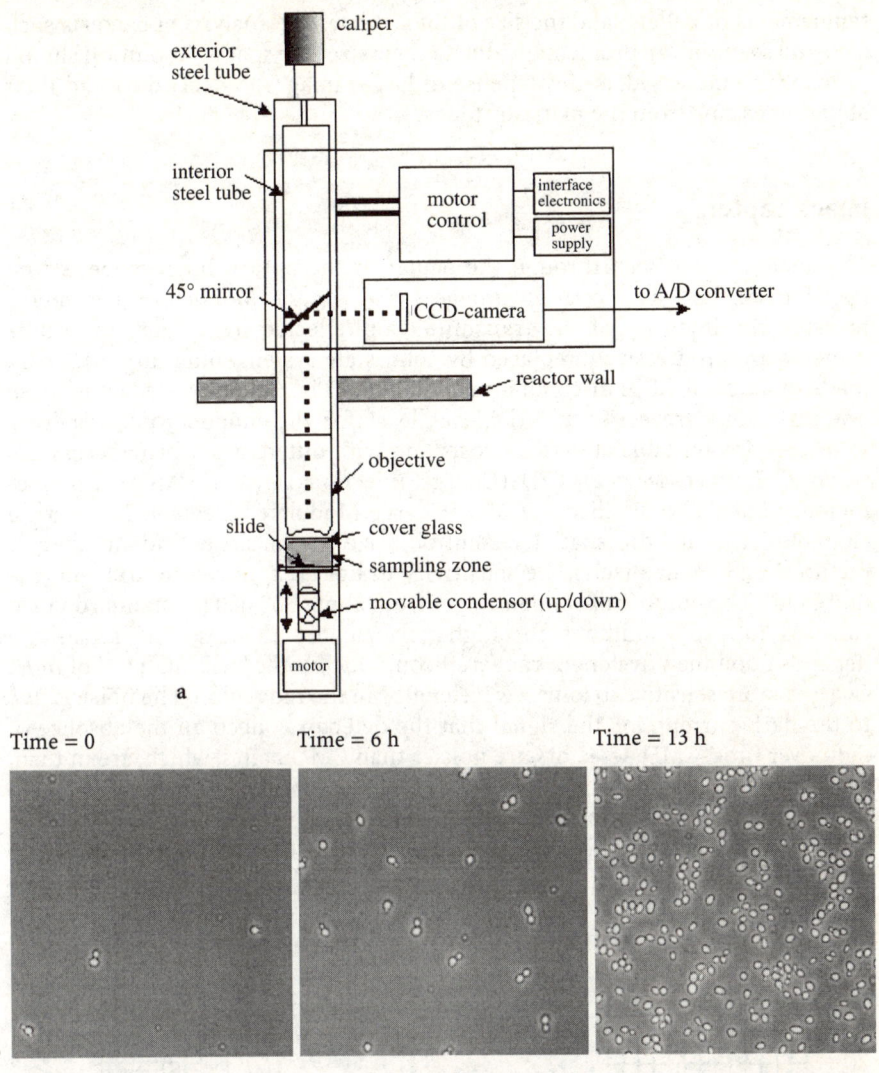

Fig. 5. a In-situ microscope proposed by Bittner et al.; **b** *Saccharomyces cerevisiae* cells observed by in-situ microscopy in function of culture time (by permission of T. Schepper)

analyze than in the previous system. The cell number can vary between 10^6 and 10^9 cells/ml, which corresponds to a cell concentration of 0.01 to 12 g/l. The sample time is rather fast (20 s for 256×224-pixels images) for a response time of 13 min, due to the necessity of counting a significant number of cells. The authors promise a decrease of the response time down to 2 min if more powerful computing systems are used. This device could be extended for cells larger than the tested *Saccharomyces* cells, e.g. filamentous species, although there, a compromise should be found between the size of the object (filament, entanglements or pellets) and the size of the hyphae. The analysis of bacteria such as *E. coli* seems more problematic due to their size (the spatial resolution should probably be increased, at the expense of larger images to deal with) and their ability to escape from the plane-of-focus.

3
Image Capture

The analog image seen through the ocular of the optical microscope is first transformed into an electronic image by a camera. Monochrome cameras measure the intensity of the transmitted or reflected light. The bulky tube cameras are progressively replaced by solid-state systems built around a chip made of an array of semi-conducting detectors (Fig. 6a): they are largely used nowadays and are easily affordable at least for the monochrome versions. Detectors are available at various costs and with different performances: CCD (Charge Coupled Device), CID (Charge Injection Device), CMOS (Complementary Metal Oxide Silicon), CMD (Charge Modulated Device). The critical characteristics are the spatial resolution (size of the array and number of picture elements or pixels), the sensitivity to light (expressed in lux), and the noise [46]. The number of elements goes from about 300,000 for standard video cameras to a few millions for high performance devices. The sensitivity depends upon the wavelength range: silicon, which is the basis material of most CCDs is more sensitive to longer wavelengths in the red region. The noise is due to the dark current, i.e. the signal that the detector collects in the absence of light over time. CMD detectors are noisier than CMOS ones, which, are in turn, noisier than CCDs.

The first motivation for processing color images is that human beings are sensitive to thousands of shades and intensities, in comparison to about thirty

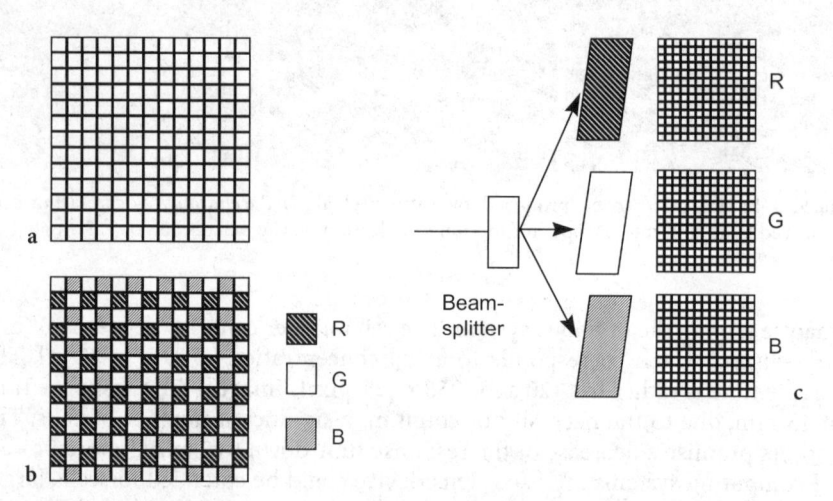

Fig. 6. Solid-state technology: **a** monochrome chip; **b** mono-chip for color imaging; **c** set-up for 3-CCD color imaging

gray-levels. The second motivation is that color, especially in physiological applications, is a powerful descriptor. A color is defined by its hue (H), saturation (S) and intensity (I) (Fig. 7a). The hue is associated with the dominant wavelength of the captured light wave. The saturation is inversely proportional to the amount of white light mixed with a hue. The HSI model of color is the closest to the human perception of color and it is usually easier to analyze color images using this model [47]. However color cameras acquire images according to the RGB model which is based on the Cartesian coordinate system formed by the primary spectral components of Red, Green and Blue (Fig. 7b). Most image analyzers easily convert color images from one model to the other.

Mono-CCD color cameras contain basically the same image sensor as monochrome CCD ones, except that the matrix elements have being covered by color micro-filters (Fig. 6b): a color image results from the combination of three primary images (Red, Green and Blue). The sensitivity is only slightly reduced but the spatial resolution is strongly decreased. Due to the high sensitivity of human beings to green, there are more green filters than red and blue ones. The filters induce a general decrease of sensitivity with respect to monochrome cameras. The much-more expensive 3-CCD color cameras contain three matrix elements (Fig. 6c). A beam splitter sends the light beam toward the three detecting units. The spatial resolution is high but the sensitivity is generally strongly reduced (filters and beam splitter). It is important to note that not all objects visualized in color need really color imaging. A image with a single hue can be captured by a monochrome camera (staining with Propidium Iodide, Neutral Red, Methylene Blue for example). However, the complex spectrum of Acridine Orange (yellow, red, green) requires a color camera.

In fluorescence and especially in bioluminescence [48, 49], the emitted light levels are usually low and normal cameras might not be sensitive enough. Intensifying cameras are available for those special applications [12]: the signal-to-noise is improved by thermoelectrically cooling the camera with a Peltier element. They are still rather expensive and cheaper solutions can be found by adjusting the integration time of normal CCD cameras for many

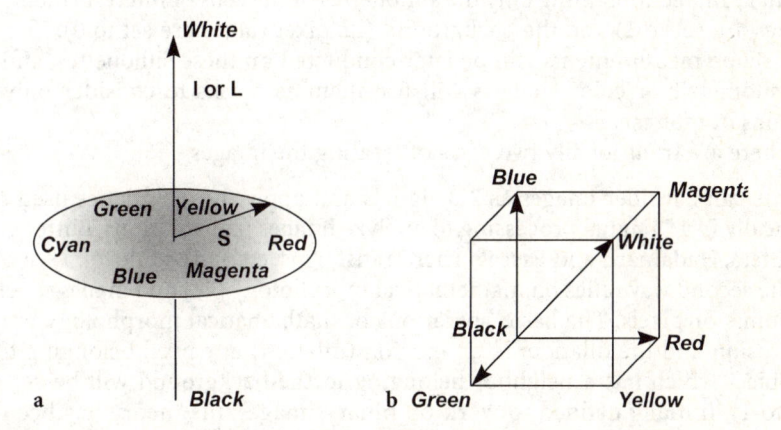

Fig. 7. Representation of color by the HSI model (**a**) and the RGB model (**b**)

routine fluorescence applications. When densitometric analysis of the cell is required (especially for physiological applications), the auto-gain of the camera should be switched off.

The video analog signal output by the tube and solid-state cameras (and by the electron microscopes) is digitized in the computer by a specialized grabbing board. The spatial resolution of the imaging system results from the combination microscope magnification-camera-grabber. Most of the actual grabbers generates rectangular images of about 586 lines of 756 pixels although some systems provide square images of 512 lines of 512 pixels. Pixels are nowadays mostly square. Older grabbers produced rectangular pixels with an aspect ratio X/Y = 1.5, requiring interpolation to restore a grid of square pixels. The intensity of the transmitted or emitted light by the surface element corresponding to the pixel on the initial analog image is nowadays transformed into a number coded on 8 bits for monochrome images and 3×8 bits for color images: the 256 gray or brightness levels goes from black (0) to white (255). Costello and Monk [50] used only 256×256 images with 64 gray-levels for one of the first routine application of image analysis to count *Saccharomyces cerevisiae* cells in the mid-1980s. A larger range may be necessary especially for fluorescence and luminescence applications: some systems use 4096 gray-levels, coded on 12 bits.

It would seem a better solution to directly output a digital signal from a solid-state camera as the pixelization is already done at the level of the detector. This is effectively the trend. Digital cameras are becoming available but the high-performance ones requested for scientific applications are still very expensive. The data readout is rather slow which can make focusing tedious.

4
Basic Treatment

A basic treatment of bio-images can be proposed. Its main objective is to reduce the total amount of information contained in the gray-level or color images to focus the quantification on the key features: the result of this treatment will be a binary image containing only the silhouettes of the cells of interest (their pixel values are set to 1) and the background (its pixel values are set to 0). The size and shape measurements will be later conducted on these silhouettes, and the densitometric or color analysis will use them as masks to consider only the regions of interest.

There are traditionally two ways of treating the images:

- one can consider images as 2D signals and apply the techniques used classically in 1D-signal processing to analyze images (convolutions, Butterworth filters, Hadamard and Fast Fourier Transforms, etc.) [51–53];
- the second way relies on mathematical morphology [54] and manages sets of points or pixels. The basic operations of mathematical morphology are the erosion and the dilation: in an erosion (dilation) any pixel belonging to an object which has a neighbor belonging to the background will be set to 0 (to 1). Initially defined to work on binary images, the theory has been extended to gray-level images, providing tools for their enhancement.

Microorganisms being individualized objects or sets of pixels, the second approach would appear more appropriate. However the initial enhancement of the gray-level image may benefit from the first approach. There is no definite solution and generally a balance between the two is used. It means also that there is no unique treatment of an image and that the final solution will depend upon the expert in charge with the development of the procedure.

Halos, light inhomogeneities, necessity to enhance the contours of cells are among the reasons for applying different algorithms to improve the quality of images. One has always to keep in mind that good quality of the initial gray-level or color images is essential: any enhancement by software will introduced some bias in the final results. Due to the halos around the cells, to inhomogeneities of staining within the cells, fluorescence images are generally delicate. It is even more true when the cells are small as for non-filamentous bacteria. Schröder et al. [55] and Viles et Sieracki [12] suggest improving the image by convolution with the Mexican-hat type matrix proposed by Marr and Hildreth [56].

The aim of segmentation is to separate the objects (cells, including most of the debris) from the background. The obtaining of the binary image (pixel value =1 for object and =0 for background) is a critical step, which should be automated as much as possible to avoid the operator "subjectivity" (differences of appreciation between operators, fatigue). Different algorithms are possible and a good overview can be found in Ref. [57].

The global segmentation methods are based on the analysis of the gray-level image (or on the Intensity image for the HSI color model) histogram, i.e. the distribution in number of the gray-levels, which is approximated by a weighted sum of two (or more) probability densities with normal distribution, one for the background region and one for the object region. Figure 8 presents two typical cases. The histogram of the *Streptomyces* image presents two regions separated by a large valley: one at low gray-level for the pellet and one at high gray-levels for the background. The image of *Saccharomyces cerevisiae* shown in Fig. 2b has been focused so that the gray-levels of the cell wall are very low and those of the cytoplasm very high. The background pixels have intermediate values. The iterative threshold optimal selection can be based on different characteristics: first moments [58], variance [59], entropy [60] and can be applied for segmentation into two or more regions.

Let p_i be the probability for a pixel to have a gray-level equal to i:

$$p_i = N_i/N \tag{1}$$

where N_i is the number of pixels of gray-level i and N the total number of pixels in the image. Let us consider the case of a segmentation into two regions: w_0 and w_1 the probabilities for a pixel to have a gray-level, respectively, less than or equal to the threshold h and greater than h:

$$w_0 = \sum_{i=0}^{h} p_i \quad \text{and} \quad w_1 = \sum_{i=h+1}^{255} p_i \tag{2}$$

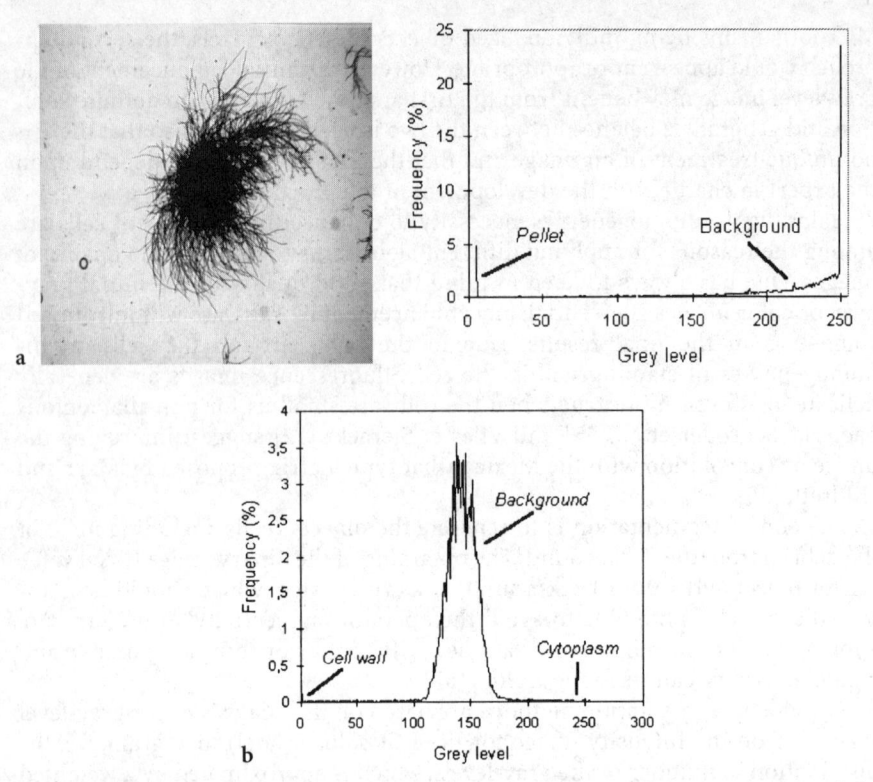

Fig. 8. Gray-level image of a *Streptomyces* pellet and its histogram (**a**) – Histogram of the *Saccharomyces cerevisiae* image (Fig. 2a) (**b**)

The first moments of the probability functions corresponding to the background and the objects will be:

$$\mu_0 = \frac{1}{w_0} \sum_{i=0}^{h} i \cdot p_i \quad \text{and} \quad \mu_1 = \frac{1}{w_1} \sum_{i=h+1}^{255} i \cdot p_i \qquad (3)$$

Ridley and Calvard [58] propose to update the initial estimate of the threshold h:

$$h_{new} = \frac{\mu_0 + \mu_1}{2} \quad \text{until} \quad h_{new} = h \qquad (4)$$

when Otsu [59] maximizes the separability between the probability functions by maximizing the interclass variance

$$w_0 w_1 (\mu_1 - \mu_0)^2 \qquad (5)$$

Kanpur [60] suggests to maximize the total entropy

$$\phi = \phi_0 + \phi_1 \tag{6}$$

where

$$\phi_0 = \left(-\sum_{i=0}^{i=h} \frac{p_i}{w_0} \cdot \text{Log}\left(\frac{p_i}{w_0}\right) \right) \quad \text{and} \quad \phi_1 = \left(-\sum_{i=h+1}^{i=255} \frac{p_i}{w_1} \cdot \text{Log}\left(\frac{p_i}{w_1}\right) \right) \tag{7}$$

Examples of applications are given in Fig. 9. For the *Streptomyces* pellet the threshold suggested by the entropy algorithm is higher (188) than the one suggested by the variance algorithm (141). The pellet is better recovered with the entropy algorithm although some of the background on the right side of the image due to the halo on the gray-level image has been selected as well. This artifact will be, however, easily removed in the enhancement of the binary image. For the yeast cells no significant difference can be found between the algorithms, which provide a clear segmentation of the wall membrane. Vicente

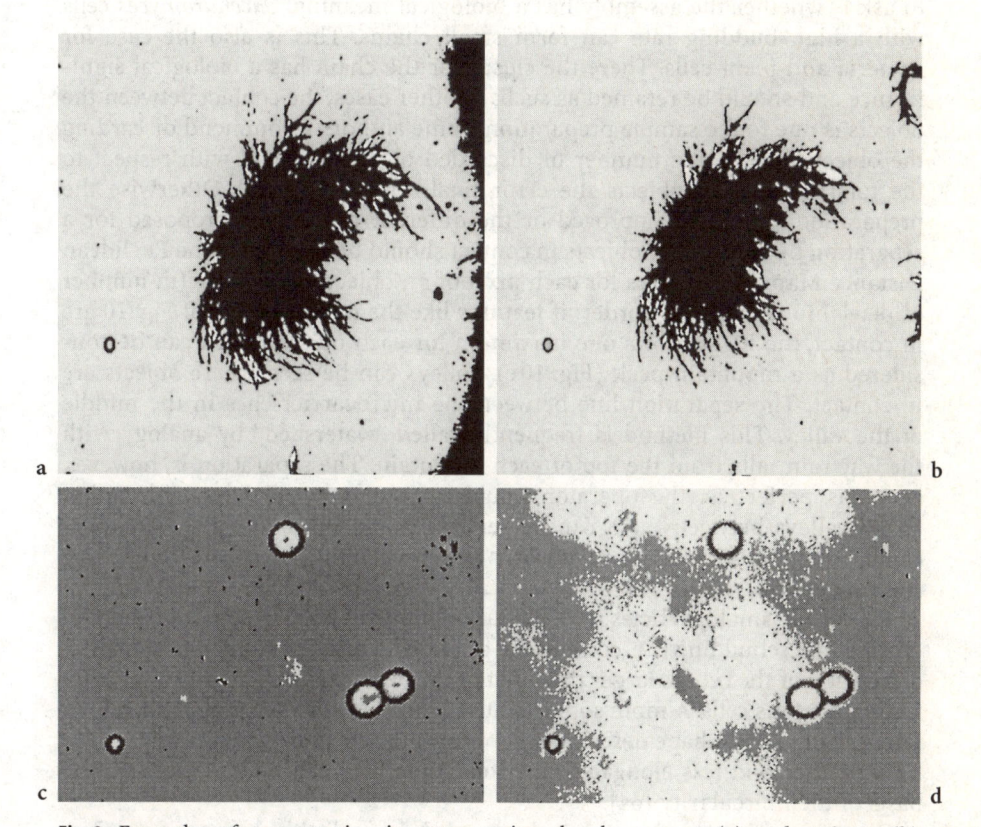

Fig. 9. Examples of segmentation into two regions by the entropy (**a**) and variance (**b**) methods and into three regions by the entropy (**c**) and variance (**d**) methods

et al. [61] obtained satisfactory results with the variance method for the segmentation of images of yeast flocs.

The local methods of segmentation consider the neighborhood of each pixel. Edge-based methods rely on discontinuities in gray level, color, texture. They are sensitive to noise which creates false edges and to bad contrast between features which gives weak edges (i.e. difficult to detect). Region-growing methods work better on noisy images. Neighboring pixels, which have reasonably similar neighborhoods in terms of gray-levels are considered to belong to the same object or background zone.

In applications dealing with individualized cells, only those fully in view can be analyzed. The largest objects have, proportionally, a higher risk in contact with the image frame and to be discarded, which will ultimately introduce a bias in the population statistics. Some authors suggest a measuring frame be inserted into the visible field [41] or a correction factor, function of the size of the feature relative to the size of the field of view, be applied to the measured size or shape parameter [62]. Border-killing is also used to remove artifacts such as the halo of Fig. 9a.

It may happen that two or more objects are in contact. The first question to ask is whether the assembly has a biological meaning. *Saccharomyces* cells with a high budding rate can form small chains. This is also the case for bacteria and plant cells. There the cluster or the chain has a biological significance and should be retained as such. In other cases, the contact between the objects is due to the sample preparation. Some authors recommend discarding the objects [63]. If the number of discarded objects is small with respect to the total number of objects the error finally made is small. Otherwise the preparation should be improved or the different procedures proposed for a separation by software of objects in contact should be used [53]. The Euclidean Distance Map (EDM) gives for each pixel of an object its distance (in number of pixels) to the nearest border: if features like the microcarriers of Fig. 10 are in contact, the EDM shows one maximum for each object, which can be considered as a mountain peak (Fig. 10e). Valleys can be seen where objects are in contact. The separation line between the microcarriers lies in the middle of the valley. This method is frequently called "watershed" by analogy with the way rain falls from the top of each mountain. The separation is, however, never as perfect as the operator would expect: only particles with rather broad valleys between them can be separated and noise along the contours can introduce false separations. It works better when the objects are in the same size range.

Classically small particles of debris are removed by a two-step procedure of erosion and rebuilding (i.e. conditional dilation). This implies than the smallest dimension of the cells is larger than that of the debris. This is usually a problem for filamentous cells. A more sophisticated procedure should be applied, taking into account some shape descriptor. For example debris are often considered to be smoother and less elongated than filaments, and can be separated on the basis of their circularity [64].

Due to the transparency of the microorganisms and with absence of staining, the visualization focuses often on the membrane of the cell. A hole-filling step

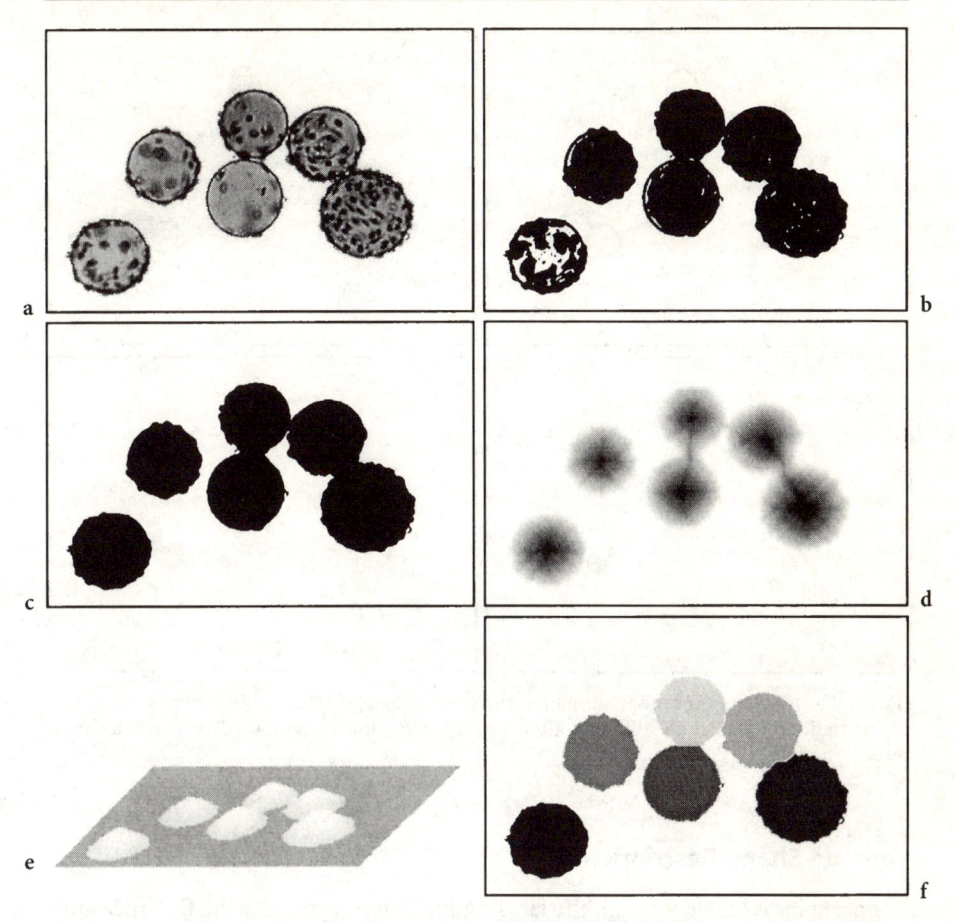

Fig. 10. Separation of objects in contact: **a** Original gray-level image, **b** Binary image, **c** Enhanced binary image (hole-filling), **d** EDM image, **e** Perpective view of the EDM image, **f** Separated microcarriers

is necessary to obtain a complete silhouette. This is simple for a yeast cell but requires more work for filaments for which holes do not mean necessary cytoplasmic zones in the case of hyphae curling.

Figure 11 summarizes the main steps of the treatment of the *Saccharomyces cerevisiae* image (Fig. 2a). There is usually one final procedure before quantitative characterization of each object or cell: the purpose of the discrete labeling is to identify each individual and assign a number to it [52].

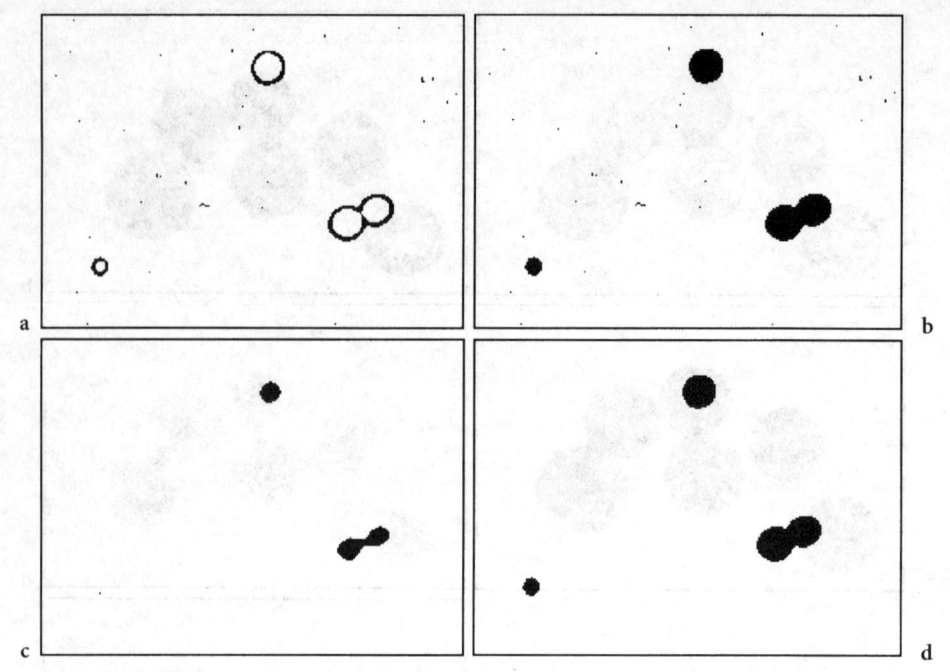

Fig. 11. The main steps of the treatment of the yeast image (Fig. 2a): **a** Binary image just after segmentation; **b** after hole-filling; **c** after a 4-order erosion to remove debris; **d** after final reconstruction

5
Size and Shape Description

Counting individuals through the labeling procedure is usually the first measurement performed on an image [65]. No problem is encountered when the objects in contact have been properly separated by software and the debris removed. Bacterial detection for growth or contamination assessment is based on the counting of cells or microcolonies, often after staining [66–69]. Manual counts are tedious and labor-intensive [70, 71]. Automation of colony counting has been proposed with epi-illumination or a transillumination [72, 73]. Sometimes a full separation of all the objects in view is difficult or not desired when one is interested in the number of elements of a cluster [74, 75]. Assuming the cells are convex the EDM provides an easy way to count the number of objects as illustrated in Fig. 12: the number of maxima found in the EDM images corresponds to the number of cells.

The primary size measurement is based on the area of the silhouette. A simple count of the pixels and the consideration of the calibration (μm^2/pixel for example) give the projected area, A. This is valid for any cell, even elongated ones [76]. Geometrical correction might be necessary to take into account the shape of the support, like in the case of spherical microcarriers [77]. An equivalent diameter (D_{eq}) is subsequently calculated:

$$D_{eq} = 2\sqrt{A/\pi} \qquad (8)$$

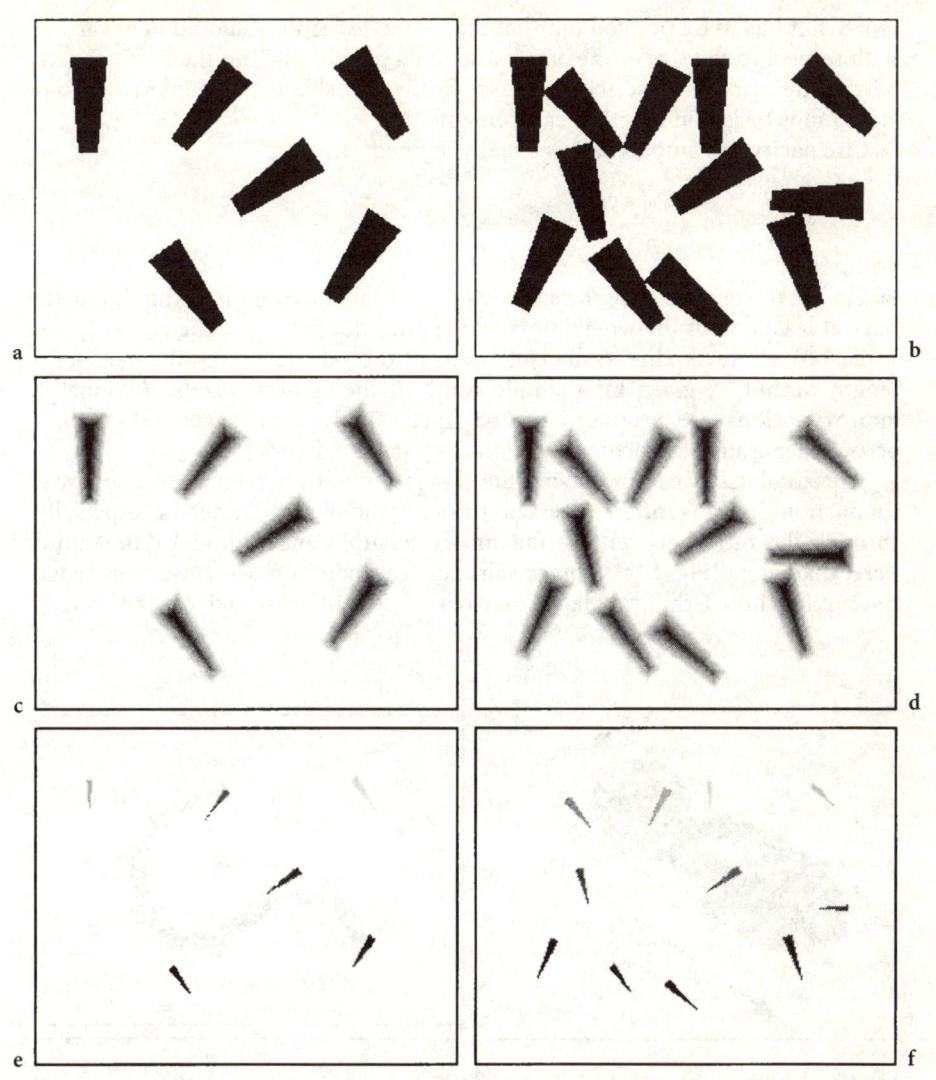

Fig. 12. Counting cells without separation: **a** and **b** Initial binary images; **c** and **d** Corresponding EDM images; **e** and **f** Maxima of EDM

If this can be applied to rather "circular" shapes (*Saccharomyces cerevisiae* [50], mammalian cells, either isolated or aggregated [26, 78, 79], microcolonies [80], etc.), it seems more difficult to characterize filamentous species by this size parameter. As mentioned before size and shape quantifications are intimately connected.

Many general shape factors exist but the nomenclature depending upon authors varies and it is suggested to always verify the mathematical definition

[53, 81]. It has to be pointed out that the shape descriptors should be invariant with respect to the size of the object and to its position within the field of view. Several descriptors describe macroscopically the shape by geometrical considerations based on length measurements.

Circularity or compactness is usually defined as

$$\frac{P^2}{4\pi A} \quad \text{or} \quad \frac{4\pi A}{P^2} \tag{9}$$

where P is the perimeter of the silhouette. Pixels on the contour of the silhouette have at least one of their neighbors which does not belong to the object (pixel value = 0). However, due to the constrains of discrete geometry, the perimeter length cannot be given by a simple count of the contour pixels. Reasonable approximations are proposed by the image analysis packages, taking into account separately the orthogonal and diagonal neighbors [53].

The circularity varies both with the elongation of the object and its rugosity. Elongation itself is often assessed through the Feret diameters, especially through the ratio between the maximal ("length") and minimal ("breadth") Feret diameters (Fig. 13a). This is valid for convex or almost convex elements, but not for those exhibiting large concavities (Fig. 13b). In such cases a rectan-

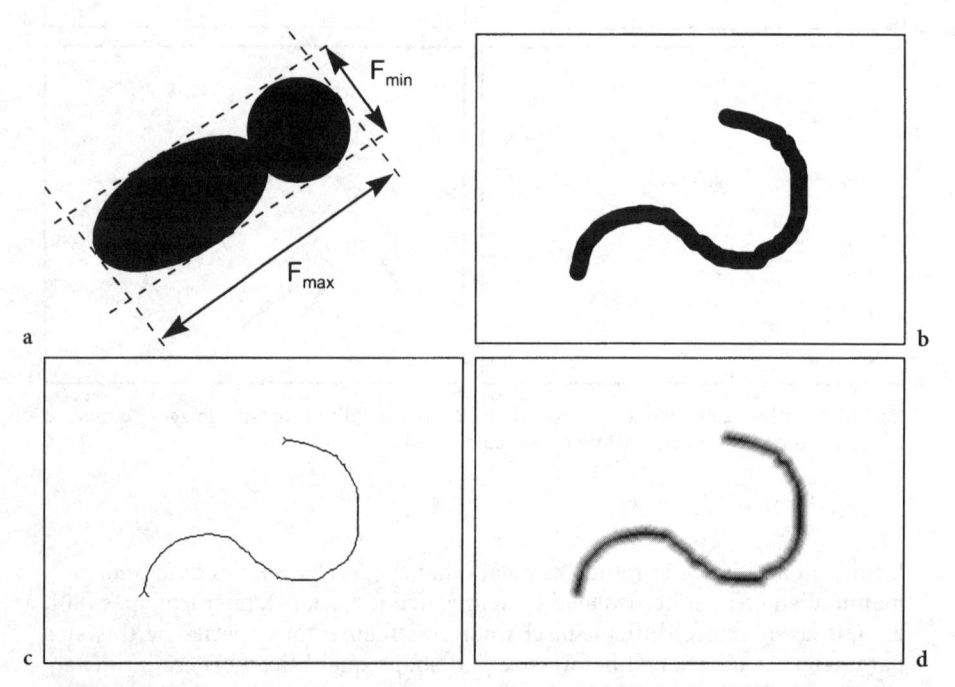

Fig. 13. a Maximal (F_{max}) and minimal (F_{min}) Feret diameters of a *Saccharomyces cerevisiae* budding cell; Schematic representation of an unbranched filament (**b**), its skeleton (**c**) and its EDM image (**d**)

gular assumption can be made about the feature shape: assuming a constant width W and a length F then

$$A = F \cdot W \tag{10}$$

and

$$P = 2 \cdot (F + W) \tag{11}$$

If the boundary is relatively smooth the perimeter can be measured with reasonable accuracy, and F and W are given by the following equations:

$$F = \frac{P + \sqrt{P^2 - 16 \cdot A}}{4} \tag{12}$$

$$W = \frac{A}{F} \tag{13}$$

Another way to proceed is to calculate the skeleton of the feature of interest (Fig. 13c) and its Euclidian Distance Map (Fig. 13d). The skeleton is obtained by successive erosions until a set of points with maximal thickness of one or two is found: the area of the skeleton gives the length of the object. Using the skeleton to select only the points which lie along the axis of the feature's EDM and retaining their gray values, an average or maximal width is obtained. In the example of Fig. 13, $F = 347$ pixels and $W = 13$ pixels when calculated from the rectangular assumption, and $F = 352$ pixels and $W = 13$ pixels when calculated from the skeleton length and the EDM.

Cazzulino et al. [82] and Hirano [83] calculate a "roughness" factor:

$$R = \frac{\pi(F + W)}{2P} \tag{14}$$

to characterize somatic embryos and budding *Saccharomyces cerevisiae* cells respectively. R is closed to 1 for a disk or an ellipse. For a disk $F = W$ and $P = \pi F$.

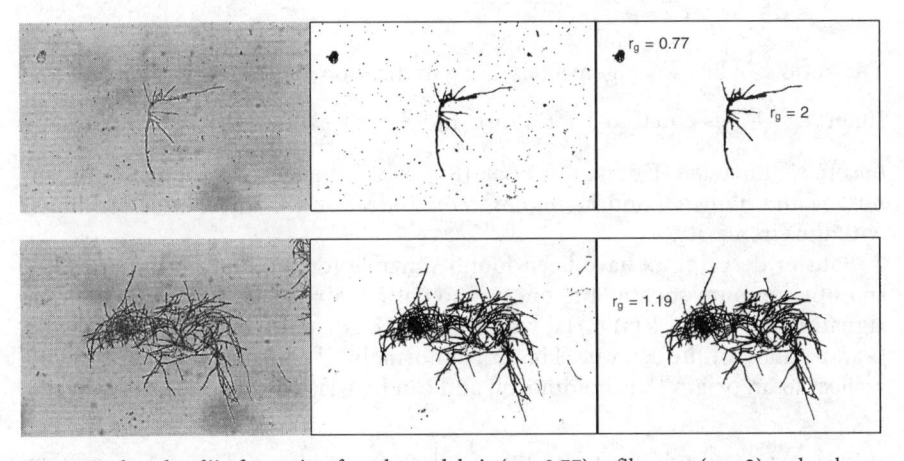

Fig. 14. Reduced radii of gyration for a large debris ($r_g = 0.77$), a filament ($r_g = 2$) and a clump ($r_g = 1.19$)

R becomes smaller than one ($0.7 \leq R < 1$) when the shape becomes irregular like a budding yeast. For Fig. 13a $R = 0.87$.

The reduced radius of gyration (r_g) is a global descriptor useful to discriminate large globular features (debris or pellets) from filamentous objects ($r_g = 0.7$ for a disc) (Fig. 14). It is based on the moments of the feature:

$$r_g = \frac{\sqrt{M_{2X} + M_{2Y}}}{D_{eq}/2} \qquad (15)$$

with

$$M_{2X} = \frac{\sum_{i=1}^{N}(x_i - x_g)^2}{N}, \quad M_{2Y} = \frac{\sum_{i=1}^{N}(y_i - y_g)^2}{N}, \qquad (16)$$

$$x_g = \frac{\sum_{i=1}^{N} x_i}{N} \quad \text{and} \quad y_g = \frac{\sum_{i=1}^{N} y_i}{N}$$

(x_i, y_i) defines the position of pixel i of an object of N pixels according to the columns and lines of the image. The inertia matrix Σ_{XY}, also based on the moments, is real and positive semi-definite, with two non-zero real eigenvalues, λ_{min} and λ_{max}

$$\Sigma_{XY} = \begin{bmatrix} M_{2X} & M_{XY} \\ M_{XY} & M_{2Y} \end{bmatrix} \qquad (17)$$

with

$$M_{XY} = \frac{\sum_{i=1}^{N}(x_i - x_g) \cdot (y_i - y_g)^2}{N} \qquad (18)$$

The ratio of the two eigenvalues is an indication of the eccentricity of the shape, as $\frac{\lambda_{min}}{\lambda_{max}}$ is equal to 1 for a disc and decreases toward 0 when the shape becomes elongated. Dubuisson et al. [84] based their classification of *Methanospirillum hungatei* and *Methanosarcina mazei* on this descriptor combined with the circularity.

Fourier descriptors have been found valuable for the description of rather smooth contours such as the ones of somatic embryos [85–87]. The contour signature defined as $R(s) = f(s)$ where $R(s)$ is the distance between the centroid G and a point of the contour (Fig. 15), of normalized curvilinear abscissa s with respect to an origin O, is holomorph and can be written as:

$$R(s) = \sum_{i=1}^{\infty} C_n e^{-jns} \quad \text{with} \quad C_n R(s)) = \frac{1}{2\pi} \int_{0}^{2\pi} R(s) e^{-jns} ds \quad \text{and} \quad s = 2\pi \frac{s'}{P} \qquad (19)$$

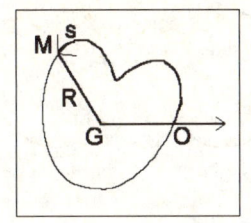

Fig. 15. Definition of the signature

The first coefficients ($2 \leq n < 10$) describe the general shape of the object when the coefficients of higher order are related to its rugosity. C_0 gives the average radius of the particle and C_1 the error in the definition of the position of the centroid [81, 88].

The comparison to a reference shape, which is generally the convex bounding polygon provides another set of descriptors, at a mesoscopic level of details. The index of concavity in length or in surface measures globally how the object departs from a convex shape:

$$IC_A = A/A_c \quad \text{and} \quad IC_P = P/P_c \tag{20}$$

where A_c and P_c are respectively the area and the perimeter of the convex bounding polygon. For the somatic embryo (torpedo) schematized in Fig. 16 $IC_A = 0.95$ and $IC_P = 1.07$. Variations of the index of concavity can be due to multiple small concavities or to a single one. The size of the largest concavity can be estimated by applying a series of erosions to the residual set of the object with respect to the convex bounding polygon [89, 90]. The residual set (Fig. 16c) disappears completely after $\omega_2 = 9$ erosions and the reduced size of its largest concavity is expressed by:

$$\Omega_2 = \frac{2\omega_2}{\sqrt{A}} = 0.14 \tag{21}$$

Fractal dimension is used to give information on rugosity at the microscopic level of details. Different algorithms are available. In the Richardson's walk [91],

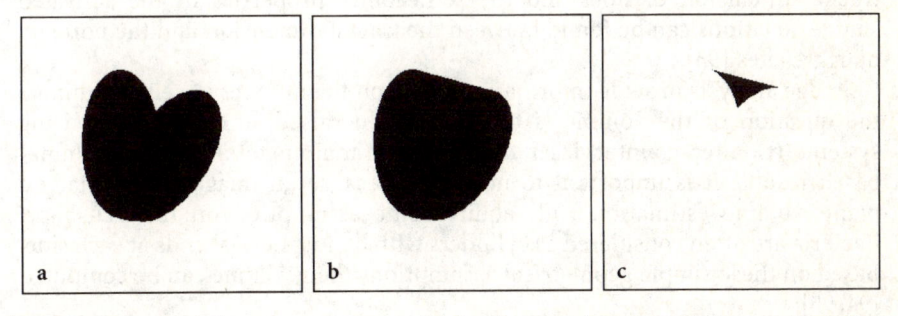

Fig. 16. Concavity index measurement: **a** Schematic somatic embryo **b** Convex bounding polygon, **c** Residual set

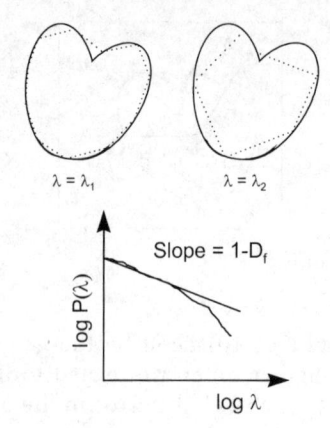

Fig. 17. Computation of the fractal dimension (below) by the Richardson's walk (above)

a step of increasing length (λ) is uses to measure the length of the contour (P) (Fig. 17 above). The slope of the $\log P$ versus $\log \lambda$ plot gives the desired fractal dimension (Fig. 17 below). As mentioned previously measuring lengths is not easy on images: many discuss the reliability of the results in function of the way the contour is defined. The results of the Richardson's walk depends also upon the starting point.

A faster and more robust way to obtain the fractal dimension is to use the EDM [53]. The brightness histogram $b(v)$ gives directly the number of pixels at each distance λ: the perimeter $P(\lambda)$ is calculated by dividing the number of pixels with a brightness less than or equal to λ by λ:

$$P(\lambda) = \frac{\sum_{v=1}^{\lambda} b(v)}{\lambda} \qquad (22)$$

It has to be recognized that the interpretation of the plot is not always straightforward since it might be difficult to find a single straight line. Nevertheless Grijspeerdt and Verstraete [92] have established a relationship between the fractal dimension of flocs and the settleability properties of the activated sludge. Relations can be found between the fractal dimension and the porosity of aggregates [93].

Image analysis provide information based on the number of cells examined: the question of the volume is therefore often raised, as some other sizing systems (Coulter counter, laser diffraction granulometers) provide volume-based results. It is important to note that, because an image is essentially a plane, volume estimation will require some assumption on the 3D-shape. Bacteria are often considered as cylinders with hemispherical ends at each side. Based on these simple geometrical assumptions their volume can be computed [94, 95]:

$$V = [(W^2 - \pi/4) \cdot (F - W)] + (\pi \cdot W^3/6) \qquad (23)$$

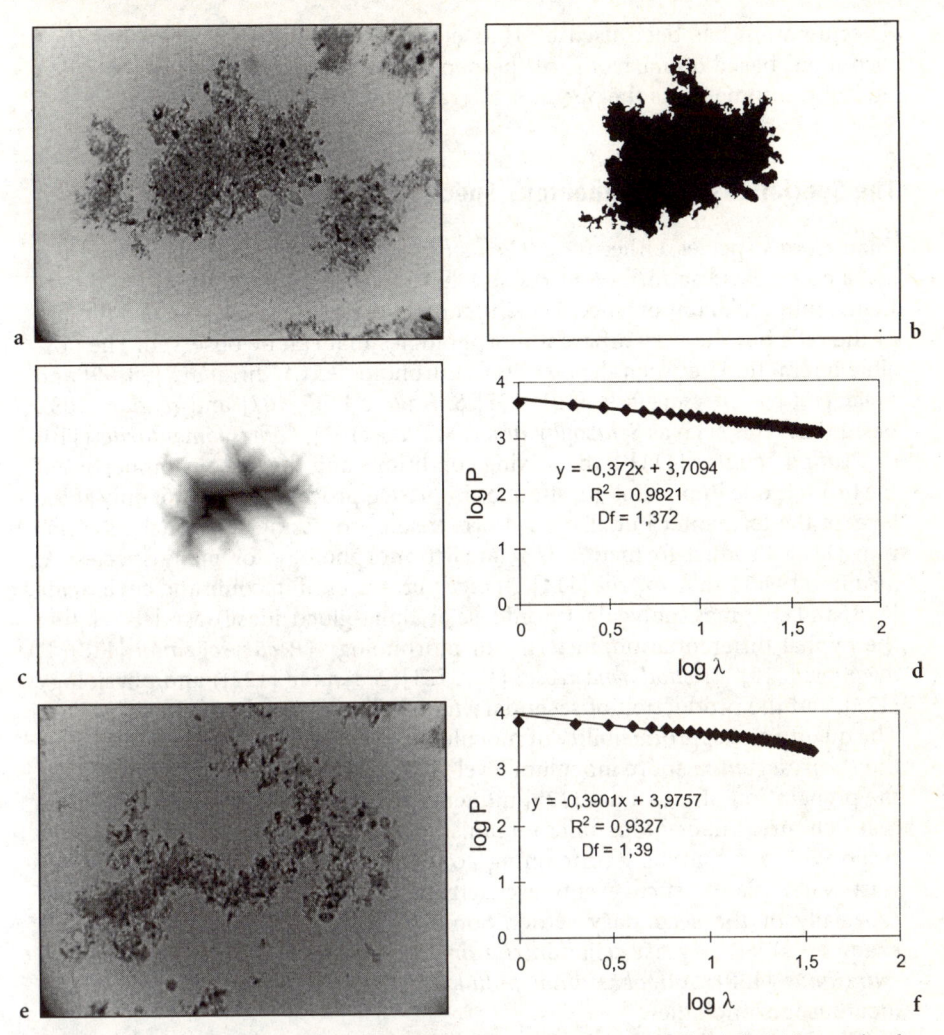

Fig. 18. Fractal dimension of activated sludge flocs "1" (**a**) and "2" (**e**) (initial gray-level images); Binary image (**b**) and EDM image (**c**) of floc "1"; Calculation of the fractal dimension of flocs "1" (**d**) and "2" (**f**)

This reduces to:

$$V = \pi \cdot W^3/6 \qquad (24)$$

when the bacteria is a coccus. *Saccharomyces cerevisiae* yeast cells are mostly assumed to be prolate ellipsoids [74]

$$V = \frac{4\pi W^2 F}{3} \qquad (25)$$

or even spheres [37, 96]. Vaija et al. [97] assume that *Saccharomyces cerevisiae* mother cell is a prolate ellipsoid with a spherical bud. A more detailed volume determination has been used by Huls et al. [29] for the same yeast but their technique, based on narrow slices perpendicular to the main axis of the cell, is not fully automated as the operator interactively selects the cell to analyze.

6
The Special Case of Filamentous Species

Filamentous species, either fungi (*Penicillium* sp., *Trichoderma* sp., *Aspergillus* sp., etc.) or bacteria (*Streptomyces* sp.), have received a lot of attention due to their commercial importance. The difficulty in their case is not so much the size of the cells but the very different morphologies that can be observed. The possible interaction between the observed morphologies (*A. awamori*, [98, 99] and *niger* [100] *P. chrysogenum* [101–105] *S. fradiae* [106, 107] and *tendae* [108], basidiomycetes such as *Schizophyllum commune* [109], *Fomes fomentarius* [110] or *Cyathus striatus* [111]), the mixing conditions and the viscous properties of the broth is one important question for improved process design, not only at the level of the fermenter but also for downstream processing steps such as filtration [112]. Product formation is related to morphology for many species: *A. awamori* [99, 113], *A. oryzae* [114], *A. niger*, producer of itaconic and citric acids [115] and enzymes (polygalacturonidase or alpha-glucosidase). Models relating the hyphal differentiation, in terms of morphology (*P. chrysogenum* [118], *A awamori* [119], *Trichoderma reesei* [120, 121], *S. tendae* [122]) and physiology [123], and the production of secondary metabolites is another field of interest. The quality and reproducibility of inoculum is a everyday problem in industry: the *P. chrysogenum* spore inoculum level [124] and the type of reactor used for the preparation of inoculum [125] influence the morphology found during the main culture. Filamentous microorganisms such as *Microthrix parvicella* are responsible for bulking and foaming accidents [126, 127] in activated sludge wastewater plants: such events are detrimental to the control of the plant, especially of the secondary settler. Some plant cells (*Morinda citrifolia* for example) [128] or yeast cells (*Saccharomyces cerevisiae* [129], *Kluveyromyces marxianus* [130], *Aureobasidium pullulans* [131, 132]) can also exhibit filamentous morphologies.

Whether growing in suspension or on a solid substrate, the life cycle of filamentous bacteria and fungi starts with spores. Spores are usually considered as being spherical particles [133] although they can probably be best represented as ellipsoids. Actinomycetes spores were found to have some degree of elongation [134]. Mitchell et al. [135] have examined a collection of 13 types of basidiospores (*Agrocybe parasitica*, *Leucoagaricus leucothites* and 11 *Agaricus* species) which were distinguished one from another by five features (length, width, elongation, circularity and area). An automated method has been developed by Paul et al. [136] to monitor the swelling, the emission of germ tubes and their subsequent elongation: the method has been applied to *P. chrysogenum* spores for different germination conditions. The shape of the germ tube depends upon the species. A general automated procedure has been devised by Oh et al. [137] to study the germination of circular (*Aspergillus fumigatus*), oval (*Curvularia lunata*) and banana-shaped (*Fusarium solani*) spores: it uses a curvature-based shape representation technique after edge-smoothing by Fast Fourier Transform. Once the germ tube has grown sufficiently, lateral branches develop as filaments [138].

Figure 19 shows the main steps of a method of characterization of spores and generated filaments which is based on the Euclidean Distance Map. The central zone of the EDM (i.e. the zone with the highest distance value R_m) is dilated R_m times. The size of the object in Fig. 20e gives the size of the spore. A logical subtraction of Fig. 19e from Fig. 19c leaves only the branches visible in Fig. 19f.

The important features of filaments in absence of entanglement of their branches are the total length of the filament, the length of the main hypha, the number of extremities or tips, the number of branching points and the distance between two branching points and between branching points and tips.

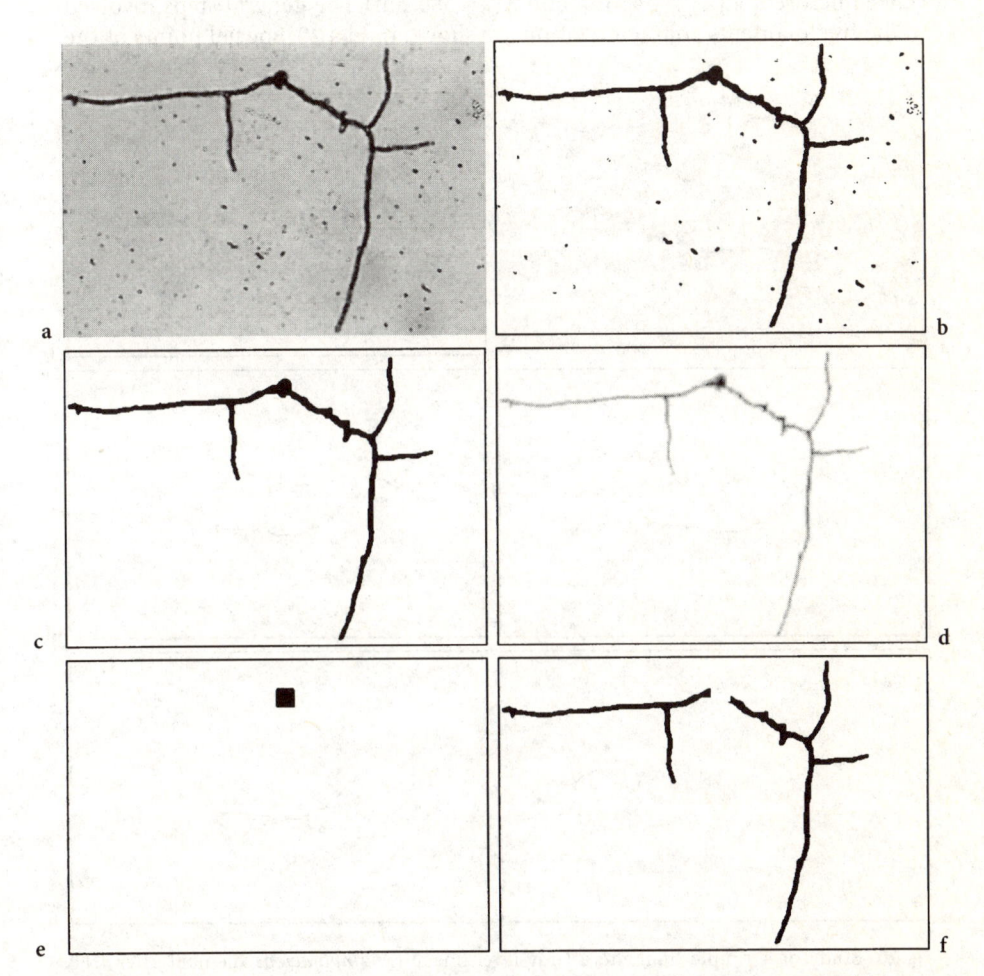

Fig. 19. Study of a germinated spore (*Trichoderma reesei*): **a** Initial gray-level image; **b** binary image; **c** binary image after cleaning; **d** EDM image; **e** Spore mask; **f** Filamentous parts

The first methods of their quantitative morphological analysis were based on a digitizing table, on which the image of filaments was projected. The filaments were spread carefully to avoid entanglements [139]. The operator, by means of a cursor, captured the hyphal coordinates, the tips and the branching points positions. This method has been used to study the influence of operation conditions on the morphology of filamentous molds [140] but it reveals itself to be very laborious, as it is necessary to study a large numbers of objects (from 100 to 2000 per sample) to have statistically valid results. This encouraged C. Thomas and his research team to develop automated image analysis methods for free filaments [141–143] initially for *S. clavuligerus* and *P. chrysogenum*. Similar methods were applied by Daniel et al. [144] for the quantification of fungi on decomposing leaves and by Walsby and Avery [145] to cyanobacteria. In the latter case, the microorganisms can be observed directly by epifluorescence microscopy (λ_{ex} = 546 nm and λ_{em} = 580 nm) The general steps involved in the free filaments characterization are shown in Fig. 20. Special points of the

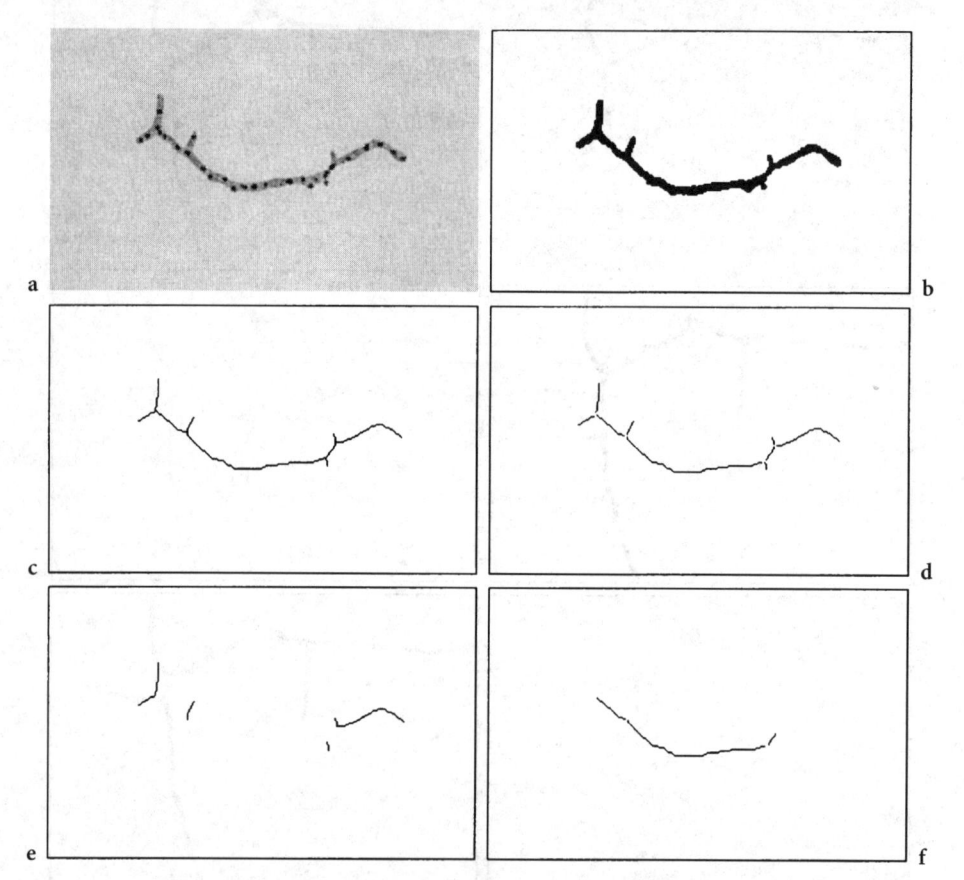

Fig. 20. Study of a simple filament: **a** Gray-level image (*S. ambofaciens* filament after INT staining); **b** binary image; **c** skeleton; **d** skeleton after removal of branching points; **e** terminal branches; **f** internal segments

skeleton (Fig. 20c) are detected: end-points to mark the apices and triple-points to mark the branching points. The triple-points are removed from the skeleton and the apices are reconstructed from the end-points (Fig. 20e). The internal segments are obtained by difference (Fig. 20f).

A similar procedure can be applied to the branches of the germinated spores of Fig. 19. The only measurement listed above which cannot be really obtained is the length of the main hypha. Its definition is somewhat unclear. In the above approach it is usually stated to be the longest path between two tips. In the example of Fig. 19, it is certainly not true as the spore from which the filaments originated is almost at the center of the structure. For a more developed filament the spore is often not clearly visible any more due to the spatial resolution.

In order to monitor closely the early steps of germination and development of hyphae, and to track the development of the main hypha, a flow-through cell has been designed by Spohr et al. [133]. The adhesion of *Aspergillus oryzae*

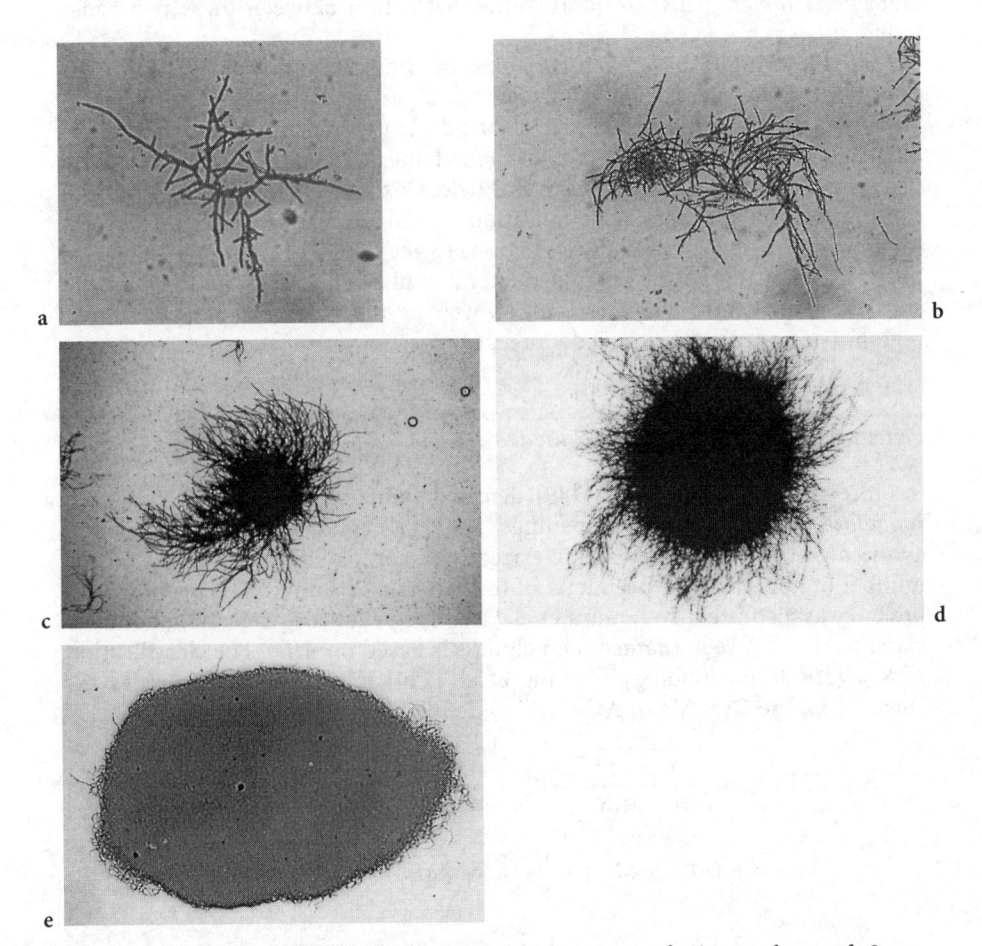

Fig. 21. Different complex filamentous morphologies: **a** *P. roqueforti* entanglement; **b** *S. ambofaciens* clump; **c** and **d** *S. ambofaciens* pellets; **e** *S. ambofaciens* bald pellet

spores was facilitated by the presence of poly-D-lysine. The spore swelling, the rate of extension of the germ tube, the branching frequency, the growth kinetics of all branches and of the main hypha could be monitored for each of the 35 hyphal elements which could be observed in a single experiment When the filament network become too complicated as in the experiment of Hitchcock et al. [146] on growth of *Trichoderma viride* on Cellophane, a fractal approach can be applied to estimate the relative importance of feeding and foraging.

In suspension also, progressive elongation of the filaments provokes their entanglements and structures of various complexity can be observed. The ultimate form is the pellet, which can itself present various morphologies, with a large or small core, with respect to the filamentous zone as illustrated in Fig. 21. The segmentation between the compact core and the filamentous zone is generally based on image analysis concepts. A staining procedure has been tested by Durant et al. on two basidiomycetes (*Schizophyllum commune* and *Fomes fomentarius*) to improve the distinction between these two zones [109, 110].

Many descriptors have been proposed by different researchers to describe the complexity of pellets and clumps and comparison between authors may be difficult [147]. This abundance reflects the difficulty to characterize these complex objects and also the diversity of their shape. It can confuse the newcomer to applications of image analysis in this field. Cox and Thomas [148] detect the central core by ultimate skeletonisation combined with series of openings (opening = erosion followed by a dilation) to get rid of the loops which appear in the skeleton due to the entangled lateral hyphae (Fig. 22). The shape of the core is subsequently characterized, as well as the hairiness of the annular region. The annular fullness is defined as:

$$\frac{(Pellet\ area - Core\ area) \cdot 100}{Pellet\ convex\ area - Core\ convex\ area} \quad (41\%\ \text{for the pellet of Fig. 22a}) \quad (26)$$

More simply Pichon et al. [149] detected the core of *S. ambofaciens* et *P. roqueforti* pellets by series of openings. The hairiness of annular zones, as well as the own of clumps was characterized by a complexity index based on the number of holes and the perimeter of the pellet and its holes. The openings approach was also used by Tamura et al. [106] for pellet core detection, but their classification between filament and clumps is based on A/P_c^2. The classification of *S. virginiae* morphology by Yang et al. [150] into three classes relies exclusively on the Oval Major Axis (OMA) defined by:

$$OMA = 2\sqrt{\frac{16 \cdot Max^3}{\pi^2 \cdot Min}} \quad \text{with} \quad (27)$$

$$Max = \frac{1}{2}(M_{2X} + M_{2Y}) + \frac{1}{2}\sqrt{(M_{2X} - M_{2Y})^2 + 4M_{XY}^2} \quad \text{and} \quad (28)$$

$$Min = \frac{1}{2}(M_{2X} + M_{2Y}) - \frac{1}{2}\sqrt{(M_{2X} - M_{2Y})^2 + 4M_{XY}^2} \quad (29)$$

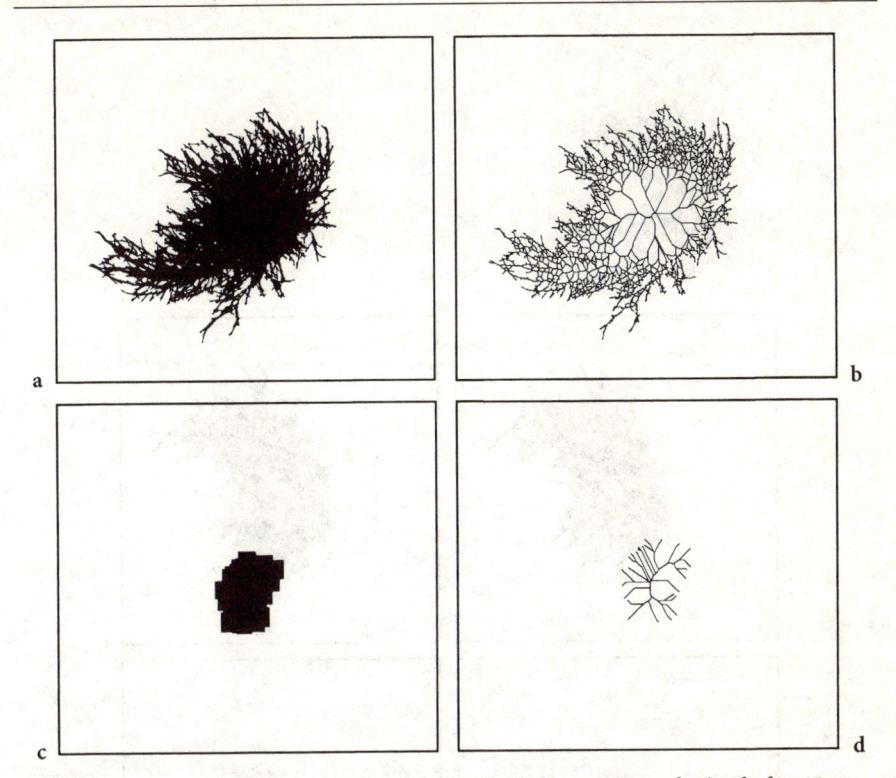

Fig. 22. Detection of a pellet core: **a** Binary image; **b** Skeleton; **c** Core obtained after 15 openings; and **d** its skeleton

In this approach an ellipse is fitted to the object, taking into account its moments. Series of openings helped Treskatis et al. [41] to characterize *S. tendae* pellets by progressive filling of the holes in the structure.

The EDM brightness histograms computed on the object before and after hole-filling bring also information on pellets and entanglements without having to detect specifically the presence of a core. The differences in the EDM images of a pellet before (object, of equivalent diameter D_{eq1}, in Fig. 23b and d) and after (mask, of equivalent diameter D_{eq2}, in Fig. 23c and e) hole-filling are seen in the graph of Fig. 23d. The maximal distance of the object gives the radius of the core: R_{c1} [151].

The filament thickness can be another interesting feature. Production of cephalosporin by *Cephalosporium acremonium* has been observed in conjunction with some swelling of the hyphae [152]. It has been often reported that, when cultivated on solid substrate, apical parts of filamentous species tend to become thinner, apparently due to substrate limitation, which induces the cell to increase its surface to volume ratio. Similar behavior has been observed in suspended media on fungi (*Trichoderma reesei* [151] and *S. ambofaciens* [153, 154]) (Fig. 24). Hyphae are often regarded as solid cylinders, and, in order to

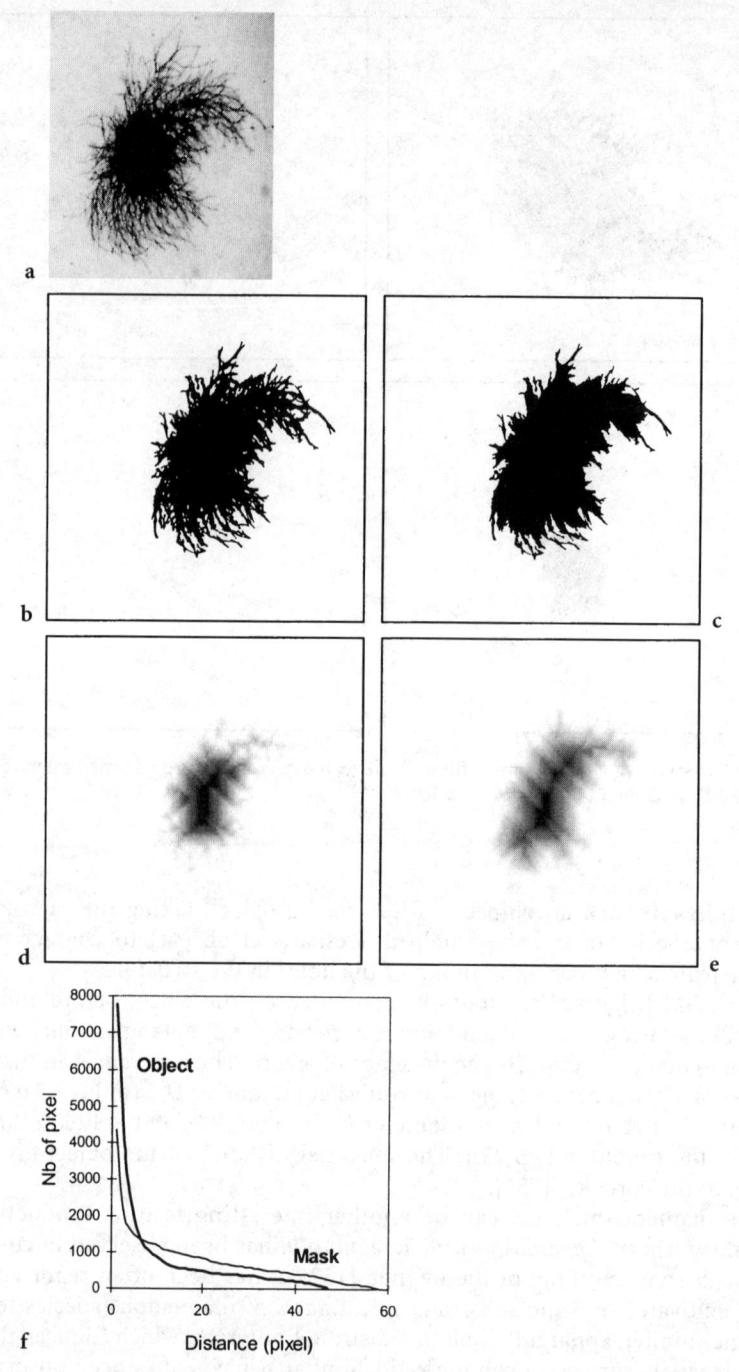

Fig. 23. *Streptomyces* pellet (**a**), its silhouette (**b** normal, **c** filled) and the EDM images of its normal (**d**) and filled (**e**) silhouettes; **f** EDM histograms for the pellet (Object) and its mask

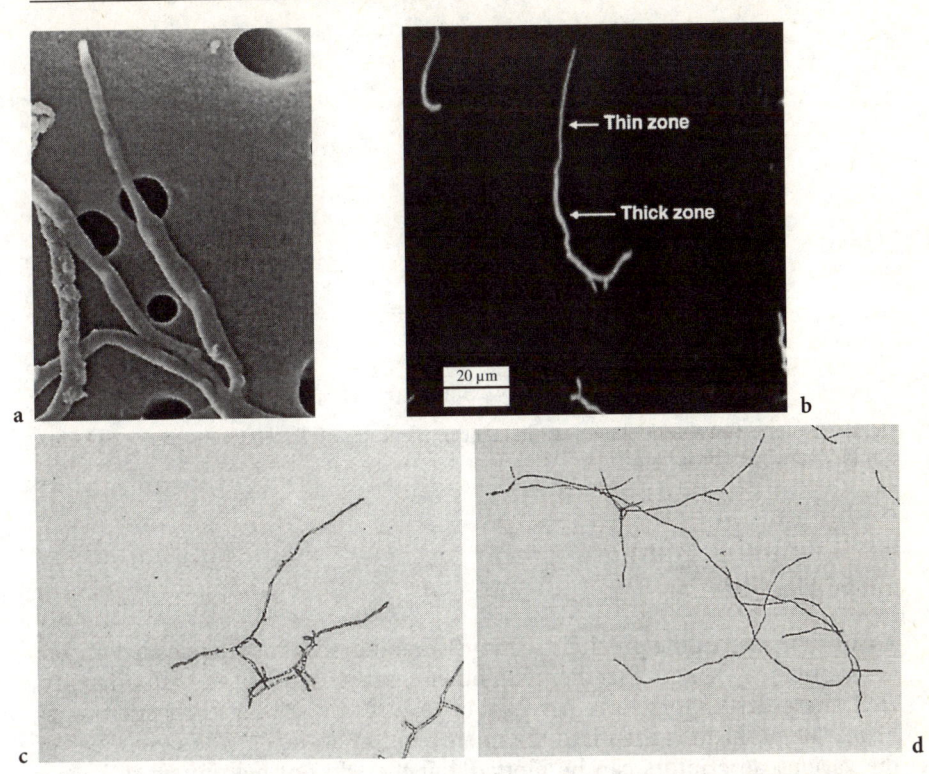

Fig. 24. Filament diameter: *S. ambofaciens* with SEM (**a**) and epifluroescence microscopy (**b**); *T. reesei* with optical microscopy at the beginning (**c**) and the end (**d**) of a culture

calculate the volume occupied by filaments, one needs to know the hyphal diameter (D_h) [155]:

$$V = \frac{\pi D_h A}{4} \tag{30}$$

Manual estimation can be obtained using a cursor-editor but a more general quantification of the filament thickness can be simply made by computing the brightness level histogram of the EDM along the skeleton, as a generalization of the procedure described in Fig. 13b. A simpler characterization in three thickness classes is proposed in Fig. 25 for *Trichoderma reesei* [151] by performing series of erosions of predetermined order to the silhouette and measuring the length of the skeleton masked by the eroded silhouette. Another morphological characteristic, the distance between septa, can be put in light by fluorescent dyes as described by Reichl et al. [156] and Drouin et al. [152] on *Streptomyces* sp. (Fig. 2b).

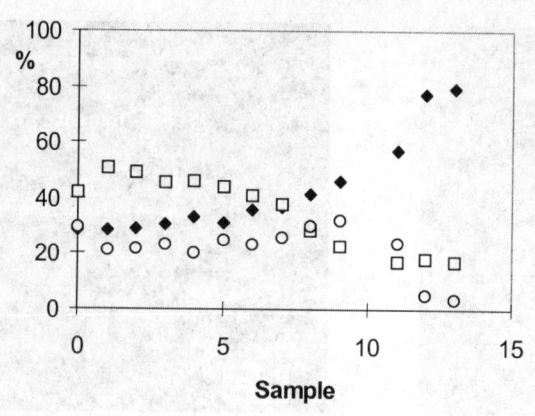

Fig. 25. Evolution of the thickness of filament in a culture of *Trichoderma reesei*: (♦) Fine, (□) Medium, (○) Thick filaments

7
Shape Classification

As we have seen in the previous sections several descriptors are needed for the description of a shape. Together with other characteristics of the cell (viability, etc.), these descriptors form a vector of features describing each microorganism. The problem is then into the interpretation of the data. Distributions of the various descriptors can be plotted, but they do not help much to have a global assessment except when the number of descriptors is limited as for the characterization of *Saccharomyces cerevisiae* budding by elongation [74, 83] or when one is specifically interested in a descriptor such as hairiness. The situation is even more complex when Fourier descriptors are used as it is difficult to find a real physical significance to them. Data-mining tools such as Principal Component Analysis [90, 135, 157, 158], Discriminant Analysis [82], fuzzy logic and Artificial Neural Nets [41, 72, 95, 132, 133, 159] are useful for clarifying the situation.

The general framework of these data interpretation techniques is:

- a learning phase: a training population is classified manually by one or several experts. The algorithm is trained to classify the corresponding data set.
- a validation step: a population, which has not been used for the training phase, is automatically classified. The performance index, i.e. the rate of misclassification with respect to the "ideal" classification which would have been done by the expert(s) is calculated.

If the agreement is satisfactory, the classification procedure can be used on a routine bias. If not, some of the classification algorithm parameters, the composition of the training population, the class characteristics or the selection of the shape descriptors should be modified.

Many applications dealing with non filamentous bacteria are not so much related to process control than to health and hygiene problems or characterization of the biopopulation in a natural system (fresh and sea waters, soils). Meijer et al. [157] have classified different bacterial strains automatically: *Streptococcus puogenes*, *Escherichia coli*, *Streptococcus* group D, *Klebsiella pneumonia*, *Proteus mirabilis*, *Pseudomonas aeroginosa* and *Staphylococcus aureus*. Descriptors of the size, the irregularity of the contour, the concavity and the global circularity of the cells were considered in a multivariate analysis of variance to distinguish the different species one from another.

Filamentous morphologies can be classified with a similar approach. Going back to the characterization based on the EDM, the maximal distance of the mask gives the radius of the core plus the entangled zone, without the lateral isolated filaments (R_{c2}). Together with the reduced radius of gyration ($r_g = 0.84$) four descriptors are defined: the ratio of the object perimeter to the mask perimeter, the ratio of the pellet diameter to the core diameter, the ratio of the filled pellet diameter to the core + entanglement diameter and the ratio of the core radius to the mean value of the pellet EDM. A general map, in the plane of the two first principal components, can be obtained, with different locations for the various morphologies (Fig. 26). Recently a similar method has been proposed to help to the automated recognition of protozoa in activated sludge [160]. The protozoa are characterized by their silhouette circularity, elongation, eccentricity, length and area. *V. microstoma* sp. and *Opercularia* sp., which are indicators of a poor efficiency of a wastewater treatment plant are well recognized.

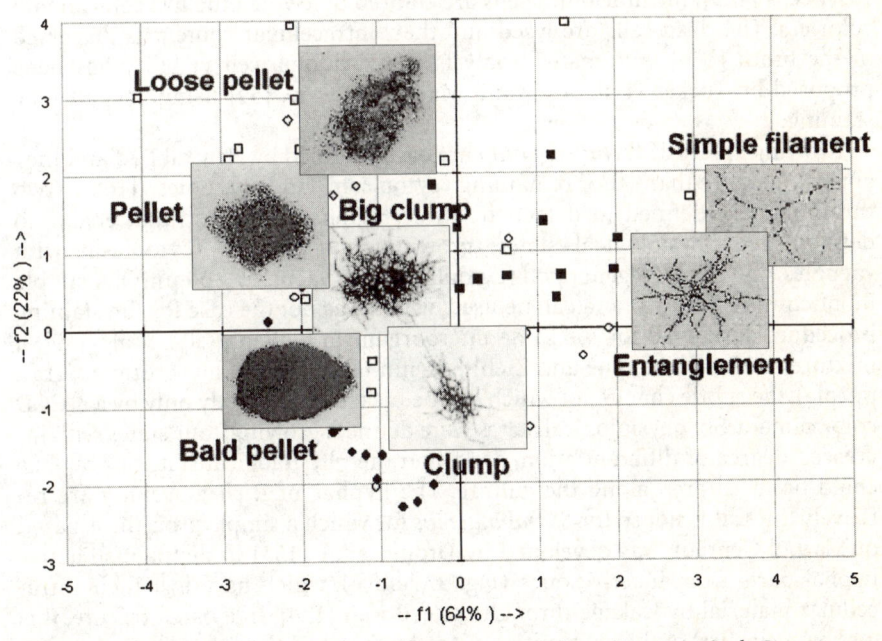

Fig. 26. Global shape classification of filamentous species using Principal Components Analysis

8
Densitometry and Visual Texture

8.1
Physiology Assessment

Saffron was used by A. van Leeuwenhoek to stain muscular fibers but it was the development of aniline stains for the textile industry in the mid 19th century that changed the face of sample preparation for optical microscopy. Orchil appears to be the first synthetic dye, prepared as early as 1300. Some of the now classical stains are Methylene Blue, Neutral Red, Hematoxylin, Carbol Gentian, Trypan Blue and Fuschin. They are almost all synthetic. These stains are first used to enhanced the contrast between the cells and the background but physiological features can also be detected due to changes in hue or intensity related to intracellular reactions between the dye and biological species. Different mechanisms are responsible for these reactions [161]. Basic Fuchsin is the main component of Schiff's reagent, often used for aldehyde detection. In 1924, Feulgen and his coworkers [162, 163] used this reagent for the first time to stain nuclei. In Feulgen's reaction, after acid hydrolysis of nucleic acids, the produced aldehydes are detected by Schiff's reagent. Crystal Violet and Safranin are the basic components of the Gram's stain: Gram-positive micro-organisms are deep-violet when the Gram-negative ones are red. Image analysis can provide morphological as well as densitometric data to monitor the cell cycle [164]. Methylene Blue is used extensively in a viability test for yeast cells [165]: the moribund cells are colored blue when the live cells remain colorless. The dead cells are lysed and their intracellular content is dispersed in the broth [166]. Automated analysis of hybridoma cell viability has been proposed by Tucker et al. [167] and Maruhashi et al. [39] using Trypan Blue staining.

Growing apices of *P. chrysogenum* are easily stained by Neutral Red and they appear darker than the remaining cytoplasm and vacuoles [168]. Two thresholds are defined to detect the three regions of interest, that enable to define three physiological states: growing, non-growing (cytoplasm with vacuoles < 30 μm^3), degenerated (cytoplasm with vacuoles > 30 μm^3). A simple monochrome CCD-camera can be used, which was not the case for the staining procedure developed for the same microorganism by Vanhoutte et al. [169]: a mixture of Methylene Blue and Ziehl Fuschin takes various hues (orange, gray, purple), the subtle changes of which can be picked up correctly only by a 3-CCD color camera. Six physiological states were defined: growing, four states with increased degree of differentiation, dead, and the physiological status of hyphae could be monitored along the culture. The hyphae of *P. chrysogenum* are relatively thick: it is not so for *S. ambofaciens* for which a simple procedure, based on Carbol Gentian, was developed by Drouin et al. [153] to distinguish active hyphal parts from inactive ones (Fig. 27 above), which have lost their intracellular material by leakage through the cell wall [170]. In a batch culture, this loss happens first at the extremities of the hyphae and then propagates into the internal parts of the filament as shown graphically in Fig. 27 below. The ability

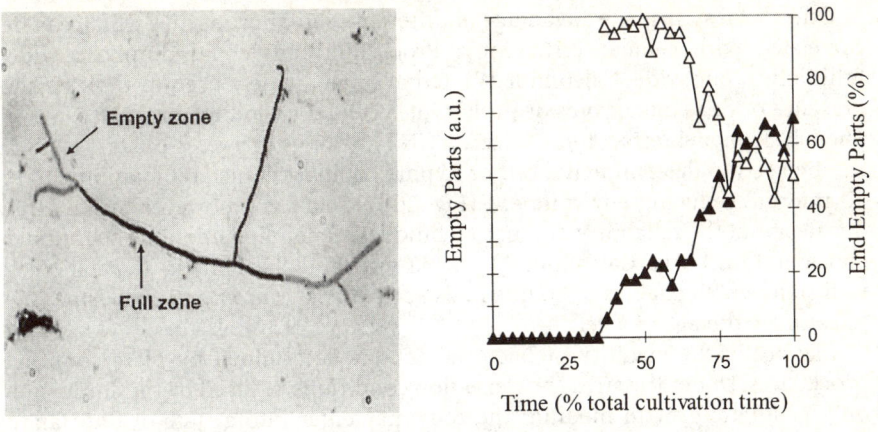

Fig. 27. *above* Full and empty parts in hyphae; *below* Proportion of empty parts in hyphae (▲) (in arbitrary units) and proportion of terminal empty parts (△)

offered by automated image analysis to examine a large number of images have facilitated investigations on the effects of operation conditions and metabolism and in function of the culture age [171].

More recent staining procedures largely use fluorescent dyes to characterize the physiological and biochemical states of cells. Fluorescein Diacetate (FDA), a non-polar substance which crosses the membrane and is hydrolyzed by intracellular esterases in viable cells to produce fluorescein, exhibits yellow-green fluorescence when excited at 490 nm. Damaged or non-viable cells in general are unable to hydrolyze FDA or to retain fluorescein within the cell [172, 173]. In combination with Ethidium Homodimer or Propidium Iodide, a similar esterase substrate, calcein acetoxy methyl ester (CAM) has been found to be reliable for viability assessment of protozoans, but not on *Candida* yeast, neither on bacteria such as *Bacillus cereus* and *Escherichia coli* [174].

Acridine Orange (3,6-bis[dimethylamino]acridinium chloride) is a well-known metachromatic stain which fluoresces orange when bound to single-stranded nucleic acids and green when bound to double-stranded nucleic acids. The color, which ranges from orange to yellow and green, can give an indication of the relative proportions of RNA (mostly single-stranded) and of DNA (mostly double-stranded). When RNA:DNA is large, such as in fast growing cells, the color is orange and tends to green in slowly growing cells [69]. A color CCD camera should be used for Acridine Orange densitometry: monochrome cameras are more sensitive to red than to green and therefore detection of cells stained in green may be difficult. It can be applied to cells cultivated in suspension or on solid substrate, without having to detach them from their support [69, 175–177]. DAPI (4'6-diamino-2-phenylindole) is another vital stain, DNA specific [178]: however, its blue fluorescence under UV light can again cause problem for detection by a CCD camera [179] although other researchers obtain a good correlation between the most-probable-number (MPN) obtained by DAPI staining and direct counts [67].

Other vital stains take advantage of different cellular properties which can be correlated with cellular physiology: Propidium Iodide, Ethidium Bromide, Ethidium Monoazide, Calcofluor White have been widely used to indicate the presence of dead eukaryotes or prokaryotes cells. 2-(p-iodophenyl-)3)(p-nitrophenyl)-5-phenyl tetrazolium chloride (INT) belongs to a class of stains which can be used to determine if a cell or hyphal compartments [180] can maintain an internal reducing environment (Fig. 20a). There are, however, still a large debate about the reliability of those techniques, depending upon the cells under consideration [181]. Calcofluor ($\lambda_{ex} \approx 380$ nm, $\lambda_{em} \approx 420$ nm) is a specific cell wall stain which enables to counts buds scars on *Saccharomyces cerevisiae* [29] to estimate the age of a cell.

Classically the viability of bacterial cells is determined by plate counting procedures. Under the stress or starvation conditions induced by the cultivation on a synthetic growth medium the cells may enter into a state of dormancy, which introduces a bias in the results. A method of direct estimation of viable bacteria has been proposed by Kogure et al. [182] for Gram-negative bacteria: the samples are incubated in presence of nutrients and of a low concentration of nalidixic acid, which acts as a specific inhibitor of DNA synthesis. The growing cells cannot divide and the enlarged, i.e. viable, cells are stained with Acridine Orange. Singh et al. [183] have automated the enumeration of bacteria on such preparations. Barcina et al. [184] have proposed another inhibitor, ciprofloxacin, for both Gram-positive and Gram-negative bacteria.

The vital stains can have an effect on the shape of the cells. It is especially true for cells which have a rather flexible membrane such as *Saccharomyces cerevisiae*: the buffers (reactants) necessary to a proper reaction of the stain affect the cell. Viable cells observed with Methylene Blue staining are larger and dying cells smaller than unstained cells [74]. Potential problems are mentioned by Huls et al. [29], with dansylchloride and Calcofluor staining procedures of *Saccharomyces cerevisiae*. Sometimes a simple difference in gray level can permit, without staining but with a phase contrast microscope, to visualize internal structures such as vacuoles in *Saccharomyces cerevisiae* (Fig. 28) [40]. Billon et al. [185] have monitored the germination of single spores of *Clostridium botulinum* by observing them with a phase-contrast microscope and simply

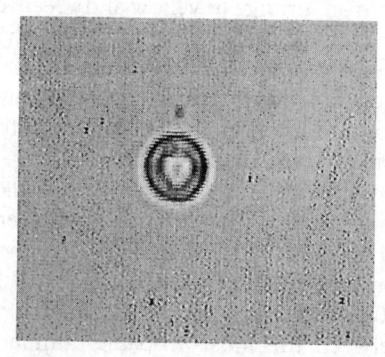

Fig. 28. Vacuole of *Saccharomyces cerevisiae* cell observed by phase-contrast microscopy

recording the change in refractivity. Jones et al. [186] have observed also a change in refractivity when they developed their automated method for *Phanerochaete chrysosporium*, but they have suggested it may be due to water evaporation. Recombinant bacteria such as *E. coli* can store the foreign proteins they produce as inclusion bodies [53, 187] or crystals [188] which do not refract light as the rest of the cell. The number of cells containing one or more inclusion bodies as well as the number of inclusion bodies per cell can be easily monitored [53].

Images obtained under different light conditions can be combined. Binnerup et al. [189] have proposed a protocol to determine rapidly the Gram characteristics of single cells or microcolonies: the polycarbonate membrane filters on which the bacteria have been deposited, immobilized with a thin layer of highly viscous silicone (which does not affect either cell growth or Acridine Orange subsequent staining), stained for Gram characterization and by Acridine Orange are first observed by light microscopy then by epifluoresence. The viability of *Saccharomyces cerevisiae* cells subjected to osmotic stresses and stained with Acridine Orange was assessed by visualization under white light, to detect all cells, followed by a visualization of the same field by epifluorescence, to detect the viable cells [172]. Multispectral imaging [190], by simultaneous or sequential capture of images using a range of excitation and emission wavelengths, has been tested by Lawrence et al. [191] for the quantification of algal (by far red autofluorescence, $\lambda_{ex} = 647$ nm, $\lambda_{em} = 680 - 632$ nm), bacterial (by a fluorescent nucleic acid stain SYTO 9, $\lambda_{ex} = 488$ nm, $\lambda_{em} = 522 - 532$ nm) and exopolymer components (by a lectin probe, *Triticium vulgaris*-TRITC, $\lambda_{ex} = 568$ nm, $\lambda_{em} = 605 - 632$ nm) of microbial biofilms observed by CLSM.

Ratio imaging takes a large advantage of the ion-sensitive fluorescent dyes [192, 193] which have been proposed for the mesurement of local concentrations of various ions, H^+ but also Na^+, K^+, Ca^{2+}, Mg^{2+}, Cl^-. Two types of dyes are available for the measurement of intracellular pH: dyes that cross the cell wall in response to a pH gradient or dyes which stayed trapped in cell compartments and whose excitation or emission spectra depend on the local pH. Acridine Orange belongs to the first type. The spectrum of fluorescein and its derivatives is pH sensitive. Figure 29a represents schematically the action of BCECF-AM (2′,7′-bis(carboxyethyl)-5-carboxyfluorescein-acetoxyl methyl) when it is loaded in a cell. The ester group AM protects the polar groups of the dye and facilitate the crossing of the cell membrane. The esterases present in the cell cleave the ester, liberating BCECF, which can be detected by emission (Fig. 29b) or excitation (Fig. 29c) ratio imaging.

The basic principle of ratio imaging is presented in Fig. 30. In the pH image, the value taken by each pixel is calculated as $F^{520\,nm}/F^{620\,nm}$ where $F^{520\,nm}$ and $F^{620\,nm}$ are respectively the fluorescence signals at 520 nm and 620 nm. Emission ratio is potentially interesting but is more difficult to apply than excitation ratio since rapid changes in the emission filters or two detectors are required. CF (5- (and 6)-carboxyfluorescein) exhibits a strong change in its excitation spectrum with respect to pH when its emission spectrum is almost unchanged: intracellular pH is measured by ratio excitation imaging [194] with $\lambda_{ex} = 518$ and 464 nm and $\lambda_{em} \approx 520$ nm. Due to the importance of calcium in cellular signal

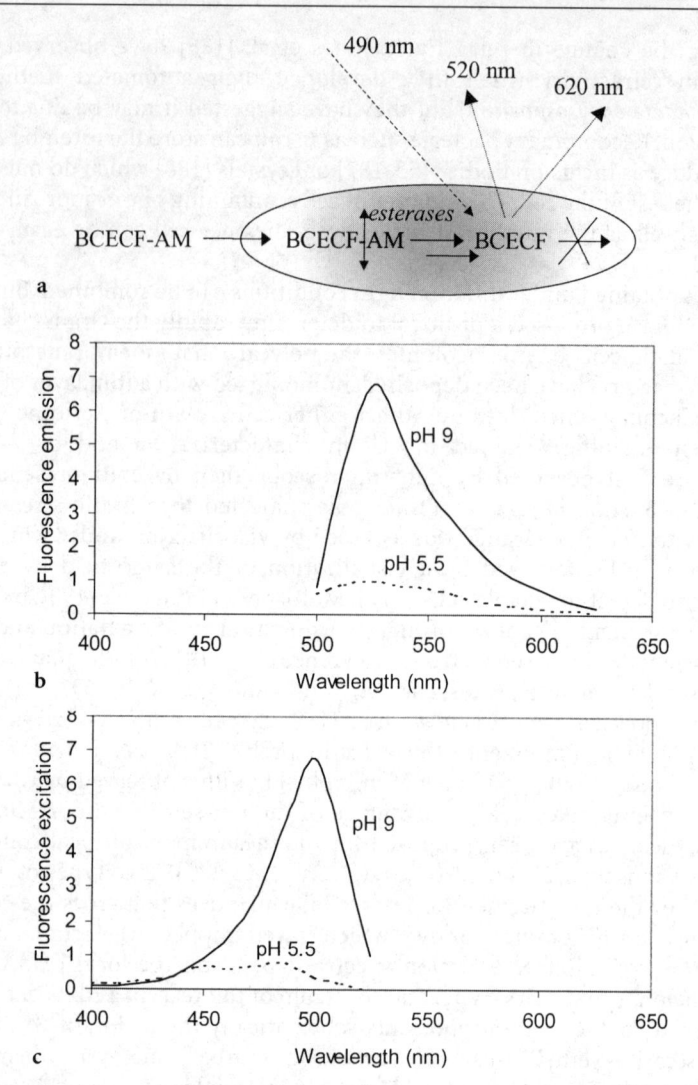

Fig. 29. Action of BCECF-AM illustrated for the emission fluorescence case (**a**); **b** Emission (excited at 490 nm) and **c** excitation (detected at 535 nm) spectra of BCECF at pH 5.5 at pH 9

transduction, many calcium indicators have been proposed. The most popular Ca^{2+} indicator for individual cells is certainly Fura-2: when the dye binds to Ca^{2+}, the excitation spectrum shift to shorter wavelengths but the emission spectrum is unchanged. An excitation pair of 340/380 nm or 350/385 nm gives a good measure of intracellular Ca^{2+}, with a reading at 505–520 nm.

In ratio imaging the quality of the calibration curve is essential for a successful quantitative method. The curve can be established in vitro (i.e. using solutions of the stain in a series of buffers at different pH) or in vivo (using labeled

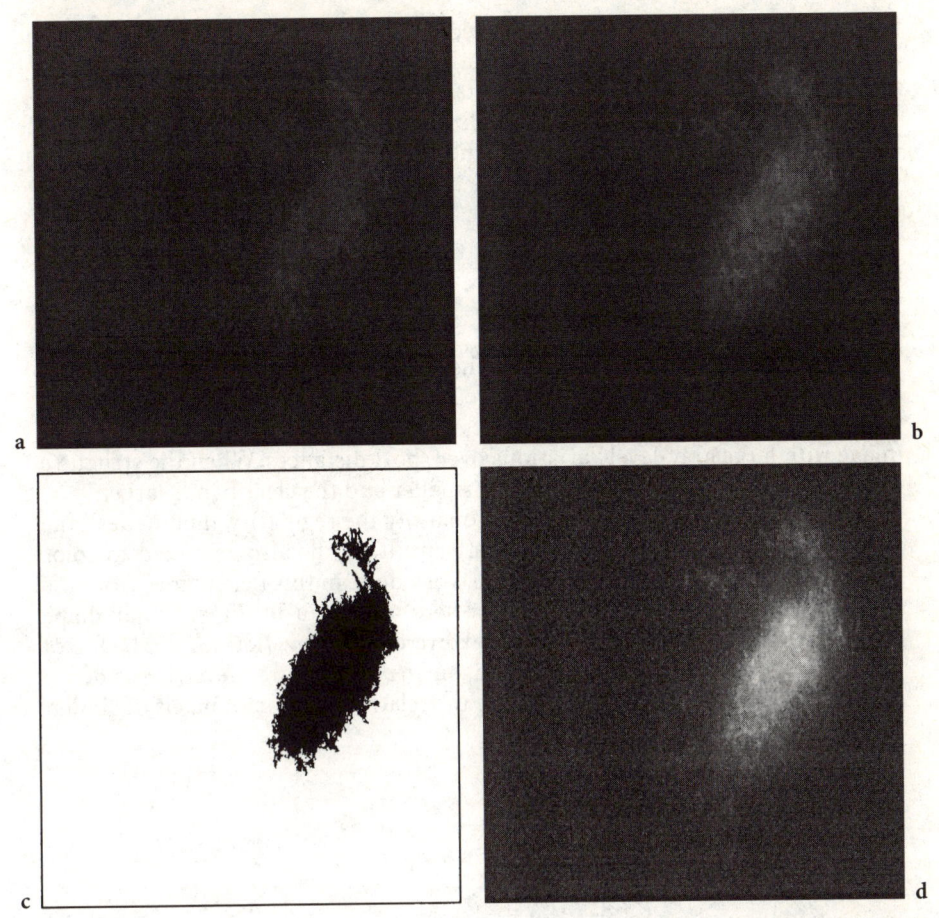

Fig. 30. Ratio imaging of a *Streptomyces* pellet, **a** Red fluorescence; **b** Green fluorescence ($\lambda_{ex} = 480$ nm); **c** Pellet mask; **d** pH image

cells). The problem with in vitro standards is that the dye does not behave exactly in the same manner in vitro and in vivo: BCECF exhibits a red-shift of about 5 nm when it is introduced in a cell. For in vivo calibration ionophores such as nigericin (for pH) or ionomycin (for Ca^{2+}) which allow alteration of the intracellular ion concentration by changing the extracellular one are used. For calcium the in vitro calibration approach proposed by Grynkiewiez et al. [195] is generally adopted [197–198]:

$$[Ca^{2+}] = K_d \left(\frac{R_m - R_{min}}{R_{max} - R_m} \right) \left(\frac{F_{free}^{\lambda_{ex}}}{F_{sat}^{\lambda_{ex}}} \right) \tag{31}$$

where K_d is the in vitro dissociation constant, R_m is the measured ratio, R_{min} the fluorescence ratio at zero calcium concentration, R_{max} the fluorescence ratio at maximal calcium concentration, $F_{\text{free}}^{\lambda_{ex}}$ and $F_{\text{sat}}^{\lambda_{ex}}$ the fluorescence signals at zero and maximal calcium concentrations respectively.

For any type of application the staining procedure should be carefully designed in connection with the visualization problem [55, 181]. Overstaining saturates the images, the background fluoresces and debris are also stained. On the contrary, when the concentration is too low, cells cannot be detected out of the background noise.

8.2
Visual Texture

Texture is here used in the particular sense of vision: it is related to the idea that a lateral lighting of an object by an electromagnetic wave will reveal the topography of the object by shading effects (Fig. 31). A rough structure gives an image with large gray-level variations over short distances. When the structure is smooth, the gray-level variations are smaller and the clear bands larger.

On these images, the challenge is to quantify the regularity, the fineness, the homogeneity, the orientation of the patterns. It can be also extended to color spatial distribution. Most of the general work done in this field comes from the analysis of aerial or satellite images. There are, as in bi-dimensional shape characterization, several methods of texture characterization [51–53]. One of the most successful of them in based on the spatial gray-level dependence matrix (SGLDM) [199] which examines the relative position of pixels of similar

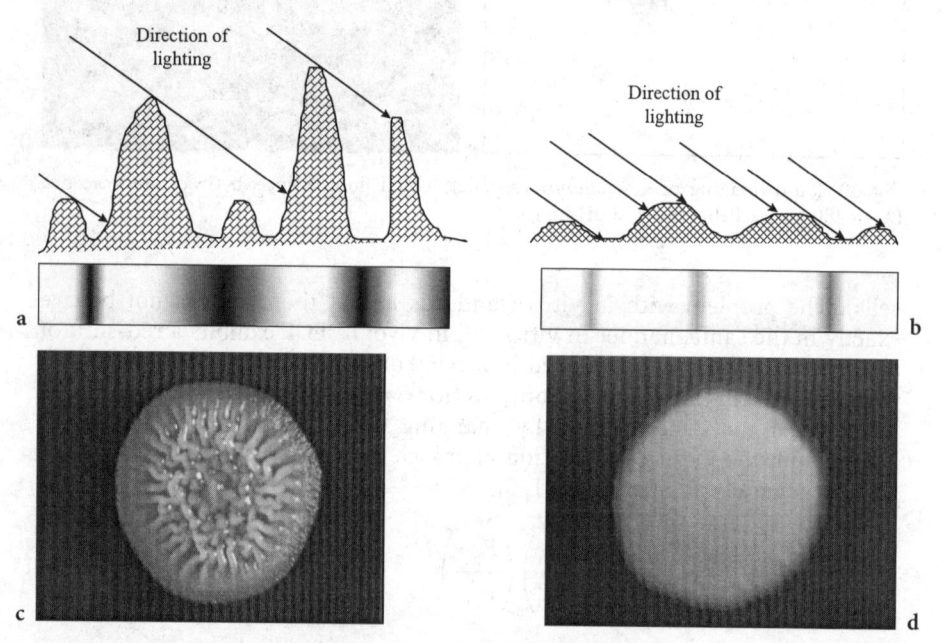

Fig. 31. Visualization of 3D shape by visual texture: rugose (**a**) and smooth (**b**) surfaces; **c** young and **d** old *P. roqueforti* colonies

gray level. The co-occurrence gray-level matrix $C(c_{ij})$ of dimension $N_g \times N_g$, where N_g is the maximal number of possible gray levels (usually 256) is defined as:

c_{ij} = frequency of occurrence of having a pixel of gray level j at a distance d and angle α of a pixel of gray level i. The C matrix is itself characterized by descriptors such as:

$$\text{Inertia} = \sum_{i=1}^{N_g} \sum_{j=1}^{N_g} (i-j)^2 c_{ij} \tag{32}$$

$$\text{Entropy} = \sum_{i=1}^{N_g} \sum_{j=1}^{N_g} \{c_{ij} \log(c_{ij} + 10^{-6})\} \tag{33}$$

$$\text{Energy} = \sum_{i=1}^{N_g} \sum_{j=1}^{N_g} c_{ij}^2 \tag{34}$$

Inertia takes small values when the largest values of C are close to its diagonal, where $(i-j)$ is small, i.e. when the image is homogeneous. Entropy describes the randomness of C: it is maximal when all the c_{ij} are equal, on the contrary to energy which will be minimal in this case. Other descriptors have been proposed by Haralick but these are the most commonly used. They are computed for the four directions of the grid ($\alpha = 0°, 45°, 90°$ and $135°$) and may be averaged with respect to direction. Several values of d are tested to characterize the texture on different ranges.

The SGLDM method can be applied to global images as well as individual objects (seen through the mask constituted by the silhouette) such as the colonies of Fig. 32. In this special case however, the obvious circular symmetry

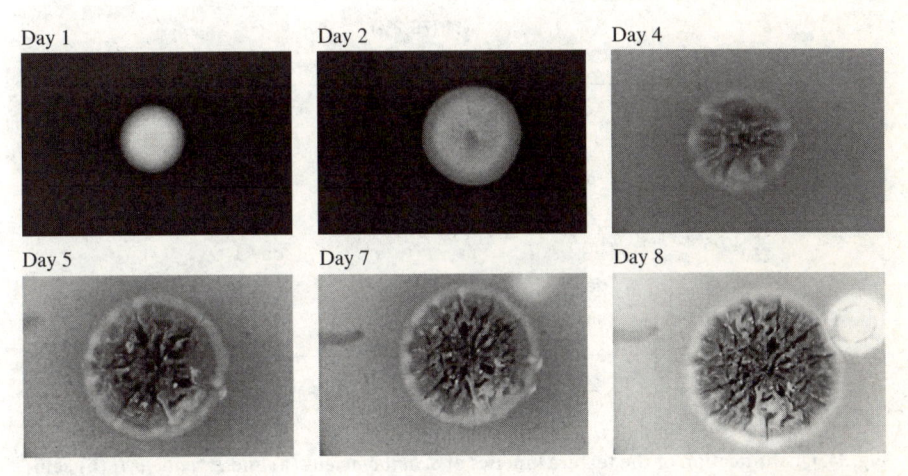

Fig. 32. Texture kinetics of a *S. ambofaciens* colony over an eight-day period (the magnification is not constant)

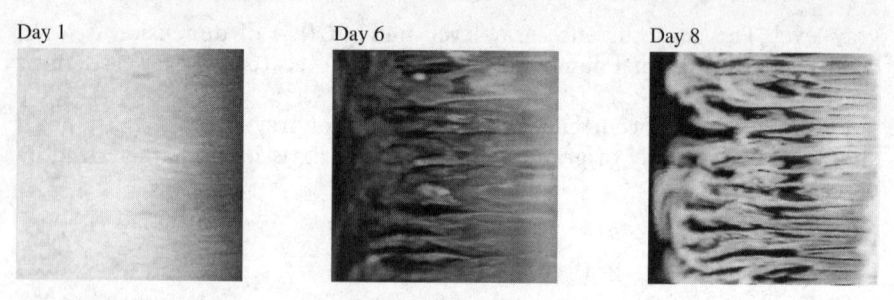

Fig. 33. Transformation of a circular colony into a square image

of the colony would be badly characterized by a matrix calculated on rectangular grid assumptions. To circumvent this problem a geometrical transformation is applied which transforms the circular colony into a square image (Fig. 33) [200].

It has to be noted that the direct analysis of the texture descriptors is generally difficult: the descriptors help to compare one structure to another, in

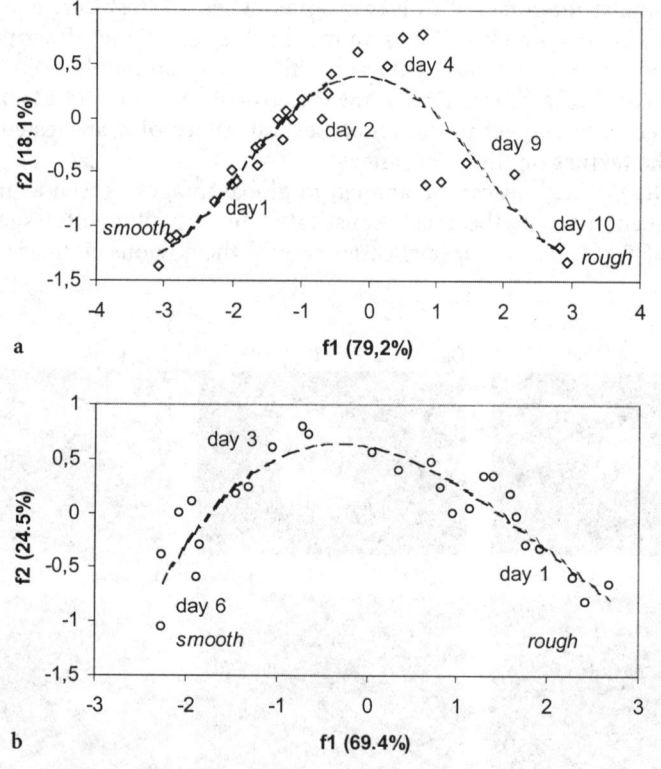

Fig. 34. Quantification of the texture kinetics of *S. ambofaciens* (a) and *P. roqueforti* (b) colonies by Principal Components Analysis. The percentages of total variability explained by the principal components are given in parentheses

similar conditions of visualization, but do not give an absolute grading of the texture. In that domain also Principal Components Analysis has been found useful for summarizing the information. The growth on agar of colonies of *S. ambofaciens* and *P. roqueforti* have been monitored over several days (one or two images/colony/day): *S. ambofaciens* colonies are initially bright and smooth and becomes rougher with aging. In contrast, *P. roqueforti* are initially rather rough and becomes smooth. The texture descriptors previously presented have been calculated for each colony and sample time and a Principal Component Analysis conducted. It is possible to represent 95% of the variability by considering the first two principal components (f_1 and f_2). In Fig. 34 the texture trajectories of the two types of colonies in the plane $f_1 - f_2$ have been plotted. The modification of the texture between smooth and rough colonies for *S. ambofaciens* (Fig. 34a) and between rough and smooth colonies for *P. roqueforti* (Fig. 34b) is clearly visible.

9
Motility

The characteristics of the motion of cells in the vicinity of solid surfaces can help to improve the design and control of bioremediation processes, filtration membranes, surface biofouling, biofilm reactors for wastewater treatment Bacterial cells respond to heat or toxic stresses by modifying their motility characteristics [201]. The motility characteristics of some protozoa make their identification easier than on still images. The chemotactic responses of protozoa or daphnia are used to monitor toxic components in wastewater. In the early seventies Berg [202] and Berg and Brown [203] developed a tracking microscope able to monitor the full spatial motion of a single cell for a long period of time. It is a very sophisticated system, still in use nowadays [204] but many cells should be monitored to get reliable statistics on the motion characteristics. Image analysis is more restrictive as only the motion in the plane of focus is observed [205]. The advantage is that several cells are monitored simultaneously. Data of various types are collected: velocities, shape of trajectories, tumbling, reversal of motion, waggling, etc. [206]. Until recently the on-line analysis of motion of cells or just dynamical investigation such as cell division [7–9] was restricted by the time required by the treatment of images. Images were stored on films or video-tapes, which were analyzed manually or automatically after the experiment. Such storage is often detrimental to the quality of images. Digital video-recorders or computers with real-time image capture and storage are now available, which permit to capture a large number of frames, at a rate of 25 frames/s or 33 frames/s. When higher grabbing frequencies are necessary, to analyze the motion of cilia or flagella for example, high-speed cameras are available, up to 4000 frames/sec.

When the dynamics are not too fast, real-time image capture and analysis is easily performed on a personal computer: Gerin et al. [207] have monitored the adhesion of *Phanerochaete chrysosporium* spores with a capture rate of one image/4 min. Bailleul and Schleunders [208] describe a method to reconstruct in real time the trajectories of moving cells, by logical addition in the frame memory (which can contain 4 images simultaneously) of the threshold images. Sequences of 25 images captured at a rate of 25 frames/s are collected and analyzed.

The basic analysis steps previously described are applicable to each individual image [206]. The localization of the cell, generally by the position of its centroid, is important here to be able to reconstruct the trajectories of the various cells from one frame to the next one. The magnification should be careful selected: high magnification will make the cell identification easier but the cell will remain very shortly in the field of view. More frames will be necessary and some data such as the distance between reversal or tumbling events will be biased.

10
Conclusions and Perspectives

The background for the application of image analysis in bioengineering is now well set: users are no more afraid by a mysterious tool and more and more examples of applications to different microorganisms can be found in the literature. Growing cells under extreme conditions [209, 210] will probably induce the development of new visualization systems. The user-friendliness of image analysis software has increased, making a routine use of it in a research environment quite possible. The main challenge resides now in the extension to industrial applications. As for many sophisticated analysis techniques, off-line means human labor, which can be incompatible with productivity constrains. Attempts for on-line development are still scarce and on a prototype scale. However it may not be necessary to dispose of the morphological information at every minute of a culture. Low-frequency sampling and analysis may be sufficient to recalibrate a model, which will use easily available measurements such as respiration data or pH variations for high-frequency estimation of the process state.

There is always a debate on the use of either image analysis or "classical" sizing systems such as Coulter counter or laser diffraction granulometry for biomass quantification. It is certain that the number of cells examined by image analysis is much lower than what is analyzed by the other systems. Physiology can be also assessed by flow cytometry, with a much higher rate. This is a problem to be considered in order to obtain results which are statistically valid over a cell population. There are also types of cells which are difficult or even impossible to analyze by laser diffraction granulometry or flow cytometry: it is the case for filamentous species and cells growing on a support. For the later, a detachment is necessary, with risks on the physiological state of the cells and of partial detachment. Even for yeast cells the agreement between the different methods is difficult to reach, due to shape factor [29, 97].

Following the trend to cultivate more and more complex cells such as human ones for gene therapy or regeneration of body parts (skin [211], nerve [212], bone [213], cartilage [214]) or simply for an easier study of cell physiology [13, 215], applications have yet to be developed, in the field of supported growth in particular, where descriptors of visual texture, to check cell confluence for example, might be interesting.

Other fields of application will be important in the future for the control of the quality of the bioproducts: different systems are already proposed on a commercial basis for the quantification of electrophoresis gels, but developments are under way also for automatic DNA fingerprinting and Fluorescence In-Situ Hybridization (FISH) [216–219].

Acknowledgements. The authors are thankful to P Corvini, JF Drouin, A Maazi, P Mauss, D Pichon, whose work helps to provide most of the illustrations.

References

1. Descartes R (1637) In: La Dioptrique, Dixième Discours
2. Ball BA, Mohammed HO (1995) Theriogenology 44:367
3. Larpent JP (1991) Biotechnologie des levures, Masson, Paris
4. Chalfie M, Tu Y et al. (1994) Science 263:802
5. Poppenborg L, Friehs K, Flaschel E (1997) J Biotechnol 58:79
6. Kuehn M, Hausner M, Bungartz HJ, Wagner M, Wilderer PA, Wuertz S (1998) Appl Environ Microbiol 64:4115
7. Lord PG, Wheals AE (1980) J Bacteriol 142:808
8. Lord PG, Wheals AE (1981) J Cell Sci 50:361
9. Wheals AE (1982) Mol Cell. Biol. 2:361
10. Pons MN, Vivier H, Voignier L, Bernard-Michel B, Rolland T (1996) Microsc Microanal Microstruct 7:467
11. Wilkinson MHF (1998) Optical systems for image analysed microscopy In: Wilkinson MHF, Schut F (eds) Digital Image Analysis of Microbes. Wiley, New York, pp 431
12. Viles CL, Sieracki ME (1992) Appl Environ Microbiol 58:584
13. Young JC, DiGiusto D, Backer MP (1996) Biotechnol Bioeng 50:465
14. Lawrence JR, Wolfaardt GM, Neu TR (1998) The study of biofilms using confocal laser scanning microscopy In: Wilkinson MHF, Schut F (eds) Digital Image Analysis of Microbes. Wiley, New York, p 431
15. Walsh PK, Isdell FV, Noone SM, O'Donovan MG, Malone DM (1996) Enz Microb Technol 18:366
16. Bloem J, Veninga M, Shepperd J (1995) Appl Environ Microbiol 61:926
17. DeLeo PC, Baveye P, Ghiorse WC (1997) J Microbiol Methods 30:193
18. Morris CE, Monier JM, Jacques MA (1997) Appl Environ Microbiol 63:1570
19. Caldwell DE, Korber DR, Lawrence JR (1992) J Microbiol Methods 15:249
20. Swope KL, Flickinger MC (1996) Biotechnol Bioeng 52:340
21. Sanford BA, de Fejiter AW, Wade MH, Thomas VL (1996) J Indus Microbiol 16:48
22. Wentland EJ, Stewart PS, Huang CT, McFeters GA (1996) Biotechnol Prog 12:316
23. Bancel S., Hu WS (1996) J Ferment Bioeng 81:437–444
24. Bancel S, Hu WS (1996) Biotechnol Prog 12:398–402
25. Coppen SR, Newsam R, Bull AT, Baines AJ (1995) Biotechnol Bioeng 46:147
26. Wu FJ, Friend JR, Hsiao CC, Zilliox MJ, Ko WJ, Cerra FB, Hu WS (1996) Biotechnol Bioeng 50:404
27. Hartke A, Giard JC, Laplace JM, Auffray Y (1998) Appl Environ Microbiol 64:4238
28. Jiang SC, Kellogg CA, Paul JH (1998) Appl Environ Microbiol 64:535
29. Huls PG, Nanninga N, van Spronsen EA, Valkenburg JAC, Vischer NOE, Woldringh CL (1992) Biotechnol Bioeng 39:343
30. Pons MN, Wagner A, Vivier H, Marc A (1992) Biotechnol Bioeng 40:187
31. Fry JC, Davies AR (1985) J Applied Bacteriol 58:105
32. Woldringh CL, De Jong MA, Van den Berg W, Koppes L (1977) J Bacteriol 131:270
33. Gutsche AT, Zurlo J, Deyesu E, Leong KW (1996) Biotechnol Bioeng 49:259

34. Fowler JD, Robertson CR (1991) Appl Environ Microbiol 57:102
35. Heldal M, Norland S, Bratbak G, Riemann B (1994) J Microbiol Methods 20:255
36. Howgrave-Graham AR, Wallis FM (1993) Biotechnol Techniques 7:143
37. Srinorakutara (1998) J Ferment Bioeng 86:253
38. Lichtfield JB, Reid JF, Richburg BA (1992) Proc IFAC Modeling and Control of Biotechnical Processes, Colorado, 275
39. Maruhashi F, Murakami S, Baba K (1994) Cytotechnol 15:281
40. Zalewski K, Buchholz R (1996) J Biotechnol 48:43
41. Treskatis SK, Orgeldinger V, Wolf H, Gilles ED (1997) Biotechnol Bioeng 53:191
42. Berner JL, Gervais P (1994) Biotechnol Bioeng 43:165
43. Perrier-Cornet JM, Marechal PA, Gervais P (1995) J Biotechnol 41:49
44. Suhr H, Wehnert G, Schneider K, Bittner C, Scholz T, Geissler P, Jähne B, Scheper T (1995) Biotechnol Bioeng 47:106
45. Bittner C, Wehnert G, Scheper T (1998) Biotechnol Bioeng 60:24
46. Van Vliet LJ, Boddeke FR, Sudar D, Young IT (1998) Image detectors for digital image microscopy In: Wilkinson MHF, Schut F (eds) Digital Image Analysis of Microbes. Wiley, New York, p 37
47. Smith MAL, Reid JF, Hansen AC, Li Z, Madhavi Dl (1995) J Biotechnol 40:1
48. Prosser J, Rattray L, Silcock D, Glover A, Killham K (1994) Binary 6:49-54
49. Hooper CE, Ansarge RE, Rushbrooke JG (1994) J Biolumin Chemlumin 9:113
50. Costello PJ, Monk PR (1985) Appl Environ Microbiol 49:863
51. Gonzalez RC, Wintz P (1987) Digital image processing, 2nd edn. Addison Wesley, Reading, MA
52. Sonka M, Hlavac V, Boyle R (1993) Image Processing, Analysis and Machine Vision, Chapman & Hall Computing, London
53. Russ JC (1995) The Image Processing Handbook, 2nd edn. CRC Press, Boca Raton, FL
54. Serra J (1982) Image analysis and Mathematical Morphology, Academic Press, London
55. Schröder D, Krambeck C, Krambeck HJ (1991) Acta Stereol 10:123
56. Marr D, Hildreth E (1980) Theory of edge detection Proc R Soc London Ser B207:187
57. Wilkinson MHF (1998) Automated and manual segmentation techniques in image analysis of microbes In: Wilkinson MHF, Schut F (eds) Digital Image Analysis of Microbes. Wiley, New York, p 135
58. Ridler TW, Calvard S (1978) IEEE Trans On Syst Man Cyber 8:630
59. Otsu N (1979) IEEE Trans Syst Man Cyber 9:62
60. Kanpur JN, Sahoo PK, Wong AKC (1985) Comput Vis Graphics Image Proc 29:273
61. Vicente A, Meinders JM, Teixeira JA (1996) Biotechnol Bioeng 51:673
62. Miles RE (1974) Stochastic Geometry, Wiley, New York, p 228
63. Jeison D, Chamy R (1998) Biotechnol Techniques 9:659
64. Paul GC, Thomas CR (1998) Characterisation of mycelial morphology using image analysis. Adv Biochem Eng Biotechnol 60:1
65. Sieracki ME, Viles CL (1998) Enumeration and sizing of micro-organisms using digital image analysis In: Wilkinson MHF, Schut F (eds) Digital Image Analysis of Microbes. Wiley, New York, p 175
66. Wierzba A, Reichl U, Turner RFB, Warren RAJ, Kilburn DG (1995) Biotechnol Bioeng 46:185
67. Bianchi A, Giuliano L (1996) Appl Environ Microbiol 62:174
68. Wirtanen G, Mattila-Sandholm T (1993) J Food Protein 56:678
69. Yeh TY, Godshalk JR, Olson GJ, McKelly RM (1987) Biotechnol Bioeng 30:138
70. Jiang C, Chen P, Shan S (1995) J Microbiol Methods 23:297
71. Guerra-Flores AL, Abreu-Grobois A, Gomez-Gil B (1997) J Microbiol Methods 30:217
72. Wit P, Busscher HJ (1998) J Microbiol Methods 32:281
73. Corkidi G, Diaz-Uribe R, Foch-Mallol JL, Nieto-Sotelo (1998) Appl Environ Microbiol 64:1400
74. Pons MN, Vivier H, Rémy JF, Dodds JA (1993) Biotechnol Bioeng 42:1352
75. Robinson A, Sadr-Kazemi N, Dickason G, Harrison STL (1998) Biotechnol Techniques 12:763

76. Korber DR, Choi A, Wolfaardt GM, Caldwell DE (1996) Appl Environ Microbiol 62:3939
77. Pichon D, Vivier H, Pons MN (1992) Proc IFAC Modeling and Control of Biotechnical Processes, Colorado, p 311
78. Dai W, Saltzman WM (1996) Biotechnol Bioeng 50:349
79. Moreira JL, Alves PM, Aunins JG, Carrondo MJT (1995) Biotechnol Bioeng 46:351
80. Hammonds SJ, Adenwala F (1990) Lett. Appl Microbiol 10:27
81. Coster M, Chermant JL (1989) Précis d'Analyse d'Images, 2nd Ed., CNRS Editions, Paris
82. Cazzulino D, Pedersen H, Chin CK (1991) Bioreactors and Image Analysis for Scale-up and Plant Propagation. Cell Culture Somatic Cell Gen Plants 8:147
83. Hirano T (1990) ASBC J 48:79
84. Dubuisson MP, Jain AK, Jain MK (1994) J Microbiol Methods 19:279
85. Chi CM, Zhang C, Staba EJ, Cooke TJ, Hu WS (1996) J Ferment Bioeng 81:445
86. Huang LC, Chi CM, Vits H, Staba J, Cooke TJ, Hu WS (1193) Biotechnol Bioeng 41:811
87. Chi CM, Zhang C, Staba J, Cooke TJ, Hu WS (1996) Biotechnol Bioeng 50:65
88. Beddow JK, Melloy P (1980) Testing and characterisation of powders and fine particles, Heyden, London
89. Pons MN, Vivier H, Dodds JA (1997) Part Part Syst Charact 14:272
90. Pons MN, Vivier H (1998) Beyond filamentous species ... Adv Biochem Eng Biotechnol 60:61
91. Kaye BH (1986) Image analysis procedures for characterising the fractal dimension of fine particles. Proc Part Technol Conf Nürnberg
92. Grijspeerdt K, Verstraete W (1997) Wat Res 31:1126
93. Logan BE, Wilkinson DB (1991) Biotechnol Bioeng 38:389
94. Loferer-Krößbacher, Klima J, Psenner R (1998) Appl Environ Microbiol 64:688
95. Blackburn N, Hagström Å, Wikner J, Cuadros-Hansson R, Bjørnsen PK (1998) Appl Environ Microbiol 64:3246
96. Marechal PA, Gervais P (1994) Appl Microbiol Biotechnol 42:617
97. Vaija J, Lagaude A, Ghommidh C (1995) A. van Leeuwenhoek 67:139
98. Cui YQ, van der Lans RGJM, Luyben KCAM (1997) Biotechnol Bioeng 55:715
99. Schügerl K, Gerlach SR (1998) Influence of the process parameters on the morphology and enzyme production of *Aspergilli*. Adv Biochem Eng Biotechnol 60:195
100. Kamilakis EG, Allen DG (1995) Process Biochem 30:353
101. Makagiansar HY, Ayazi Shamlou P, Thomas CR, Lily MD (1993) Bioprocess Eng 9:83
102. Tucker KG, Thomas CR (1993) Trans IchemE 71:111
103. McIntyre M, McNeil B (1997) Enz Microb Technol 20:135
104. Jüsten P, Paul GC, Nienow AW, Thomas CR (1998) Biotechnol Bioeng 59:762
105. Olsvik E, Tucker KG, Thomas CR, Kristiansen B (1993) Biotechnol Bioeng 42:1046
106. Tamura S, Park Y, Toriyama M, Okabe M (1997) J Ferment Bioeng 83:523
107. Choi DB, Park EY, Okabe M (1998) J Ferment Bioeng 86:413
108. Vecht-Lifshitz SE, Magdassi S, Braun S (1990) Biotechnol Bioeng 35:890
109. Durant G, Cox PW, Formisyn P, Thomas CR (1994) Biotechnol Techniques 8:759
110. Durant G, Crawley G, Formisyn (1994) Biotechnol Techniques 8:395
111. Gehrig I, Bart HJ, Anke T, Germerdonk R (1998) Biotechnol Bioeng 59:525
112. McCarthy AA, O'Shea D, Murray NT, Walsh PK, Foley G (1998) Biotechnol Prog 14:279
113. Johansen CL, Coolen L, Hunik JH (1998) Biotechnol Prog 14:233
114. Carlsen M, Nielsen J, Villadsen J (1996) J Biotechnol 45:81
115. Gomez R, Schnabel I, Garrido J (1988) Enzyme Microb Technol 10:188
116. Hemmersdorfer H, Leuchtenberger A, Wardsack C, Ruttloff H (1987) J Basic Microbiol 27:309
117. Kobayashi H, Suzuki H (1977) Biotechnol Bioeng 18:37
118. Krabben P, Nielsen J (1998) Modeling the mycelium morphology of *Penicillium* species in submerged cultures. Adv Biochem Eng Biotechnol 60:125
119. Cui YQ, Okkerse WJ, van der Lans RGJM, Luyben KCAM (1998) Biotechnol Bioeng 60:216
120. Lejeune R, Nielsen J, Baron GV (1995) Biotechnol Bioeng 47:609

121. Lejeune R, Baron GV (1997) Biotechnol Bioeng 53:139
122. King R (1998) Mathematical modelling of the morphology of *Streptomyces* species. Adv Biochem Eng Biotechnol 60:95
123. Paul GC, Thomas CR (1996) Biotechnol Bioeng 51:558
124. Tucker KG, Thomas CR (1992) Biotechnol Letters 14:1071
125. Tucker KG, Thomas CR (1994) Biotechnol Techniques 8:153
126. Rossetti S, Christensson, Blackall LL, Tandol V (1997) J Appl Microbiol 82:405
127. Seviour EM, Blackall LL, Christensson C, Hugenholtz P, Cunningham MA, Bradford D, Stratton HM, Seviour RJ (1997) J Appl Microbiol 82:411
128. Kieran PM, O'Donnel HJ, Malone DM, MacLoughlin PF (1995) Biotechnol Bioeng 45:415
129. Shimosaka M, Masui S, Togawa Y, Okazaki M (1991) J Ferment Bioeng 72:485
130. O'Shea DG, Walsh PK (1996) Biotechnol Bioeng 51:679
131. Shabtai Y, Ronen M, Mukmenev I, Guterman H (1996) Comp Chem Engng 20:S321
132. Guterman H, Shabtai Y (1996) Biotechnol Bioeng 51:501
133. Spohr A, Dam-Mikkelsen C, Carlsen M, Nielsen J (1998) Biotechnol Bioeng 58:541
134. Reponen TA, Gazenko SV, Grinshpun SA, Willeke K, Cole EC (1998) Appl Environ Microbiol 64:3807
135. Mitchell AD, Walter M, Gaunt RE (1997) Biotechnol Techniques 11:801
136. Paul GC, Kent CA, Thomas CR (1993) Biotechnol Bioeng 42:11
137. Oh KB, Chen Y, Matsuoka H, Yamamoto A, Kurata H (1996) J Biotechnol 45:71
138. Reichl U, Buschulte TK, Gilles ED (1990) J Microsc 158:55
139. Metz B, de Bruin EW, van Suijdam JC (1981) Biotechnol Bioeng 23:149
140. van Suijdam JC, Metz B (1991) Biotechnol Bioeng 23:111
141. Adams HL, Thomas CR (1988) Biotechnol Bioeng 32:707
142. Packer HL, Thomas CR (1990) Biotechnol Bioeng 35:870
143. Thomas C, Packer H (1990) Binary 2:47
144. Daniel O, Schönholzer F, Zeyer J (1995) Appl Environ Microbiol 61:3910
145. Walsby AE, Avery A (1996) J Microbiol Methods 26:11
146. Hitchcock D, Glasbey CA, Ritz K (1996) Biotechnol Techniques 10:205
147. McIntyre M, Berry DR, Eade JK, Cox PW, Thomas CR, McNeil B (1998) Biotechnol Techniques 12:671
148. Cox PW, Thomas CR (1992) Biotechnol Bioeng 39:945
149. Pichon DR, Vivier HL, Pons MN (1992) Acta Stereol 11:243
150. Yang YK, Morikawa M, Shimizu H, Shioya S, Suga KI, Nihira T, Yamada Y (1996) J Ferment Bioeng 81:7
151. Maazi A, Pons MN, Vivier H, Latrille E, Corrieu G, Cosson T (1998) 2nd Eur Symp Biochem Eng Science, Porto
152. Matsumura M, Imanaka T, Yoshida T, Taguchi H (1980) J Ferment Technol 58, 197
153. Drouin JF, Louvel L, Vanhoutte B, Vivier H, Pons MN, Germain P (1997) Biotechnol Techniques 11:819
154. Pons MN, Drouin JF, Louvel L, Vanhoutte B, Vivier H, Germain P (1998) J Biotechnol 65:3
155. Packer HL, Keshawarz-Moore E, Lilly MD, Thomas CR (1992) Biotechnol Bioeng 39:384
156. Reichl U, Yang H, Gilles ED (1990) FEMS Microbiol Lett 67:207
157. Meijer BC, Kootstra GJ, Wilkinson MHF (1990) Binary 2:21
158. Meijer BC, Wilkinson MHF (1998) Optimized population statistics derived from morphometry In: Wilkinson MHF, Schut F (eds) Digital Image Analysis of Microbes. Wiley, New York, p 225
159. Uozomi N, Yoshino T, Shiotani S, Suehara KI, Arai F, Fukuda T, Kobayashi T (1993) J Ferment Bioeng 76:505
160. Amaral AL, Baptiste C, Pons MN, Nicolau A, Lima N, Ferreira EC, Mota M, Vivier H (1999) Biotechnol Techniques 13:111
161. Lillie RD, JJ (1997) Conn's Biological Stains, 9th edn. Williams & Wilkins, Baltimore
162. Feulgen R, Rossenbeck H (1924) Z Physiol Chem 135:203
163. Feulgen R, Voit K (1924) Z Physiol Chem 135:249
164. Nicolini C, Kendall F (1982) Acta Med Pol 23:155–181

165. Yeast Group of EBC (1962) J Inst Brewing 68:14
166. Pons MN, Vivier H (1998) Morphometry of yeast. In: Wilkinson MHF, Schut F (eds) Digital Image Analysis of Microbes. Wiley, New York, p 199
167. Tucker KG, Chalder S, Al-Rubeai, CR Thomas (1994) Enz Microb Technol 16:29
168. Paul GC, Kent CA, Thomas CR (1994) Biotechnol Bioeng 44:655
169. Vanhoutte B, Pons MN, Thomas CR, Louvel L, Vivier H (1995) Biotechnol Bioeng 48:1
170. Tanaka H, Mizuguchi T, Ueda K (1975) Hakko Kogaku Zasshi 53:35
171. Whiteley AS, Grewal R, Hunt A, Barer MR (1998) Determining biochemical and physiological phenotypes of bacteria by cytological assays In: Wilkinson MHF, Schut F (eds) Digital Image Analysis of Microbes. Wiley, New York, p 281
172. Raynal L, Barnwell P, Gervais P (1994) J Biotechnol 36:121
173. Nikolai TJ, Peshwa MV, Goetghebeur S, Hu WS (191) Cytechnol 5:141
174. Kaneshiro ES, Wyder MA, Wu YP, Cushion MT (1993) J Microbiol Methods 17:1
175. Wirtanen G, Nissinen V, Tikkanen L, Mattila-Sandholm T (1995) Int J Food Sci Technol 30:523
176. Wirtanen G, Ahola H, Mattila-Sandholm T (1995) Trans Int. Chem. Engineers 73 (part C):9
177. Sjoberg AM, Wirtanen G, Mattila-Sandholm T (1995) Trans Int. Chem. Engineers 73 (part C) 17
178. Coleman, AW, Maguire, MJ, Coleman, JR (1981) J Histochem Cytochem 29:959
179. Davies CM (1991) Letters Appl Microbiol 13:58
180. Mauss P, Drouin JF, Pons MN, Vivier H, Germain P, Louvel L, Vanhoutte B (1997) Biotechnol Techniques, 11, 813
181. Williams SC, Hong Y, Danawal DCA, Howard-Jones MH, Gibson D, Frischer ME, Verity PG (1998) J Microbiol Methods 32:225
182. Kogure K, Simidu U, Taga N (1979) Can J Microbiol 25:415
183. Singh A, Pyle BH, McFeters GA (1989) J Microbiol Methods 10:91
184. Barcina I, Arana I, Santorini P, Iriberri J, Egea L (1995) J Microbiol Methods 22:139
185. Billon CMP, McKirgan CJ, McClure PJ, Adair C (1997) J Appl Microbiol 82:48
186. Jones CL, Lonergan GT, Mainwaring DE (1992) Biotechnol Techniques 6:417
187. Brandis JW, Ditullio DF, Lee JF, Armiger WB (1989) Process controlled temperature induction during batch fermentations for recombinant DNA products. In: Computer applications in fermentation technology. Elsevier, London, p 235
188. Srinivas G, John Vennison S, Sudha SN, Balasubramanian P, Vaithilingam Sekar (1997) Appl Environ Microbiol 63:2792
189. Binnerup SJ, Højberg O, Sørensen J (1998) J Microbiol Methods 31:185
190. Slavik J (1998) Single and multispectral parameter fluorescence microscopy. In: Wilkinson MHF, Schut F (eds) Digital Image Analysis of Microbes. Wiley, New York, p 281
191. Lawrence JR, Neu TR, Swerbone GDW (1998) J Microbiol Methods 32:253
192. Slavik J (1998) Microspectrofluorometry: measuring ion concentrations in living microbes In: Wilkinson MHF, Schut F (eds) Digital Image Analysis of Microbes. Wiley, New York, p 309
193. Tsien, RY (1988) Fluorescent indicators of ion concentrations. In: Taylor DL, Wang YlL (eds) Fluorescence Microscopy of Living Cells in Culture, part B. Academic Press, San Diego, p 127
194. Imai T, Ohno T (1995) J Biotechnol 38:165
195. Grynkiewicz C, Poenie M, Tsien RY (1985) J Biol Chem 260:3440
196. Kawai S, Yoshizawa Y, Mizutani J (1993) Biosci Biotechn Biochem 57:1115
197. Kawaii S, Yoshizawa Y, Mizutani J (1994) Biosci Biotech Biochem 58:982
198. Tsunoda Y, Yodozawa S, Tashiro Y (1988) FEBS Lett 231:29
199. Haralick RM (1978) Statistical and structural approaches to texture. Proc 4th Int Joint Conf Patt Recog, Kyoto, p 45
200. Pons MN, Mona H, Drouin JF, Vivier H (1995) Texture characterization of colonies on solid substrate. Proc CAB6, Garmisch-Partenkirchen, Elsevier, p 189

201. Tsuchido T, Takeuchi H, Kawahara H, Obata H (1994) J Ferment Bioeng 78:185
202. Berg HC (1971) Rev Sci Instrumm 42:868
203. Berg HC, Brown DA (1972) Nature 239:500
204. Frymier PD, Ford RM (1997) AIChE J 43:1341
205. Biondi SA, Quinn JA, Goldfine H (1998) AIChE J 44:1923
206. Cercignani G, Lucia S, Petracchi D (1998) Motility, chemotaxis and phototaxis measurements by image analysis. In: Wilkinson MHF, Schut F (eds) Digital Image Analysis of Microbes. Wiley, New York, p 343
207. Gerin P, Bellon-Fontaine MN, Asther M, Rouxhet PG (1995) Biotechnol Bioeng 47:677
208. Baillieul M, Scheunders P (1998) Wat Res 32:1027
209. Walther I, Bechler B, Müller O, Hunzinger E, Cogoli A (1996) J Biotechnol 113
210. Pinheiro R, Belo I, Mota M (1997) Biotechnol Lett 19:703
211. Prenosil JE, Villeneuve PE (1998) Biotechnol Bioeng 59:679
212. Tai HC, Buettner HM (1998) Biotechnology 14:364
213. Hollister SJ, Kikuchi N (1994) Biotechnol Bioeng 43:586
214. Carver SE, Heath CA (1998) Biotechnol Bioeng 62:167
215. Bhatia SN, Balis UJ, Yarmush ML, Toner M (1998) Biotechnol Prog 514:378
216. Joos S, Fink TM, Rätsch A, Lichter P (1994) J Biotechnol 35:135
217. Ramm P (1994) J Neurosci Methods 54:131
218. Tanke HJ, Florijn RJ, Wiegant J, Raap AK, Vrolijk J (1995) Histochem J 27:4
219. Pukall R, Brambilla E, Stackebrandt E (1998) J Microbiol Methods 32:55

Monitoring the Physiological Status in Bioprocesses on the Cellular Level

K. Christian Schuster

Research Group Bioprocess Engineering, Institute of Chemical Engineering,
Fuel and Environmental Technology, Vienna University of Technology, Getreidemarkt 9/159,
A-1060 Vienna, Austria. *E-mail: kcs@mail.zserv.tuwien.ac.at*

The trend in bioprocess monitoring and control is towards strategies which are based on the physiological status of the organism in the bioprocess. This requires that the measured process variables should be biologically meaningful in order to apply them in physiologically based control strategies. The on-line monitoring equipment available today mostly derives information on the physiological status indirectly, from external variables outside the cells. The complementary approach reviewed here is to analyse the microbial cells directly, in order to obtain information on the internal variables inside the cells. This overview covers methods for analysis of whole cells (as a population or as a single cell), for groups of cellular components, and for specific compounds which serve as markers for a certain physiological status. Physico-chemical separation methods (chromatography, electrophoresis) and reactive analysis can be used to analyse elemental and macromolecular composition of cells. Spectroscopic methods (mass, dielectric, nuclear magnetic, infrared, and Raman) have only recently been applied to such complex multicomponent mixtures such as microbial cells. Spectroscopy and chemical separation methods produce large amounts of data, which can often be used in the best way by applying chemometrics. Some of the methods can yield information not just on the average of the microbial cell population, but also on the distribution of sub-populations. The suitability of the methods for on-line coupling to the bioprocess is discussed. Others not suitable for on-line coupling can be established in routine off-line analysis procedures. The information gained by the methods discussed can mainly be used to establish better knowledge of the basis for monitoring and control strategies. Some are also applicable in real-time monitoring and control.

Keywords. Macromolecular compounds, Marker molecules, Chemical patterns, Infrared spectroscopy, Raman spectroscopy

1	Introduction – The Approach to Obtain Physiological Status Information on the Cellular Level	186
2	Morphology	189
3	Biomass Elemental Composition	191
4	Chemical Analysis of (Macro-) Molecular Components (Protein, Carbohydrate, Lipid, Nucleic Acids)	191
5	Intracellular Key Metabolites (ATP/ADP, NADH, cAMP, Inorganic Phosphate)	192
6	Electrical Properties: "Dielectric Spectroscopy"	193

7	Marker Molecules for Physiological Events	194
7.1	Enzymes	194
7.2	Stress Response Proteins and Other Marker Proteins	195
8	Chemical Patterns of Cellular Components or Whole Cells	196
8.1	Lipid Pattern	196
8.2	Protein Pattern	198
8.3	Chemical Pattern by Pyrolysis Mass Spectrometry	200
8.4	Chemical Pattern by Infrared (IR-) Spectroscopy	201
8.5	Chemical Pattern on the Single Cell Level by Confocal Raman Microscopy	204
9	Conclusions and Outlook	205
	References	205

List of Symbols and Abbreviations

AAS	atomic absorption spectroscopy
ADP DP	adenosine diphosphate
AMP	adenosine monophosphate
ATP	adenosine triphosphate
cAMP	cyclo-adenosine monophosphate
DNA	deoxyribonucleic acid
FIA	flow injection analysis
FT-IR	fourier transform infrared (spectroscopy)
GC	gas chromatography
IR	infrared
(m)RNA	(messenger) ribonucleic acid
NAD(P)	nicotinic acid adenine dinucleotide (phosphate)
NAD(P)H	nicotinic acid adenine dinucleotide (phosphate), reduced form
NMR	nuclear magnetic resonance (spectroscopy)
PyMS	pyrolysis mass spectrometry
SDS-PAGE	sodium dodecyl sulphate polyacrylamide gel electrophoresis
SERS	surface enhanced Raman spectroscopy

1 Introduction – The Approach to Obtain Physiological Status Information on the Cellular Level

Research in bioprocess control aims at the maximum performance of the process, which depends mostly on the performance of the living organism used as the biocatalyst. The data for setting the optimal or at least permissive environment for performance of the organism is delivered by the monitoring equipment, preferably in real-time. The design of effective controls is based on knowledge about the process and the organism which is acquired during process development and optimisation. Simple processes can be optimised and controlled on the basis of purely practical experience. The more complex a process is, the less likely is an optimisation by empirical variation of process parameters and a control strategy via fixed parameters. The advanced approach is to describe the physiological state of the organism by a set of several or many process variables with relevant information about the organism's state [1, 2], and [the chapter by Sonnleitner in this volume]. Process control then aims to keep this status within a desired range. These monitored process variables should be biologically meaningful. On the other hand they should be derived from on-line, real-time measurable variables under the restrictions of low disturbance to the organism and without compromising the sterile barrier, thus leading to non-invasive monitoring concepts [2].

A problem arising from a definition of the physiological status purely as a vector of (directly or indirectly) on-line measurable variables [1] is the loss of information available from the cells, which mostly cannot be obtained on-line. Such a definition is very method-oriented and does not a take enough notice of the basis for the measured effects, the metabolic network inside the cell. The concepts reviewed here cover a wider idea: The physiological status can be seen as a certain setting in the metabolic network activities, leading to a characteristic pattern of substrate uptake and product formation which is distinct from other settings possible in the same organism. Well-known examples for such settings and shifts between different settings include the physiological events in the yeast cell-cycle [2], the overflow metabolism at high substrate availability [2], the induction of recombinant protein production [1], or the induction of pathways of the secondary metabolism such as antibiotic production (see [3] for examples).

This chapter deals with gaining information on this actual metabolic setting at a time, called the physiological state, from analysis of the organism's cells. This approach looks directly at the (microbial) biomass in the process, measuring internal variables inside the cells. It is complementary to the non-invasive concepts to trace back this status mostly from external variables which can be measured outside the organism's cells. The questions asked determine the choice of the methods to apply (Table 1). Here, mostly "global" approaches looking at the bulk of the cell population will be discussed.

The issue of distribution in the population of a monoseptic culture is only recently getting more attention. Studies on the single-cell level by image analysis (see the chapter in this volume by M.N. Pons), and by flow cytometry (see [4, 5] and the section in the chapter by Sonnleitner in this volume), reveal that most microbial cultures are by no means a mass of identical cells, even those which do not show morphological inhomogeneity. Some of the methods discussed here can also give information on distribution of subpopulations, at least when these differ by clear-cut boundaries (for an overview see Table 2).

The discussion of the approaches to analyse microbial biomass for physiological status information starts with the better known methods and proceeds to more novel developments. Morphology has been used for a long time to

Table 1. Approaches to physiological status monitoring on the cellular level, overview

Object of the Approach	Whole Cells			Groups of Cell Components	Specific Compounds (Markers)
	Single Cells	Population			
Primary Results	Morphology / Population Distribution	Chemical Patterns		Elemental Composition / Macromolecular Composition / Lipid pattern / Protein Pattern	Small Key Metabolites / Enzyme Activities / Stress Markers
Method Group	Microscopy & Image Analysis (*) / Flow Cytometry (*)	Spectroscopy		Chemical Separation or Reactive Analysis	
Method	Microscopy & Image Analysis (*) / Flow Cytometry(*)	Raman (Micro-) Spectroscopy / Infrared Spectroscopy / Pyrolysis Mass Spectroscopy / NMR / Dielectric Spectroscopy	Specific Analysis	Specific Analysis / Chromatography / Electrophoresis	Specific Analysis / Specific Analysis or mRNA blot / Electrophoresis or mRNA blot

* Not reviewed in detail in this chapter. For image analysis, see the chapter by M.N. Pons in this volume. For some more details and references, see the paragraph in the chapter by Sonnleitner, this volume, and [4].

Table 2. Information content on sub-populations of the approaches to physiological status monitoring on the cellular level

Detailed, specific information on sub-populations	Some information can be extracted from spectra or patterns	Generally no information on sub-populations
Microscopy and Image Analysis	Infrared Spectroscopy	Macromolecular composition by specific analysis
Flow Cytometry	Pyrolysis Mass Spectroscopy	Enzyme activities by specific analysis or mRNA blot
	NMR spectroscopy	Stress markers by electrophoresis or mRNA blot
Raman (Micro-) Spectroscopy	Dielectric Spectroscopy	Small key metabolites
	Lipid pattern by GC	Elemental composition by specific analysis
	Protein pattern by Electrophoresis	NADH by fluorescence
	Special cases of enzyme activities (by specific analysis or mRNA blot)	
	Some stress markers (by electrophoresis or mRNA blot)	

Table 3. Suitability for on-line monitoring of the approaches to physiological status monitoring on the cellular level

In situ	Ex-situ, with interfaces for automatic sampling, similar to on-line chromatography	Difficult, complex automated set-ups necessary, possible with FIA and/or laboratory robots
Microscopy and Image Analysis	Microscopy and Image Analysis	Macromolecular composition by specific analysis
NMR spectroscopy	Elemental composition by specific analysis	Protein pattern by Electrophoresis
Dielectric Spectroscopy	Flow Cytometry	Enzyme activities by specific analysis or mRNA blot
NADH by fluorescence	Raman (Micro-) Spectroscopy	Stress markers by electrophoresis or mRNA blot
	Infrared Spectroscopy Pyrolysis Mass Spectroscopy NMR spectroscopy Lipid pattern by GC Some Small key metabolites Some enzyme activities	Small key metabolites

obtain qualitative information on the status of a microbial culture. Chemical analysis of the biomass for its elemental and macromolecular composition is a useful tool for process development, but can also provide information for process monitoring. Intracellular key metabolites can serve as general or very specific markers for certain physiological states. The electrical properties of the cells can not just be used for biomass quantification, but also for qualitative information. Macromolecules involved in biosynthetic pathways and signalling can serve as specific markers for physiological events.

The methods yielding chemical patterns of whole cells or of certain molecular component groups were originally developed for identification of organisms or for studying mixed populations, and have only rather recently started being used to monitor changes within a culture of a single organism during a process run.

The suitability of the mentioned methods for on-line monitoring will also be discussed. For an overview see Table 3. With some, it is straight forward and has been done, others have been used off-line but are feasible on-line. With others the effort to go on-line may be too high for the results obtained, although with flow injection systems or laboratory robots nearly anything is feasible in principle.

2
Morphology

Observing the morphology is one of the oldest methods of monitoring the physiological status of a cellular culture qualitatively and can often give valuable global and single-cell information on the state of an organism. This is especially true for fungal and higher organism cell cultures, and most applications of systematic morphology monitoring deal with these. For example, in a previous volume of this series [3] on the relationship between morphology and process performance, five out of six chapters dealt with filamentous organisms and only one with single-cell organism cultures (bacteria, yeasts, and animal cells) [6]. However, many yeasts and bacteria also display distinct morphological changes which can be related to physiological events, especially the groups of the budding yeasts, and the bacteria groups of Actinomycetes, the Coryneformes, (for examples see [7], the Lactobacilli [6] and the sporulating genera *Bacillus* and *Clostridium*. For example, with solventogenic Clostridia, the shift of metabolic pathways between solvent and acid production is closely related to morphological changes [8, 9, 10]. Distortions of morphology in otherwise rod-shaped bacteria can be an indication for trace element limitations [11].

However, the morphological information is often neglected; this is partly due to the subjectivity of microscopic observation and the difficulty to quantify the information. The approach to overcome this problem is the use of automated computer-based image analysis of micro-images (see the chapter on image analysis in this issue).

Light microscopic studies of bacteria work at the edge of the theoretical resolution of this method, so many morphological features cannot be detected clearly. Additional insight in morphological features especially of bacteria can

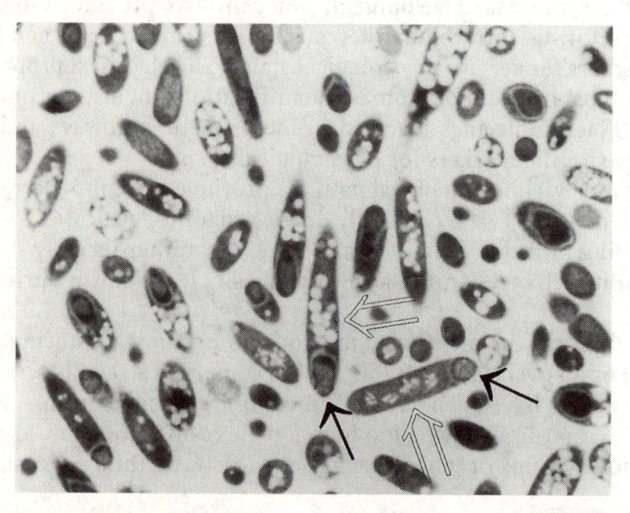

Fig. 1. Electron micrograph of cells from a continuous culture of a solventrogenic *Clostridium* (ultrathin section). In the light microscope, the cells look more or less rod-shaped with slightly different sizes, whereas under the electron microscope, additional features are detected: Examples for starting spore formation are marked with *black arrows* (→). Examples for storage carbohydrate granules (bright round spots) are marked with *double arrows* (⇒). Remark: Circular-looking shapes of cells are cross-sections of rods perpendicular to the long axis of the rod [12]

be gained by electron microscopy. Preparation methods for electron microscopy are certainly too laborious for routine process monitoring, but the detection of morphological distributions in a culture not visible in the light microscope is valuable information. Figure 1 [12] is an example.

3
Biomass Elemental Composition

Analysis of the biomass elemental composition is mainly used in bioprocess engineering as a basis for growth medium development. A recent example for assessing the effectiveness of a medium design procedure is given by Mandalam and Palsson [13]. However, elemental analysis can also be used for characterising elemental limited physiological states [14]. Determination of the major biological elements carbon, hydrogen and nitrogen (and oxygen) is generally performed by often automated standard methods for biological samples; a review on these methods and the standardised use, starting at the sampling procedure, to obtain reliable results is given by Gurakan et al. [15]. The often interesting, limiting element phosphorous can be analysed according to Ames [16]. Additional elemental analysis of minor and micro elements require specific methods for each element. Trace elements are often analysed by atomic absorption spectrometry (AAS) [17].

4
Chemical Analysis of (Macro-) Molecular Components (Protein, Carbohydrate, Lipid, Nucleic Acids)

Macromolecular biomass composition is of obvious interest when the biomass itself is the product, such as algal biomass in [18], or for production of single-cell protein, for e.g. animal feedstock. Moreover, for a precise metabolic flux analysis, changes in biomass composition should be taken in account. For example, Henriksen et al. [19] observed with *E. coli* under different growth rates, that the levels of DNA and lipids were relatively constant, whereas the proteins and stable RNA levels increased with the specific growth rate and the total amount of carbohydrates decreased.

The number of chemical methods for analysis of cell component groups is vast. See Table 4 for examples of methods and references. More methods are reviewed by Herbert et al. in [20], in [21] by Cooney, and mentioned in [22]. Results of different methods for the same components are usually difficult to compare, as chemical reactions of different selectivity are involved.

To perform the complete set of chemical analyses for macromolecular compounds is generally too labour intensive for routine process monitoring. Following the concentrations of selected compounds can, however, give interesting insights: for example, Bley et al. [23] found that a phosphate – limited population differentiated into two sub-populations of different DNA – and PHB (poly-3-hydroxybutyrate) – content. The cell-cycle of baker's yeast could be followed by a fluorimetric dye reaction for DNA content in a sample stream from the bioreactor [24].

Table 4. Some methods for analysis of cell component groups

Component	Reference	Remarks to Method
Carbohydrate	Dubois et al. [25]	
Free Amino Acids	Stewart [26]	
Lipids	Henriksen et al. [19]	gravimetry
RNA	Benthin et al. [27]	
DNA	Herbert et al. [20]	
Protein	Henriksen et al. [19]	Kjehldahl nitrogen analysis and subtracting other N-compounds
	Bradford [28]	Coomassie blue dye reaction
	Lowry [29]	Folin reagent colour reaction
All components	Herbert [20]	Review with many chemical (reactive) methods
	Cooney [21]	Review with many chemical (reactive) methods
	Grube et al. [22]	Simultaneously by infrared spectroscopy (see also Sect. 8.4) and comparison with chemical methods (with references)

An alternative, at least semi-quantitative method to follow changes in biomass composition is infrared (IR) spectroscopy [22]. From dried samples of microbial cells, IR spectra can be obtained which contain information on all major cell components. The spectra are analysed as a multi-component mixture: Characteristic bands in the spectra are identified, the extinction coefficients for each component (protein, carbohydrate, lipid, and nucleic acids) at each band are determined, and the concentrations are calculated by a system of linear equations. The method gives results on all major cell components simultaneously, and is relatively quick and easy to perform, compared to the chemical analysis methods. For details see Sect. 8.4 below.

5
Intracellular Key Metabolites (ATP/ADP, NADH, cAMP, Inorganic Phosphate)

The most generally interesting intracellular metabolites are the adenine nucleotide phosphates (ATP/ADP/AMP) and the nicotinic acid/adenine dinucleotide phosphates (NAD(P)H/NAD(P)). The pool concentrations and ratios of these can serve as markers for the physiological status of a culture.

ATP, or exactly the specific ATP concentration, and the ratios to ADP and AMP in the biomass, indicate the energy content (or energy charge) of the cells. Estimation of the fairly low pools of the adenine nucleotides can be achieved by bioluminiscence methods ([30] by Kimmich et al.); automated to a flow analyser ([31] by Reiman et al.), or by enzymatic methods ([32] by Compagnone and Guilbault; automation in FIA ([33] by Richter and Bilitewski). An application for the specific ATP pool analysis is the assessment of biomass quality in immobilised biofilms [34].

NAD(P)H concentration in biomass and its ratio to NAD(P) gives a measure for the culture reduction status [35]. NAD(P)H can, in principle, be detected directly in vivo by in situ fluorescence measurement (see the chapter by Sonnleitner in this volume – the section on culture fluorescence). This measure serves as a biomass concentration sensor if the specific NAD(P)H concentration stays constant. If not, the ratio of culture fluorescence and biomass concentration, the specific fluorescence, can be a measure for the culture reduction state, or indicate other more complex events like metabolic pathway shifts [35] or even the formation of a variant in a culture [36] (see also Figs. 2 and 3).

Methods for chemical analysis of NAD(P)H include enzymatic methods [37] and bioluminiscence methods ([38] by Grupe and Gottschalk using a commercial test kit from Boehringer Mannheim). Increased NADH levels often indicate oxygen limited states of aerobic organisms (well known for baker's yeast, and also observed with hybridoma cultures by Zupke et al. [39], where the situation is more complex due to cell compartments).

A combined application for culture energy charge and reduction state is the monitoring of events in *Clostridium acetobutylicum* at the shift from acid to solvent production [38] (Grupe and Gottschalk, 1992)

Other key metabolites analysed as physiological markers are the intracellular orthophosphate [40], and cyclic nucleotides as cyclo-AMP (review by Smith et al. [41]).

For all off-line analysis of low-concentrated key metabolites, the extremely short relaxation times in the observed biochemical pathways have to be taken in account. Special sampling and analysis techniques have to be employed to assure accurate estimations [40].

A method to assay all phosphorous-containing metabolites at once is ^{31}P-NMR spectroscopy (see the chapter by Sonnleitner in this volume – the section on NMR), the main limitation being the sensitivity for the very low-concentrated metabolites. NMR should be, in principle, also suitable for obtaining space-resolved (imaging) information [42], however, the technical difficulties have yet to be solved.

6
Electrical Properties: "Dielectric Spectroscopy"

Dielectric spectroscopy or culture capacitance measurement is used as an on-line, non-invasive method for biomass estimation (see the chapter by Sonnleitner in this issue – the section on electrical properties) and responds mainly to living cells [43, 44]. Observed difficulties in using the signal as a pure biomass concentration sensor, i.e. deviations from the simple correlation with cell density, were attributed to dependencies on the physiological state [43], and could be used to discriminate different populations in yeast cultures [45]. Connections with morphological features could be found for budding yeast

[46]. For organisms with more complex morphology, e.g. filamentous bacteria, the interpretation of dielectric signals is difficult [47]. However, in principle the method seems to hold more potential than has been exploited up to now, especially when dielectric spectra are evaluated with advanced mathematical and statistical methods (see also the chapter by A. Shaw and D. Kell in this volume).

7
Marker Molecules for Physiological Events

Physiological events which are the basis of gradual or sudden changes in product formation of a culture can be followed by analysing the biosynthetic components of the pathways involved, or of signal molecules. Examples include enzyme activities and molecules related to stress response, like heat shock proteins and other stress proteins.

7.1
Enzymes

The activities of intracellular enzymes give an indication of the activity of certain metabolic pathways. They are classically assayed by performing their specific reactions in vitro, using the conversion of natural or artificial substrates and often some additional detection reaction. The sample preparation procedure involves cell harvesting and preparation of cell extracts. A well-studied example is the metabolic shift in *Clostridium acetobutylicum* from acid to solvent production, the so-called solvent shift. Andersch et al. [48] studied the activities of 10 different enzymes involved in this shift, in batch cultivations where the shift is self-induced by the products formed, and under continuous culture conditions with an externally induced shift.

Difficulties with using enzyme activities as a physiological marker include: the general problem of comparing in vitro measured activities with the in vivo situation; the stability of enzymes under the conditions of cell extract preparation, which has to be tested and optimised for each individual enzyme, e.g. [48]; and the activity assay itself, which again has to be established for each enzyme. Thus, it has to be stated that it is a tremendous amount of work to analyse several enzymes at regular time intervals, and therefore the approach is hardly suitable for routine monitoring of a bioprocess over extended periods of time.

The upcoming alternative to assess enzymatic activities is the detection of enzyme induction on the transcription level by assaying the corresponding messenger RNA. To be precise, the assay of mRNA concentrations does not correspond directly to the enzyme activity, but to the rate of enzyme synthesis; however, a high rate of enzyme synthesis eventually leads to a high enzyme activity, due to the intracellular protein turnover. So, the two estimation approaches cover different time scales. The typical kinetic pattern of mRNA concentration for an induced enzyme shows a quick rise to a peak, followed by a drop to a steady-state level. For the above example of the solvent shift in *Clostridium*, this was shown by Sauer and Dürre [49] and Girbal and Soucaille [50] by hybridisation of mRNA on electrophoresis gels with radioactively labelled DNA probes against known enzyme gene sequences. (Instead of DNA probes, also RNA probes can be used in a similar way).

The assay of enzyme induction at the mRNA level is much easier to perform than the chemical assay of each individual enzyme and the response to a physiological event is intrinsically faster to detect. Moreover, mRNA is often more stable under cell extract preparation conditions than enzymes, if destruction by RNAses can be prevented. For analysing a number of different enzymes in parallel, only one sample preparation procedure is necessary, which produces a crude RNA extract, and only in the final step the different DNA or RNA probes are used for detection of specific enzyme RNAs. Recently, the detection of RNA is further facilitated by ready-to-use RNA isolation kits, non-radioactive DNA or RNA probes and the dot-blot and slot-blot techniques, respectively, [51, 52].

A certain limitation to the method is the fact that DNA or RNA probes against the genes involved have to be available for hybridisation. This implies that only rather well-known organisms can be studied, for which the gene sequences involved are known, or that the preparatory work to set up the method has to identify suitable sequences from related organisms or pathways.

7.2
Stress Response Proteins and Other Marker Proteins

Various proteins have been identified as markers for general stress in microorganisms or for specific stress like temperature, pH, toxic products, oxygen deficiency for strict aerobes, or oxygen toxicity for anaerobes, etc.; most of them are rather small in size. Most work was done with *E. coli* (for a review see e.g. [53] and with *B. subtilis* [54]).

Classically, the stress proteins are assayed by two-dimensional electrophoresis, with a separation according to molecular mass (in SDS) in one direction and isoelectric focusing in the other, and autoradiographic detection. An example for an application is the study of stress response and its relationship to a metabolic shift in *Clostridium* by Terraciano et al. [55] and Pich et al. [56]. This technique is not applicable to bioprocess monitoring on a routine basis, as the analysis by 2-dimensional electrophoresis is fairly laborious and needs high skill, and moreover the autoradiographic detection requires feeding of radioactive substrates. However, for basic studies and identification of a few marker proteins which can afterwards be detected e.g. by antibody reactions, it is useful. A recent development in this field is the identification of proteins directly from the spot on the gel by *N*-terminal sequencing and comparison with known sequences. In *B. subtilis*, this was shown [57]; and also in *Clostridium* for proteins involved in the solvent shift [58].

Recently, an easier, more applicable alternative was developed by Völker et al. [59], detecting the expression of "stress genes" by the messenger RNA produced. Schweder et al. [52,60] used this to monitor process-related stress in large scale bioreactors. Cells exposed to stress in certain zones of the reactor expressed general and specific stress genes within a few seconds, and the elevated mRNA levels could be detected for a few minutes after the stress event, thus serving as a "historic" marker for past physiological events.

At the moment, the procedure takes about a day for the analysis of 8 stress genes and an improved method is being developed. The approach is the DNA/RNA chip technology (T. Schweder, personal communication). For the methodical advantages and limitations, the considerations are similar as for enzyme activity analysis via mRNA – see [51, 52, 60], and Sect. 7.1 (Enzymes) above.

8
Chemical Patterns of Cellular Components or Whole Cells

Most of the following methods, yielding characteristic chemical patterns of cells, were originally developed for identification of organisms or for studying mixed populations, and are as such well established in microbiology laboratories. For organism identification, care has to be taken to grow and harvest the cells under certain, well-defined growth conditions, otherwise the patterns are not comparable. Rather recently, this problem has in some cases been turned into an opportunity for monitoring the changes within a culture of a single organism during cultivation. As the patterns contain a large amount of data, evaluation may not be straight forward but may need chemometric methods for extracting the relevant information or for making correlations with the description of the physiological status by other process variables [61, 62] (see also the chapter on chemometrics by Shaw and Kell in this volume). Also because of this high information content, it is possible to extract quantitative data on heterogeneity of cultures, provided that features can be identified in which the sub-populations differ clearly.

In principle, most of these methods can be linked to a bioreactor on-line and ex-situ, with interfaces for automatic sampling similar to those used for on-line chromatography (for interfaces, see the chapter by Sonnleitner in this volume). An exception is the analysis of protein patterns by electrophoresis, which is difficult to perform automatically.

8.1
Lipid Pattern

The analysis of lipid or fatty acid patterns of organisms by gas chromatography is a common method in taxonomy [63]. The fatty acid pattern changes with growth conditions like growth rate and phase, aeration, pH value, medium composition, etc., which is attributed to adaptations of the cell membrane. For *E. coli*, the changes were studied by Arneborg et al. [64] and Pramanik and Keasling [65]. With thermophilic bacteria, the changes due to growth temperature are especially remarkable and connected with short-term and long-term temperature adaptation [66, 67], thus also serving as a "historic" marker for conditions the organisms have experienced some time before the actual observation.

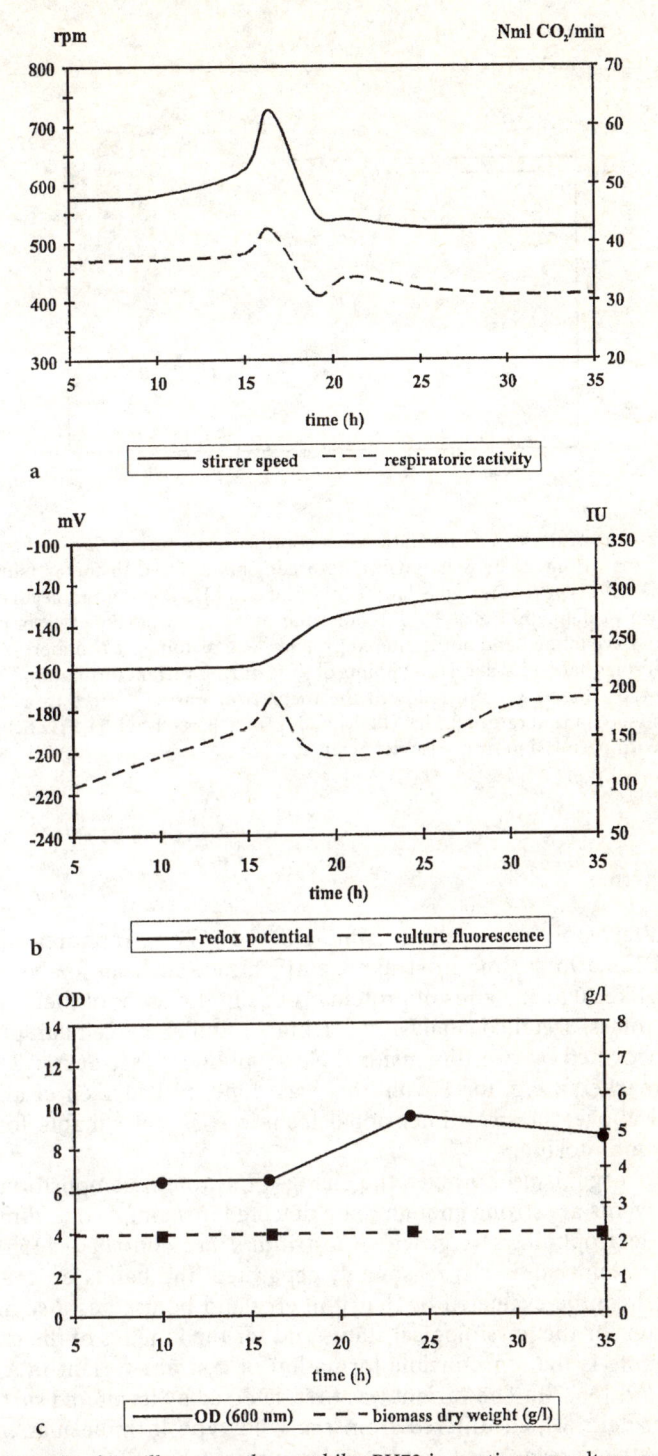

Fig. 2a–c. Growth of *Bacillus stearothermophilus* PV72 in continuous culture on a synthetic medium containing glucose (8 gl^{-1}) as the sole carbon and energy source. The dissolved oxygen concentration was controlled at 50% and the dilution rate was 0.3 h^{-1}. As derived from the measured process variables, variant formation started at about 15–16 h after inoculation. Shown are the measures for: **a** Respiration activity **b** External and internal reduction state (redox potential and culture fluorescence) **c** Cell density (Reprinted from: J. Biotechnol. 54, K.C. Schuster et al., p. 19, 1997, with permission from Elsevier Science)

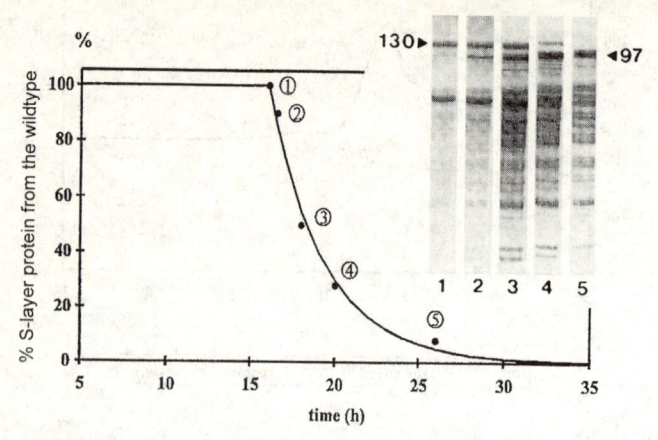

Fig. 3. Monitoring of the variant formation in a continuous culture of *Bacillus stearothermophilus* PV72 (as in Fig. 2) by SDS-PAGE. The organisms differed in the S- (surface) layer proteins. The wild type formed an S-layer protein of 130 kDa apparent molecular mass, the variant S-layer protein appears at 97 kDa molecular mass. The wild-type S-layer protein was quantified relative to the band of the altered protein by densitometry. Numbers on the curve represent samples harvested at distinct points of time during variant formation. The decrease of the wild-type S-layer protein followed the theoretical washout curve in a stirred tank reactor at the set dilution rate of 0.3 h^{-1} (Reprinted from: J. Biotechnol. 54, K. C. Schuster et al., p. 20, 1997, with permission from Elsevier Science)

8.2
Protein Pattern

Protein patterns of whole cells by simple SDS-PAG-electrophoresis [68] have been used for a long time in strain identification. Such an analysis resolves something like 20 to 50 bands of protein groups in the order of their (apparent) molecular mass. Detailed analysis of a high number of cellular proteins is usually performed by two-dimensional electrophoresis (see Sect. 7.2 on stress response markers); e.g. for *E. coli*, this was done by Pedersen et al. [69]. As mentioned above, the two-dimensional technique is not suitable for routine bioprocess monitoring.

For some organisms, however, the changes in protein composition by physiological events are strong enough to be detected in a simple one-dimensional SDS-PAG-electrophoresis (e.g. with *Clostridium acetobutylicum* [55]). If the changing protein bands are reasonably separated, the bands on the electrophoresis gel can be evaluated by densitometry and be used as a semi-quantitative marker for the physiological status and for the kinetics of the changes.

An example is the reproducible formation of a strain variant in a thermophilic *Bacillus* [36, 70]. The variant was first detected by its altered surface layer (S-layer) protein, which differed from the wild type in molecular mass and crystalline structure, as observed by electron microscopy. In continuous cultivation, the event of variant formation was accompanied by a characteristic pattern of changes in several process variables (Fig. 2), and the kinetics of variant formation could be followed by quantified SDS-PAG electrophoresis (Fig. 3).

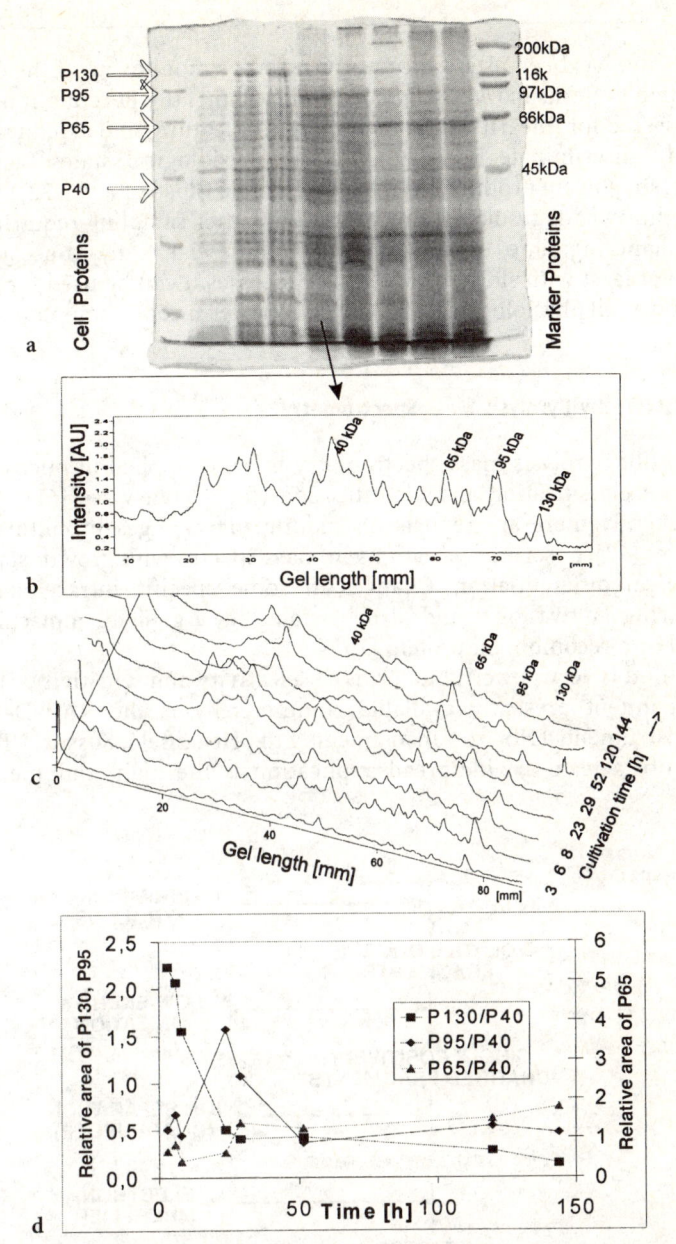

Fig. 4a–d. Protein patterns by SDS-PAGE of whole cells of *Clostridium beijerinckii* in the course of a batch cultivation. The functions of the proteins changing in concentration are not yet known. However, the results in [58] suggest that dehydrogenases of the acid and solvent producing pathways are involved. For assessing the physiological status, it is sufficient to use the quantified bands as markers. **a** Image of the electrophoresis gel, stained with Coomassie blue. The apparent molecular masses of the cell proteins changing in concentration are indicated at the left side of the gel, the marker proteins on the right side. Sampling times are as given in (**c**). **b** Densitogram of one lane (24 h cultivation time). **c** Three-dimensional representations of the protein patters with changes over the batch cultivation time. Sampling times are indicated. **d** Quantification by relative peak areas of the proteins P130, P95 and P65 against reference protein P40 (from [72])

The method has been used for semi-quantitative monitoring of the changes in the protein patterns in *Clostridium acetobutylicum* [10]. Recently, it has been further assessed for linearity and other analytical features and extended in application to discriminate between different physiological states (acid production versus solvent production) in solventogenic Clostridia [71, 72]) (Fig. 4).

The method yields results within a few hours after sampling requiring only moderate handling time, especially when commercial ready-to-use gels are applied, so at least with slowly proceeding processes it can be used for control decisions on a cell physiological basis, e.g. in long-term continuous cultivations.

8.3
Chemical Pattern by Pyrolysis Mass Spectrometry

The method of pyrolysis mass spectrometry is being applied in microbiology for organism classification and identification (for a review see [61]) (Figs. 5 and 6). Recently there are reports on monitoring changes in cultures of a *Streptomyces* [73], observing variations in mass spectra with growth stage and morphological differentiation. Changes in some specific intracellular substances during cultivation could also be detected, e.g. small molecules like indole [74] or a recombinant protein [75].

The method is very powerful in terms of selectivity and sensitivity. The only PyMS instrument presently available commercially is the RAPyD-400 by Horizon Instruments (Horizon Instruments Ltd., Heathfield, Sussex, UK) [76]. A certain obstacle to a widespread application is the high equipment cost.

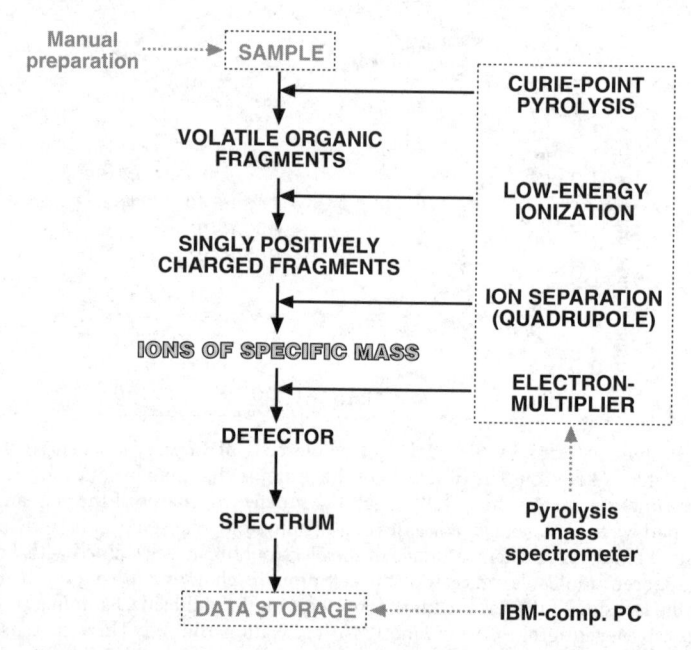

Fig. 5. Flow diagram of the main steps in pyrolysis mass spectrometry (from [76])

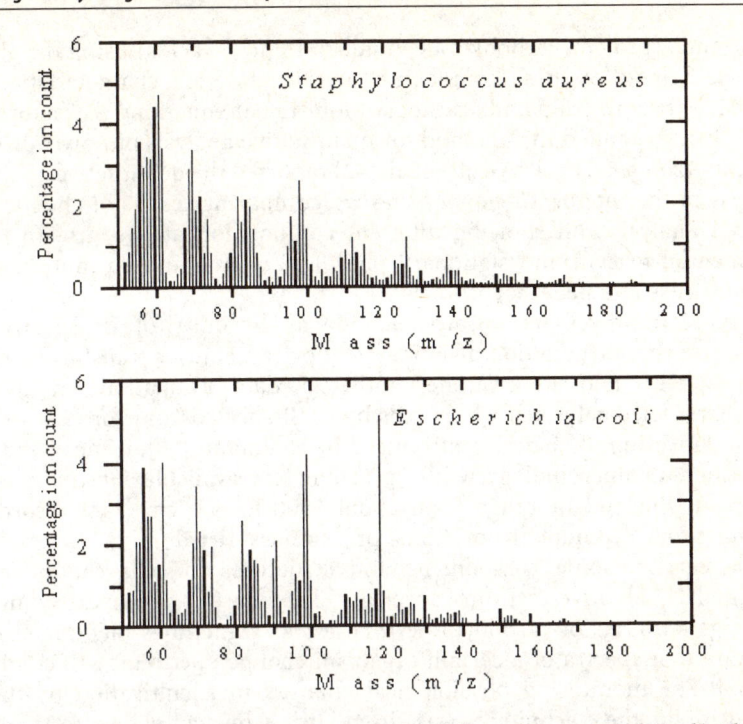

Fig. 6. Pyrolysis mass spectra of two organisms (*Staphylococcus aureus* and *E.coli*). The data from PyMS may be displayed as quantitative pyrolysis mass spectra. The abscissa represents the *m/z* (mass to electric charge) ratio whilst the ordinate contains information on the ion count for any particular *m/z* value ranging from 51–200. Below 51 *m/z*, there is no biologically relevant information in the spectra (from [76])

Spectra are very complex, and the instruments show a drift over longer times [61]. Both problems can be accounted for by data evaluation, but require sophisticated mathematical methods like supervised learning of neural networks. For a comprehensive overview and more detail, see the chapter by Shaw and Kell, this volume.

8.4
Chemical Pattern by Infrared (IR-)Spectroscopy

Reports on a chemical analysis of microorganisms by infrared spectroscopy, including ideas to use it for identification, go back to the 1950s, e.g. [77]. Due to the poor reproducibility between instruments and problems with evaluating complex spectra, not much development followed these early works for a long time.

A group in Latvia and Russia [78] analysed microbial cell mass by a fairly simple classical infrared spectroscopic method, recording spectra of dried biomass embedded in potassium bromide pellets with a dispersive IR spectrometer. They could calibrate the evaluation for four major groups of cellular compounds (protein, carbohydrate, lipid, and nucleic acids) using the absorption coefficients of these compounds at four different characteristic wavelengths of the spectrum and a system of four linear equations. Savenkova et al. [79] (1994) expanded this method for quantitative analysis of polyhydroxybutyrate in *Azotobacter* cells. Grube et al. [22] reported the quantitative analysis of biomass of the fungus *Thielavia terrestris*, comparing results of chemical and infrared analysis and showing different composition at two growth phases (exponential growth and stationary phase) as well as in two morphological states (diffuse and pellet form).

A great improvement towards a wide applicability of IR spectroscopy came from the introduction of moderately priced Fourier transform (FT-) IR spectrometers, and multivariate statistical data evaluation on personal computers. The analysis of whole cells by FTIR spectroscopy for identification and classification of bacteria, introduced by Naumann [80], is meanwhile well established in microbiology, with spectrometers available for this purpose. Spectra in the mid-IR range from about 3500 to 500 cm^{-1} are recorded in transmission technique from films of biomass dried on a zinc selenide sample carrier. Some components and structures in cells can be clearly identified qualitatively from spectra, like spores, polyhydroxybutyrate, and large amounts of carbohydrate [81]. It was sometimes mentioned in the literature that spectra of a certain organism change due to growth conditions [81, 82]. For monitoring physiological changes in a cultivation by the cell composition, the technique was used by Schuster et al. [83] during *Clostridium beijerinckii* batch and fed-batch cultures, Fig. 7. The physiological state of the organism, characterised by the formation of different products, corresponded to different clusters of similar cell composition revealed by multivariate evaluation of IR spectra. Moreover, the growing phase is clearly characterised by strong bands of nucleic acids, whereas the states at high concentrations of toxic products are connected with a high lipid content in the cells, probably reflecting a response to stress on the cell membrane [83, 84].

The technique is easy and fast to use, involving only cell harvesting, washing and drying of the biomass on an IR-transparent carrier as sample preparation, before the spectra are recorded and evaluated. An instrument dedicated for organism identification is available and has been used for the cited process monitoring experiments with complete spectrometer hardware and software, model IFS 28/B from Bruker (Milton, ON, Canada; or Bruker Optik, Karlsruhe, Germany). The whole procedure takes less than an hour.

A different method of recording FTIR spectra of microbial biomass is the diffuse reflectance-absorbance technique [62]. To my knowledge, this has not yet been used to monitor biomass properties in bioprocesses but for related purposes such as screening for overproducing mutants (for more detail see the chapter by Shaw and Kell in this volume).

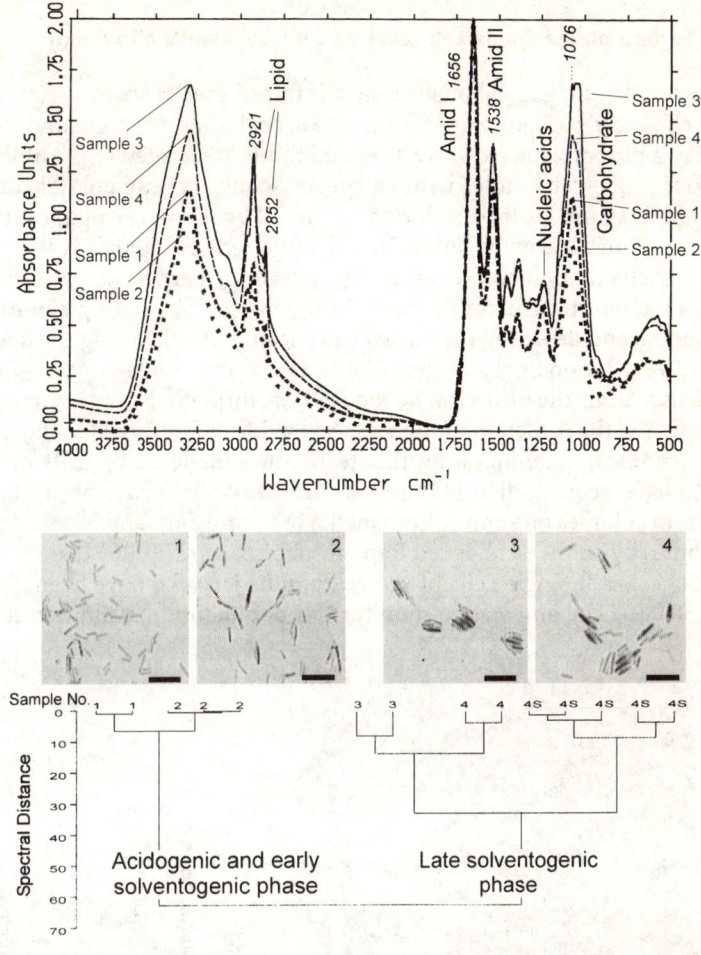

Fig. 7. Analysis of bacterial biomass by FTIR spectroscopy. Samples of *Clostridium beijerinckii* biomass were taken from different phases of batch cultivations. Sample numbering is according to the degree of differentiation visible in the light microscope: *Sample 1* is least differentiated, *sample 4* most differentiated. *Samples 1* and *2* are from the acidogenic and early solventogenic phase, *samples 3* and *4* are from the middle and late solventogenic phase. *Sample 4 S* is a flocculating fraction of *sample 4*, with cells sticking together by a slime. *Inserted images*: Light micrographs from the growth phases of *Clostridium beijerinckii*. Bright field mode, iodine stain, video images. *Dark stained spots* are from storage polysaccharide (granulose). *Bars* 5 µm. *Top*: FT-Infrared spectra of *Clostridium beijerinckii* biomass in different phases of batch cultivations. The main differences between the spectra are in the carbohydrate (1076 cm^{-1}), the nucleic acid (around 1250 cm^{-1}), and the lipid region (around 2900 cm^{-1}). Spectra were normalised to the protein band (Amid I). *Bottom*: Statistical similarities of the biomass spectra according to a hierarchical cluster analysis (Calculation of the distance matrix with "scaling to 1st range" algorithm, frequency range 1800–700 cm^{-1}). Identical numbers represent individual preparations of the same biomass sample. Each sample was prepared several times to assess the variation in the method. It can be seen that the parallel preparations of one sample are always more similar to each other than to different samples. Correlating with the physiological states, samples from the early states in the cultivation (acidogenic and early solventogenic) form one cluster, and the samples from late stages form another. (Modified from: Vibrational Spectroscopy vol. 19, K.C. Schuster et al., 1999, p. 471–2, with permission from Elsevier Science.)

8.5
Chemical Pattern on the Single Cell Level by Confocal Raman Microscopy

From the results of chemical analysis and infrared spectroscopy it can be seen that the chemical composition of microbial cells may vary greatly over the duration of a process run. However, these methods are limited to the bulk of the population. Single-cell studies with IR spectroscopy are only possible down to a size of about 10 µm, as the resolution is limited by the wavelength of the light applied. So, IR microscopy is limited to the large cells of higher organisms.

The complementary method to gain chemical information from molecular vibrations is Raman spectroscopy. This technique works by exciting vibrations by means of light, and detecting the inelastically scattered light at slightly lower frequencies. As excitation light sources, laser beams in the visible or near – infrared range are used, thus the beam can be focused e.g. through the optical set-up of a microscope, and the detection of the scattered light can be limited to the focus by pinholes (confocal technique). By this technique, single living cells of higher organisms have been studied [85] and strains of bacteria have been identified directly from colonies on solid culture media [86], or using samples of a few cells from colonies by surface-enhanced Raman spectroscopy (SERS) [87].

Lately, spectra of single cells of *Clostridium beijerinckii* have been obtained (Fig. 8) [88, 89]. The aim was to monitor the population distribution in a sol-

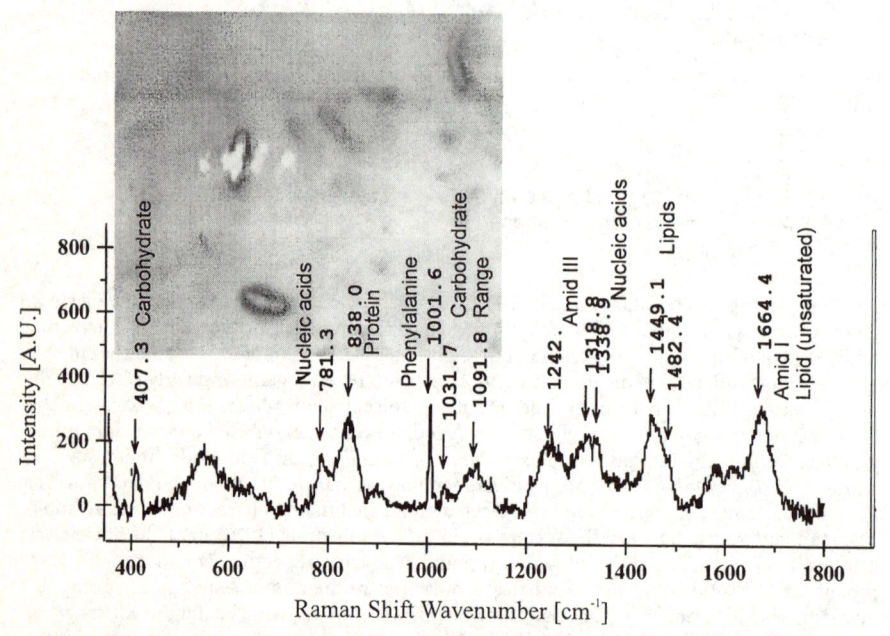

Fig. 8. Raman spectrum of a single cell of *Clostridium beijeinckii*, of a size of about 2×4 µm. Peaks are marked with the wavenumber of the Raman shift, and tentative attributions of the bands are given. *Insert*: Video image showing a *C. beijerinckii* cell in the focused laser beam of the Raman microscope. The diameter of the laser focus, which determines the sampling volume by the excitation of the Raman effect, is about the same size as the cell. (from [89])

vent-producing culture. Cell suspensions were dried on various glass-like surfaces. Raman spectra of single cells were obtained using a confocal Raman microscope (Dilor, France). The laser beam was focused on individual cells through the microscope objective (100× magnification), obtaining a spatial resolution of less than 2 µm in the horizontal plain. Raman spectra of single cells with good signal/noise ratio could be recorded with excitation in the visible range and using a calcium fluoride carrier. Cells of different morphology and cell samples from different development stages showed different spectra. These first results show that it is possible to study the chemical composition of single bacterial cells in the size of about 1 µm by confocal Raman spectroscopy. This can be used e.g. to study the heterogeneity in a microbial population.

9
Conclusions and Outlook

During the last decade, a wide range of approaches to gain information about the physiological status of a culture on the cellular level has been developed, and methods from other fields like microbial identification have been applied for this purpose. Especially the spectroscopic methods, yielding multidimensional data on the chemical composition of microbial cells, are only at the beginning of being exploited and have a bright prospect. The link with chemometrics will significantly improve the information yield.

It is expected that the systematic application of the discussed methods to bioprocesses will bring deeper understanding of the physiological events in the processes, and of the distinct phases that an organism experiences during a process, from inoculum preparation to final product recovery. Studies of culture inhomogeneity and dynamics of sub-populations in monoseptic cultures (by flow cytometry, image analysis, and some of the methods discussed here) should further improve our view of the cellular processes. Integration of the physiological information on the cellular level with non-invasive monitoring can lead to more efficient bioprocess control.

Acknowledgement. The author's work on infrared and Raman spectroscopy shown here was supported by the companies Bruker Optik (Karlsruhe, Germany, contact: Dr. Frank Mertens) and Dilor (Bensheim, Germany, contact: Dr. Eva Urlaub). Thanks are due to Doug Hopcroft, Hort Research New Zealand (Palmerston North, New Zealand) for preparing the specimen for the electron micrograph in Fig. 1. The author also acknowledges Florentina Dana's help with finalising the reference list for the manuscript.

References

1. Konstantinov KB (1996) Biotechnol Bioeng 52:271
2. Fiechter A, Sonnleitner B (1994) Adv Microbial Physiology 36:145
3. Schügerl K (vol ed) (1998) Relation between morphology and process performance. Advances in Biochemical Engineering, Biotechnology, vol. 60 (series ed. T. Scheper) Springer, Berlin Heidelberg New York

4. Shapiro HM. (1995) Practical flow cytometry, 3rd edn. Wiley-Liss, New York
5. Shapiro H, Srienc F (vol ed) (1999) J Microbiol Methods, special issue with contributions of the Int. Conference on Analysis of microbial cells on the single-cell level, Como, Italy, March 1999 (in preparation)
6. Pons MN, Vivier H (1998) Beyond filamentous species. In: Schügerl K (vol ed) Relation between morphology and process performance, Advances in Biochemical Engineering, Biotechnology, vol 60 (series ed. T. Scheper) Springer, Berlin Heidelberg New York, p 61
7. Schlegel HG (1992) Allgemeine Mikrobiologie. 7th edn. Georg Thieme Verlag, Stuttgart, for examples see pp 100–110
8. Jones DT, Woods DR (1986) Microbiol Rev 50:484
9. Dürre P, Bahl H (1996) Microbial production ofacetone/butanol/isopropanol. In: Rehm H-J, Reed G (eds) Biotechnology – a multi-volume comprehensive treatise, vol 6b, Verlag Chemie, Weinhein
10. Schuster KC, van den Heuvel R, Gutierrez NA, Maddox IS (1998) Appl Microbiol Biotechnol 49:669
11. Schuster KC (1994). Ph. D.Thesis, Vienna University of Technology, Vienna, Austria, p 73
12. Schuster KC (1995) unpublished electron micrograph
13. Mandalam RK, Palsson BO (1998) Biotechnol Bioeng 59:605
14. Rho D, Andre G (1991) Biotechnol and Bioeng 38:579
15. Gurakan T, Marison I, Stockar U, Gustafsson L, Gnaiger E (1990) Thermochimica Acta 172:251
16. Ames BN (1965) Assay of inorganic phosphate, total phosphate and phosphates. In: Neufeld EF, Ginsburg V (eds) Methods in Enzymology 8, Academic Press, New York, p 115
17. Kamnev AA, Renou-Gonnord MF, Antonyuk LP, Colina M, Chernyshev AV, Frolov I, Ignatov VV (1997) Biochem Mol Biol Int 41:123
18. Fontes AG, Moreno J, Vargas MA (1998) Biotechnol Bioeng 34:819
19. Henriksen CM, Christensen LH, Nielsen J, Villadsen J (1996) J Biotechnol 45:149
20. Herbert D, Phipps PJ, Strange RE (1971) Chemical analysis of microbial cells. In: Norris JR, Ribbons DW (eds) Methods in Microbiology 5B, Academic Press, London, UK, p 209
21. Cooney CL (1981) Growth of microorganisms. In: Rehm H-J, Reed G (eds) Biotechnology – a multi-volume comprehensive treatise, vol 1, Verlag Chemie, Weinhein, p 73
22. Grube M, Zagreba J, Gromozova E, M. Fomina M (1999) Vibrational Spectroscopy 19:301
23. Bley T, Ackerman J-U, Muller S, Loesche A, Babel W (1995) J Biotechnol 39:9
24. Sonnleitner B (1993) Experimental verification of rapid dynamics in biotechnological processes. In: Mortensen U, Noorman H (eds) Bioreactor Performance, Biotechnology Research Foundation, Lund, Sweden, p 143
25. Dubois M, Gilles KA, Hamilton JK, Rebers, Smith F (1956) Anal Chem 28:350
26. Stewart PR (1975) Analytical methods for yeast. In: Prescott DM (eds) Methods in cell biology, 12. Academic Press, New York, p 111
27. Benthin S, Nielsen J, Villadsen J (1991) Biotechnol Technol 5:39
28. Bradford MM (1976) Anal Biochem 73:248
29. Lowry OH, Rosebourgh NJ, Farr AL, Randall RJ (1951) J Biol Chem 193:265
30. Kimmich GA, Randles J, Brand JS (1975) Anal Biochem 69:187
31. Reimann A, Bigge F, Hall (1997) Direkte Bestimmung der Stoffwechselaktivität von Zellen: Die Luciferase-Reaktion mit automatisierten Biosensorsystemen (FH Furtwangen, Fand E Verfahrens- und Biotechnik, D-78054 Villingen-Schwenningen), Poster at ACHEMA fair and conference, Frankfurt/Main, Germany
32. Compagnone D, Guilbault GG (1997) Analytica Chimica Acta 340:109
33. Richter T, Biliteswki U (1997) A flow injection analysis system using a bi-enzyme sensor for the determination of ATP, (GBF, Braunschweig, D), 8th European Biotechnology Conference, Budapest, Hungary
34. Gikas P, Livingston A (1997) Biotechnol Bioeng 55:660
35. Srivastava AK, Volesky B (1991) Biotechnol Bioeng 38:181
36. Schuster KC, Pink T, Mayer HF, Hampel WA, Sára M (1997) J Biotechnol 54:15

37. Veech R (1987) Pyridine nucleotides and control of metabolic processes. In: Dolphin D, Avromovic O, Poulson R (eds) Pyridine nucleotide coenzymes. Wiley, New York, p 79
38. Grupe H, Gottschalk G (1992) Appl Environm Microbiol 58:3896
39. Zupke C, Sinskey A, Stephanopoulos G (1995) Appl Microbiol Biotechnol 44:27
40. Theobald U, Mohns J, Rizzi M (1996) Biotechnol Tech 10:297
41. Smith BJ, Wales MR, Perry MJ (1993) Appl Biochem Biotechnol 41:189
42. Pörtner R (1997) Festbettsysteme für die Kultivierung tierischer Zellen (TU Hamburg-Harburg, Bioprozeß- und Bioverfahrenstechnik, Hamburg, D), Poster at ACHEMA fair and conference, Frankfurt/Main, Germany
43. Matanguihan R, Konstantinov K, Yoshida T (1994) Bioprocess Engineering 11:213
44. Markx G, Kell D (1995) Biotechnology Progress 11:64
45. Woodward A, Kell D (1991) Bioelectrochemistry Bioenergetics 25:395
46. Gheorghiu E (1996) Bioelectrochemistry Bioenergetics 40:133
47. Crespel E, Altaba S, Vivier H, Pons MN, Latrille E, Corrieu G, Engasser JM (1998) Physiological characterisation of Streptomyces submerged cultures by image analysis and electrical capacitance (CNRS-ENSIC-INPL, Nancy, France), Poster at ESBES-2–2nd European Symposium on Biochemical Engineering Science, Porto, Portugal
48. Andersch W, Bahl H, Gottschalk G (1983) Europ J Appl Microbiol Biotechnol 18:327
49. Sauer U, Dürre P (1995) FEMS Microbiol Letters 125:115
50. Girbal L, Soucaille P (1998) Trends Biotechnol 16:11
51. Treuner-Lange A, Kuhn A, Dürre P (1997) J Bacteriol 179:4501
52. Schweder T, Krüger E, Xu B, Jürgen B, Blomsten G, Enfors S-O Biotechnol Bioeng (submitted)
53. VanBogelen RA, Neidhardt FC (1990) FEMS Microbiol Ecol 74:121
54. Hecker M, Schumann W, Volker U (1996) Molecular Microbiol 19:417
55. Terracciano JS, Rapaport E, Kashket ER (1988) Appl Environm Microbiol 54:1989
56. Pich A, Narberhaus F, Bahl H (1990) Appl Microbiol Biotechnol 33:697
57. Antelmann H, Bernhardt J, Schmid R, Mach H, Volker U, Hecker M (1997) Electrophoresis 18:1451
58. Schaffer S (1998) Ph D Thesis, University of Ulm, Ulm, Germany
59. Völker U, Engelmann S, Maul B, Riethdorf S, Völker A, Schmid R, Mach H, Hecker M (1994) Microbiology 140:741
60. Schweder T, Krüger E, Xu B, Jürgen B, Mosterz J, Enfors S-O, Hecker M. 1998. Monitoring of genes that respond to process related stress in large-scale fermentations, Poster, ESBES-2–2nd European Symposium on Biochemical Engineering Science, Porto, Portugal
61. Goodacre R, Kell DB (1996) Curr Opinion Biotechnol 7:20
62. Goodacre R, Timmins.ÉM., Rooney P, Rowland J, Kell D (1996) FEMS Microbiol Letters 140:233
63. Moss CW (1981) J Chromatography 203:337
64. Arneborg N, Salskov-Iversen A, Mathiasen T (1993) Appl Microbiol Biotechnol 39:353
65. Pramanik J, Keasling JD (1998) Biotechnol Bioeng 60:230
66. Langworthy TA, Pond JL (1986) Membranes and lipids of thermophiles. In: Brock TD (eds) Thermophiles: General, molecular and applied microbiology. Wiley Interscience, New York, p 107
67. Nordstrom KM (1992) Arch Microbiol 158:452
68. Laemmli UK (1970) Nature (London) 227:680
69. Pedersen S, Bloch PL, Reeh S, Neidhardt FC (1978) Cell 14:179
70. Sára M, Kuen B, Mayer HF, Mandl F, Schuster KC, Sleytr UB (1996) J Bacteriol 178:2208
71. Raidl M (1999) Masters thesis, Vienna University of Technology, Vienna, Austria
72. Raidl M, Schuster KC, Buchinger W, Gapes J, Hampel WA (1999) Proteinmuster von Clostridien zur Prozeßüberwachung in der Aceton-Butanol- ABE-Fermentation. Jahrestagung der Österreichischen Gesellschaft für Biotechnologie, Seggau, Austria
73. Kang SG, Kenyon RGW, Ward AC, Lee KJ (1998) J Biotechnol 62:1
74. Goodacre R, Kell DB (1993) Anal Chim Acta 279:17
75. Goodacre R, Karim A, Kaderbhai MA, Kell DB (1994) J Biotechnol 34:185

76. Goodacre, R (1998) World Wide Web-page: http://gepasi.dbs.aber.ac.uk/ROY/pyms-home.htm
77. Norris KP, Greenstreet JES (1958) J Gen Microbiol 19:566
78. Zagreba ED, Savenkov V, Ginovska M (1990) IR-spectrophotometric control of chemical composition of microbial biomass. In: Microbial conversion – fundamental aspects. Zinatne, Riga, Latvia, p 139
79. Savenkova L, Bonartseva G, Gertsberg Z, Beskina N, Zagreba J, Grube M (1994) Proc of the Latvian Academy of Sciences Section B, 562/563:89
80. Naumann D, Fialja V, Labischinski H, Giesbrecht P (1988) J Molecular Structure 174:165
81. Helm D, Naumann D (1995) FEMS Microbiol Letters 126, 75
82. Zeroual W, Choisy C, Doglia SM, Bobichon H, Angiboust J-F, Manfait M (1994) Biochimica Biophysica Acta 1222:171
83. Schuster KC, Mertens F, Gapes JR (1999) Vibrational Spectroscopy 19:467
84. Schuster KC, Mertens F, Grube M, Gapes JR (1998) Bioprocess monitoring by infrared spectroscopy of bacterial cells, 2nd Workshop "FT-IR Spectroscopy in Microbial and Medical Diagnostics," Berlin
85. Puppels GJ, deMul FFM, Otto C (1990) Nature 347:301
86. Puppels GJ (1999) Europ J Histochemistry 43/S1:35 (Abstract)
87. Sockalingum GD, Lamfarraj H, Pina P, Beljebbar A, Allouch P, Manfait M (1999) Europ J Histochemistry 43/S1:39 (Abstract)
88. Schuster KC, Urlaub E, Gapes JR (1999) Europ J Histochemistry 43/S1:37 (Abstract)
89. Schuster KC, Urlaub E, Gapes JR (1999) Analysis of single bacterial cells by confocal Raman cpectroscopy, Conference "Analysis of Microbial Cells at the Single-Cell Level," Como, Italy

Metabolic Network Analysis
A Powerful Tool in Metabolic Engineering

Bjarke Christensen, Jens Nielsen*

Center for Process Biotechnology, Department of Biotechnology, Building 223, Technical University of Denmark, DK-2800 Lingby, Denmark
* E-mail: jn@ibt.dtu.dk

Metabolic network analysis is a tool for investigating the features that identify the topology of a metabolic network and the relative activities of its individual branches. The pillars of metabolic network analysis are mathematical modeling, allowing for a quantitative analysis, biochemical knowledge of, for example, reaction stoichiometry, and the experimental techniques, providing input for the modeling part. The modeling part includes metabolite balancing, usually the basis for metabolic flux analysis, and isotope balancing. Isotope balancing can be used for both identification of active pathways and for estimation of the relative fluxes through two pathways leading to the same metabolite, aspects that are difficult to investigate using metabolite balancing. The combination of metabolite and isotope balancing is very powerful and constitutes the basis of metabolic network analysis. With the main focus being on investigating the metabolic network structure, this review describes how central metabolic features, for example, pathway identification, flux distribution, and compartmentation, can be addressed using a combination of metabolite balancing and labeling experiments.

Keywords. Metabolic flux analysis, Flux estimation, Compartmentation, Metabolic channeling, ^{13}C tracer experiments

1	Introduction	210
2	Metabolic Flux Analysis	210
2.1	Metabolite Balancing	210
2.2	Isotope Balancing	212
2.3	Mathematical Modeling	214
3	Metabolic Network Analysis	217
3.1	Pathway Identification	217
3.2	Metabolic Channeling	221
3.3	Compartmentation	223
3.4	Bidirectional Fluxes	225
4	Conclusions and Future Directions	229
	References	230

1
Introduction

A directed approach to optimization of cellular processes requires an in-depth understanding of the reactions supporting growth and product formation. Identification and quantification of metabolic reactions in an organism is therefore a central element in *Metabolic Engineering* [1, 2]. As metabolic reactions are subject to complex regulation, it is very likely that an attempt to direct the metabolism towards a product of interest will result in an unexpected metabolic response. While one part of the optimization process is focused on elucidation of regulation, another part deals with analyzing the metabolic response resulting from changes made to the biological system. It is important to realize that the metabolic response may manifest itself in both changes in the magnitudes of the metabolic fluxes and in changes in the network topology, and both fluxes and network structure therefore have to be investigated. Based on mass balances for all the metabolites in the network, the metabolic fluxes may be quantified from measurements of a few fluxes, e.g., the glucose uptake rate. This approach is normally referred to as *Metabolic Flux Analysis* or metabolite balancing and it is used for estimating net fluxes in a pre-defined metabolic network. However, when it comes to discriminating between pathways, e.g., in networks containing two different pathways leading to the same metabolite, metabolite balancing is often not adequate. Through the use of ^{13}C-labeled substrates and subsequent measurement of the labeling profile in intracellular metabolites, additional balances can be set up for estimation of the fluxes. Furthermore, the use of labeled substrates enables discrimination of different pathways leading to the same metabolite. The combination of metabolite balancing and labeling experiments enables a more detailed analysis that includes flux estimation, identification of the network topology, metabolic channeling, and subcellular compartmentation. It is therefore more appropriate to use the term *Metabolic Network Analysis* for this combination of techniques

This review aims at giving an overview of the potential and the versatility of the results that can be obtained using metabolite balancing and labeling analyses, and at describing the methodologies applied. Special emphasis is put on cases where a combination of the two techniques has been applied, but illustrative examples are also taken from literature where only one of the techniques was used. A detailed description of the mathematical modeling and techniques for measuring labeling patterns has been reported by Wiechert and de Graaf [3] and Szyperski [4].

2
Metabolic Flux Analysis

2.1
Metabolite Balancing

The flux distribution in the entire central metabolism can in some cases be obtained from calculations based on knowledge of the stoichiometry of the individual biochemical reactions and a set of measured fluxes [5, 6]. The fundamental topics of metabolite balancing such as system identifiability, flux estimates in over- and under-determined systems, and identification of gross measurement errors have already been covered in several papers and reviews, see e.g., [7, 8]. Here, the focus will be on how metabolite balancing can be exploited as one of the pillars in metabolic network analysis.

Besides the flux constraints that arise from metabolite balances over intermediates, i.e., the sum of fluxes into an intermediate has to equal the sum of fluxes leaving the intermediate, a number of constraints arise from balances over cofactors like NADH, NADPH, and ATP. These cofactors are all produced and consumed in several pathways, and the balances over these metabolites therefore impose constraints on the activities of pathways that are not necessarily related. The NADPH and NADH balances, in particular, have been used for making a link between catabolic reactions, biomass forming reactions, and product formation [9, 10]. Based on these balances Nissen et al. [10] were able to describe and, more importantly, to improve the yield of ethanol on glucose in anaerobic cultivations of *Saccharomyces cerevisiae*. The improvement was obtained by reducing the production of glycerol and thereby increasing the fraction of glucose carbon atoms available for ethanol production. The conceptually important point in this work is the understanding of the role of glycerol, which is produced as a consequence of the need for oxidation of NADH produced in the biomass-forming reactions. The ethanol biosynthesis from glucose is redox-neutral (Fig. 1), and the reason for the need for an NADH-oxidizing reaction is that the biomass-producing reactions give a net contribution of NADH which has to be oxidized by other reactions, e.g., the reactions in the glycerol pathway.

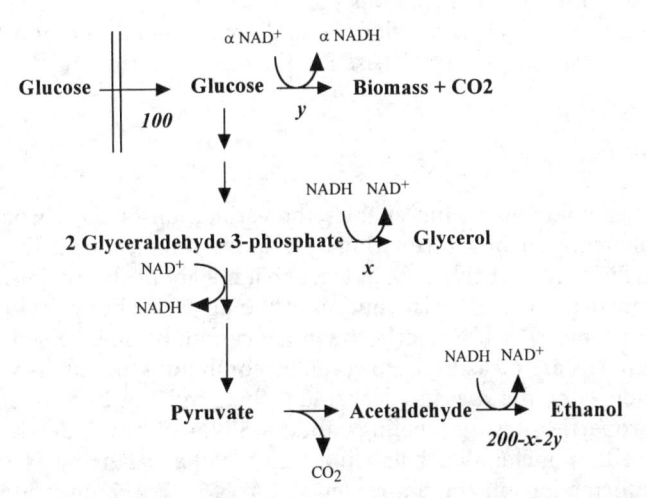

Fig. 1. Redox metabolism in *Saccharomyces cerevisiae* during anaerobic growth on glucose. The ethanol yield is lowered by the production of biomass and glycerol. The glycerol flux, x, can be decreased, and the ethanol yield thereby increased if the stoichiometric coefficient α for biomass formation is reduced, e.g., by having nitrogen assimilation via an NADH-dependent glutamate dehydrogenase [10]

By replacing the NADH-dependent glutamate dehydrogenase, which is the major nitrogen assimilation route in *S. cerevisiae*, the NADH formation associated with biomass synthesis was reduced, i.e., the stoichiometric coefficient α in Fig. 1 was reduced.

Like the examples mentioned above, most examples of metabolic flux analysis by metabolite balancing have redox balances as a central constraint used in the determination of the flux distribution. However, the redox balance is, especially under aerobic conditions, subject to uncertainties which make it less suitable for estimation of the fluxes. Part of the reason for this is to be found in futile cycles, e.g., oxidation of sulfides to disulfides, where reductive power is needed to reduce the disulfides. The net result of this reaction is reduction of molecular oxygen to water, and oxidation of NADPH to $NADP^+$. Since the consumption rate of oxygen of these specific reactions is impossible to measure, the result may be that the NADPH consumption is underestimated. This is in accordance with the finding that when the NADPH-producing reactions are estimated independently of the NADPH-consuming reactions, there is usually a large excess of NADPH that needs to be oxidized by reactions not included in the network, e.g., futile cycles [11–13].

Reactions where the net reaction is hydrolysis of ATP represent another example of reactions that are difficult to investigate using metabolite balancing. It is well known that the consumption of ATP needed for the biomass- and product-forming reactions cannot account for the entire production of ATP by the catabolic reactions, and an ATPase reaction therefore has to be included in the metabolic network [9]. Because there are many combinations of reactions that have ATP hydrolysis as the net result, the ATP balance cannot be used for estimating fluxes in single futile cycles. Since both the redox and the energy balances are central to the metabolism, it is of interest to study these balances independently of the carbon balances, and this can only be done by using isotope-labeled substrates as discussed in the following section.

2.2
Isotope Balancing

Whereas metabolite balancing enables the estimation of the absolute fluxes, isotope balancing enables estimation of relative fluxes, as illustrated by the example in Fig. 2. It can be seen from Fig. 2 that metabolite balancing alone does not give information on the distribution of the fluxes in the system if only one of the fluxes is measured. Similarly, the fluxes cannot be determined if only the labeling patterns are measured. However, by combining the extensive property of the system, i.e., the measured absolute flux, and the information on the intensive properties, i.e., the labeling patterns, all fluxes can be determined. The fact that neither metabolite balancing nor isotope balancing is capable of yielding information on the entire set of fluxes in Fig. 2 illustrates that information obtained using the two techniques is complementary.

Isotope balancing can be carried out at two levels, atom-wise or molecule-wise. In the case of isotope balancing on the level of carbon atoms, fractional enrichments are used for estimating the fluxes. Thus, the sum of labeled carbon

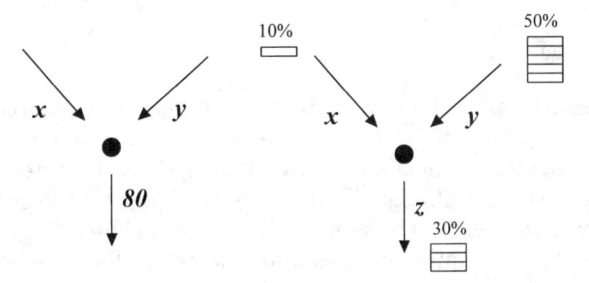

Fig. 2. Principle of metabolite (*left*) and isotope balances (*right*). Two reactions, with fluxes x and y, lead to the intermediate which is produced with a flux z, that can be measured. The enrichments of the metabolites are indicated by the boxes and the values given in percent. Using the isotope balance, x and y are determined as being of equal magnitude. Using the metabolite balance, the absolute magnitudes of x and y are determined to be 40

atoms entering a given position of a metabolite has to equal the sum of labeled carbon atoms leaving this position. This type of balance is similar to the metabolite balances, the only difference being that the fluxes in and out of a carbon position in a metabolite are now weighed with respect to the fractional enrichments of the carbon atoms involved.

The picture becomes more complicated when isotope balancing is carried out on a molecular basis. Instead of looking at the fraction of labeled carbon atoms in a given position of a metabolite, the fraction of the molecules having a certain combination of labeled carbon atoms is considered. Thus, in a C_4-molecule there are $2^4 = 16$ combinations, or positional isotopomers, which are to be compared with a total of only four fractional enrichments of the carbon positions in the metabolite. The positional isotopomer distribution contains the complete information on the labeling pattern of a metabolite, and fractional enrichments can therefore be calculated from the positional isotopomer distribution by adding the abundances of the appropriate positional isotopomers. Similarly, all other types of measurements of labeling property can be calculated from the positional isotopomer distribution. This is an important feature of mathematical models based on positional isotopomers since all types of labeling information, e.g., mass isotopomers and NMR coupling patterns, can be used for the flux estimations. Moreover, giving a complete description of the labeling pattern of a metabolite, positional isotopomer-based analyses are likely to give the most detailed description of the metabolism and the best estimates of the fluxes. However, besides the complexity arising from the increase in variables compared with fractional enrichments, the use of positional isotopomers is particularly complicated in bimolecular reactions, as the positional isotopomer generated by such a reaction is a function of the positional isotopomer distribution of both substrates. Due to the complexity, the use of positional isotopomers has primarily been limited to studies where primarily unimolecular reactions are encountered, e.g., Krebs cycle with associated reactions.

2.3
Mathematical Modeling

A distinct feature of metabolic network analysis, as compared with classical labeling experiments, is the mathematical modeling approach. The mathematical method in metabolic network analysis is characterized by being general in a way that all metabolic features, e.g., reversible fluxes, compartmentation and metabolic channeling, can be included [14]. At the same time the modeling is readily amenable to changes in the metabolic network. Martin et al. [15] used 81 equations for deducing the fluxes in five cycles involved in the glutamate metabolism. While their results are very valuable from a biological point of view, the complexity of their calculations makes it difficult to modify the equations if the metabolic network has to be slightly modified. It should be mentioned that the modeling carried out by Martin et al. [15] did not require that steady state was obtained, which is normally the case [13, 14, 16], leading to more complex equations. In the following, modeling of isotopic steady state data will be considered, as the steady state condition simplifies the modeling significantly. The methodology is therefore readily applicable to chemostat cultivations, where steady state is achieved with a labeled substrate, since this is a special case where all metabolites, whether or not they are directly included in the model, are in isotopic steady state. However, the modeling also applies to situations where only the metabolites included in the model are in a steady state, thereby neglecting the labeling state of, e.g., the macromolecules.

As the number of equations needed for isotope balances is large, it is important to use a condensed notation to describe the carbon transitions in a reaction. In the case of isotope balancing based on fractional enrichments, the balances can be expressed using so-called atom-mapping matrices [17]. The principle of using atom-mapping matrices is illustrated by considering an example where the carbon transitions in the first reactions in the Krebs cycle are shown (Fig. 3).

Citrate is an intermediate in this pathway, and it is therefore possible to set up metabolite and isotope balances over citrate. Since citrate is a C_6-molecule,

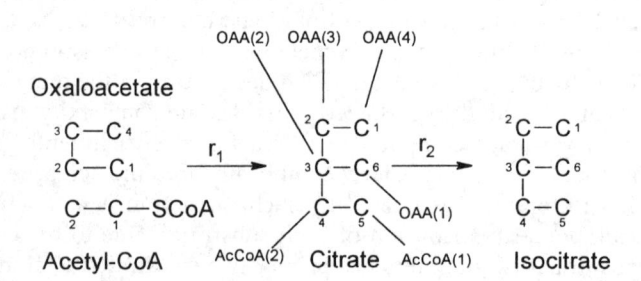

Fig. 3. Condensation of acetyl-CoA and oxaloacetate to citrate and isomerization of citrate into isocitrate. The numbers next to the carbon atoms indicate the enumeration of the metabolite carbon atoms. *OAA* oxaloacetate; *AcCoA* acetyl-coenzyme A

there are a total of six equations deriving from isotope balances over the carbon atoms:

$$r_1 x_{OAA(4)} - r_2 x_{Citrate(1)} = 0$$

$$r_1 x_{OAA(3)} - r_2 x_{Citrate(2)} = 0$$

$$r_1 x_{OAA(2)} - r_2 x_{Citrate(3)} = 0 \qquad (1)$$

$$r_1 x_{AcCoA(2)} - r_2 x_{Citrate(4)} = 0$$

$$r_1 x_{AcCoA(1)} - r_2 x_{Citrate(5)} = 0$$

$$r_1 x_{OAA(1)} - r_2 x_{Citrate(6)} = 0$$

where r_i is the reaction rate of the i'th reaction and $x_{\text{metabolite}(y)}$ is the fractional enrichment of the y'th carbon atom in the *metabolite*.

This set of equations is a bilinear system which is not as simple as the linear system of metabolite balancing, but neither as complex as a completely nonlinear system. The equation system can be written in a matrix form, giving only one matrix equation:

$$r_1 \begin{bmatrix} 0 & 0 \\ 0 & 0 \\ 0 & 0 \\ 0 & 1 \\ 1 & 0 \\ 0 & 0 \end{bmatrix} \begin{bmatrix} x_{AcCoA(1)} \\ x_{AcCoA(2)} \end{bmatrix} + r_1 \begin{bmatrix} 0 & 0 & 0 & 1 \\ 0 & 0 & 1 & 0 \\ 0 & 1 & 0 & 0 \\ 0 & 0 & 0 & 0 \\ 0 & 0 & 0 & 0 \\ 1 & 0 & 0 & 0 \end{bmatrix} \begin{bmatrix} x_{OAA(1)} \\ x_{OAA(2)} \\ x_{OAA(3)} \\ x_{OAA(4)} \end{bmatrix} - r_2 \begin{bmatrix} 1 & 0 & 0 & 0 & 0 & 0 \\ 0 & 1 & 0 & 0 & 0 & 0 \\ 0 & 0 & 1 & 0 & 0 & 0 \\ 0 & 0 & 0 & 1 & 0 & 0 \\ 0 & 0 & 0 & 0 & 1 & 0 \\ 0 & 0 & 0 & 0 & 0 & 1 \end{bmatrix} \begin{bmatrix} x_{Citrate(1)} \\ x_{Citrate(2)} \\ x_{Citrate(3)} \\ x_{Citrate(4)} \\ x_{Citrate(5)} \\ x_{Citrate(6)} \end{bmatrix} = \begin{bmatrix} 0 \\ 0 \\ 0 \\ 0 \\ 0 \\ 0 \end{bmatrix}$$

(2)

It is important to note that if all the reaction rates are known, the equation system becomes linear, and equation systems from several reactions can be combined to give a linear system. In this case, an analytical solution to the fractional enrichments can be found. This result is valid for all systems containing any number of reactions. Similarly, if all the fractional enrichments are known, the system becomes linear, but it is not certain that there is a unique solution for the reaction rates. At present, it is quite hypothetical to assume that all fractional enrichments are measured, and estimates of the complete set of reaction rates are therefore, in general, not accessible using analytical expressions. Instead, an iterative approach is used, where a guess is made of the fluxes and the corresponding set of fractional enrichments is calculated. The fractional enrichments are then compared with the measured values and the deviation is calculated. Methods for minimizing the error for a general metabolic network have been described in recent papers and reviews [3, 14, 18, 19].

As mentioned above, the calculations based on positional isotopomers are more complicated than the calculations based on fractional enrichments. The reason for this is to be found in the nonlinear structure of the equations as a consequence of some metabolites being composed of carbon atoms from several molecules. If the reactions in Fig. 3 are considered again, the positional isotopomer equations arising from the isotope balancing over citrate give rise to $2^6 = 64$ equations that all contain terms that are nonlinear in the isotopomer abundances:

$$r_1 I_{AcCoA(00)} I_{OAA(0000)} - r_2 I_{Citrate(000000)} = 0$$

$$r_1 I_{AcCoA(01)} I_{OAA(0000)} - r_2 I_{Citrate(000100)} = 0$$

$$r_1 I_{AcCoA(10)} I_{OAA(0000)} - r_2 I_{Citrate(000010)} = 0 \qquad (3)$$

$$r_1 I_{AcCoA(11)} I_{OAA(0000)} - r_2 I_{Citrate(000110)} = 0$$

$$r_1 I_{AcCoA(00)} I_{OAA(0001)} - r_2 I_{Citrate(100000)} = 0$$

$$\vdots$$

$$r_1 I_{AcCoA(11)} I_{OAA(1111)} - r_2 I_{Citrate(111111)} = 0$$

where, e.g., $I_{AcCoA(00)}$ is the fraction of the acetyl-CoA molecules that are unlabeled in both positions, and $I_{Citrate(000110)}$ is the fraction of the citrate molecules that are labeled in positions 4 and 5.

Even if all the reaction rates are known, the system is still not linear. As the number of reactions in the system increases, an analytical solution to the flux distribution becomes difficult to obtain, perhaps even impossible. Furthermore, the iterative approach also becomes difficult because the isotopomers are not linear functions of the isotopomer abundances. A rapid-equilibrium approximation has been suggested as a way of simplifying the mathematical model. The approximation applies whenever a reaction is near equilibrium, and states that the substrates and products of such reactions can be pooled in a single pool [20]. An example of this is the pentose phosphates in the pentose phosphate pathway, which are assumed to be in near equilibrium and these metabolites are therefore combined in a single metabolite pool [21]. Isotopomer-mapping matrices have been introduced to facilitate the calculations [22, 23]. Matrix-based methods have been applied for flux determination of the metabolism of *Escherichia coli* [24] and in various metabolic cycles [25, 26]. Even though modeling using isotopomers is very complex (Table 1) and, therefore, more time consuming than the fractional enrichment based method, positional isotopomer analysis opens up the possibility for all types of labeling input to be used in the flux estimation, and this type of analysis is therefore the method with the most potential.

Table 1. Mathematical structure of metabolite and isotope balances

Balance	Mathematical structure
Metabolite balance	Linear
Fractional enrichment balance	Bilinear
Positional isotopomer balance	Nonlinear

3
Metabolic Network Analysis

Metabolic networks contain a number of interesting features that can be investigated using a combination of metabolite balancing and labeling experiments. Pathway identification naturally occupies a central position among these features, but other interesting aspects, including compartmentation, futile cycles, and metabolic channeling, may also play a significant role in the overall metabolic picture. The following examples are primarily taken from studies where these features were investigated separately. However, it is important to keep in mind that the integrated approach, i.e., metabolite balancing over the entire central metabolism combined with data from labeling experiments, is in many cases capable of investigating several or all of these features based on a single experiment.

3.1.
Pathway Identification

A quantitative description of the fluxes through the individual branches in a metabolic network relies on knowledge of the pathways that are active in the primary metabolism. Labeling experiments give a unique possibility for determining the set of pathways that are active in vivo. While several other techniques, such as genome analysis, mRNA measurements, and in vitro enzyme assays, are able to give information on the possible activity of a pathway, metabolite balancing and labeling experiments, preferably in combination, are the only tools that allow investigation of actual in vivo activities.

Partly because of the high information content of the labeling pattern of glutamate, and partly because of the ease of analysis due to the high intracellular concentration of glutamate in most organisms, the labeling pattern of glutamate is often used for investigating the activities of various central pathways. Glutamate is formed from 2-ketoglutarate which, in turn, is derived indirectly from oxaloacetate and pyruvate. Glutamate therefore contains information on both the labeling pattern of pyruvate, giving information on the activities of the glucose-degrading pathways: Embden-Meyerhof-Parnas pathway, Entner-Doudoroff pathway, and the pentose phosphate pathway, and on the labeling pattern of oxaloacetate, which is involved in, e.g., the anaplerotic reactions and the TCA cycle. Clearly, the glutamate-labeling pattern may yield valuable information on many of the central pathways in the metabolism. This is illustrated by a study of the glucose-degrading pathways in *Corynebacterium melassecola*, where the labeling pattern of glutamate was studied when the microorganism was grown on $1-^{13}C$-glucose and $6-^{13}C$-glucose, respectively [27]. Measurements of the fractional enrichments of C1, C2, C3, and C4 in glutamate lead to a quantification of the relative activities of the Embden-Meyerhof-Parnas pathway and the pentose phosphate pathway, and the relative activities of the Krebs cycle, the glyoxylate shunt, and the anaplerotic reactions yielding C_4-units by carboxylation of C_3-units [27].

There are, however, cases where the glutamate-labeling pattern does not contain sufficient information for identification of the active pathways. One such example is the anaplerotic reactions generating oxaloacetate by carboxylation of phosphoenolpyruvate catalyzed by phosphoenolpyruvate carboxylase, and by carboxylation of pyruvate catalyzed by pyruvate carboxylase. When labeled glucose is used, the labeling patterns of the oxaloacetate formed in these reactions are indistinguishable. The anaplerotic reactions in *C. glutamicum*, in particular, have been studied with some interest because these reactions lead to the oxaloacetate needed for production of amino acids derived from Krebs cycle intermediates, in particular lysine and glutamate. Phosphoenolpyruvate carboxylase is present in *C. glutamicum*; however, the study of phosphoenolpyruvate-negative mutants gave the surprising result that lysine production was independent of the presence of phosphoenolpyruvate carboxylase [28].

The potential candidates for the generation of C_4-units in the Krebs cycle are: phosphoenolpyruvate carboxykinase, the glyoxylate shunt, and pyruvate carboxylase. The first two possibilities are unlikely since these activities are normally not found when glucose is present in the medium. Attempts to find activity of pyruvate carboxylase were not successful. Thus, none of these reactions qualified for being responsible for the anaplerotic reactions. Using $H^{13}CO_3^-$ in a minimal medium with glucose as the sole carbon source, it was shown that threonine was highly labeled in C-4 and, to a much lesser extent, in C-1 [29]. Threonine is derived from oxaloacetate and has therefore the same labeling pattern as oxaloacetate. Since the Krebs cycle produces symmetrically labeled oxaloacetate, the labeling pattern could only be explained if the anaplerotic reaction forming oxaloacetate was taking place by carboxylation of a C_3-unit [29]. Using gene-disruption techniques, the possibility of phosphoenolpyruvate carboxylase was excluded. This result, in combination with the labeling experiments, therefore pointed towards pyruvate carboxylase as the enzyme responsible for the formation of oxaloacetate, despite the fact that no in vitro activity of pyruvate carboxylase was found. It was later shown that pyruvate carboxylase is present in *C. glutamicum* [30], and the gene coding for the enzyme was sequenced [31].

Labeling experiments are particularly useful in cases where pathways having the same substrates and the same products are considered. Especially when metabolic studies are undertaken of microorganisms which are not well described, labeling experiments serve as a valuable tool for investigating the primary metabolism. An example of this is the degradation pathways of glucose in *Helicobacter pylori* which were identified using ^{13}C-labeled glucose as taking place via the Entner-Doudoroff pathway [32]. Labeling experiments provide an easy way of distinguishing between the Entner-Doudoroff pathway and the Embden-Meyerhof-Parnas pathway since the labeling patterns of pyruvate produced by the two pathways are completely different (Fig. 4). The same method was applied in a study of hyperthermophilic archaea, where it was shown that the Entner-Doudoroff pathway and the Embden-Meyerhof-Parnas pathway were active at the same time [33]. The labeling experiments allowed for a quantitative determination of the relative in vivo activities of the two pathways.

According to the reaction scheme shown in Fig. 4, pyruvate and, therefore, alanine and lactate, formed from degradation of 1-^{13}C-glucose should only be

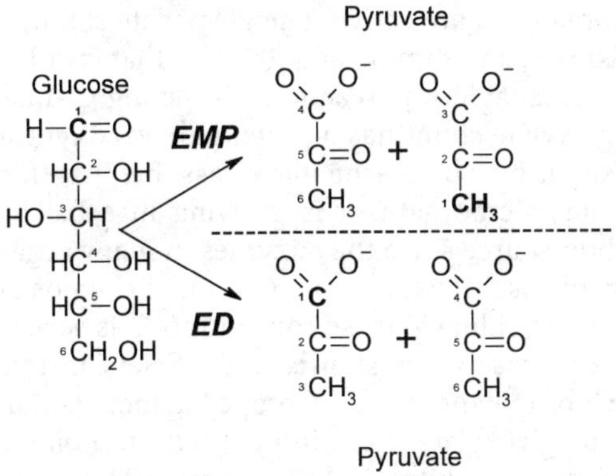

Fig. 4. Carbon transitions in the Embden-Meyerhof-Parnas (EMP) and the Entner-Doudoroff (ED) pathway. In the EMP pathway, C1-labeled glucose is converted into 50% unlabeled and 50% C3-labeled pyruvate, while the ED pathway produces 50% unlabeled and 50% C1-labeled pyruvate

labeled in positions 1 and 3 if the Entner-Doudoroff and Embden-Meyerhof-Parnas pathways are the only pathways producing pyruvate. In the study of the glucose degradation by the archaeon *Thermoproteus tenax*, position 2 in alanine also turned out to carry a significant labeling that could not be explained by the Entner-Doudoroff or the Embden-Meyerhof-Parnas pathway [34]. A possible explanation for this could be activity of the pentose phosphate pathway. However, since the reasons for the enrichment in C2 is not clear, a correction for the influence of this pathway on the enrichments of C1 and C3 cannot be made, and the estimates of the relative activities of the Entner-Doudoroff and the Embden-Meyerhof-Parnas pathways are therefore subject to some uncertainty. Even though the qualitative finding of activities of the two glucose-degrading pathways is indisputable, a more complete picture of the glucose degradation could be obtained by measuring the labeling patterns of more metabolites that may be able to explain the enrichment of C2. The advantage of analyzing the labeling patterns of several metabolites is illustrated by a study of the propionate metabolism in *Escherichia coli* [35]. Analyzing the labeling patterns of alanine, glutamate, and aspartate in the biomass of *E. coli* grown on labeled propionate as the sole carbon source, the major anabolic reaction producing C_4-units for the Krebs cycle was shown to be the glyoxylate shunt rather than the phosphoenolpyruvate carboxylase catalyzed reaction [35]. Furthermore, by studying the labeling patterns, it was possible to reduce the number of possible propionate-degrading pathways from seven to only two. This result is remarkable because the products of six of the seven pathways were either pyruvate or acetyl-CoA, rendering analysis of the pathways using metabolite balancing very difficult or even impossible.

The examples above primarily consider the identification of the pathways that are available for a given microorganism. An equally interesting study is the investigation of the conditions under which the pathways, already identified in the studied microorganism, are active. The glyoxylate shunt, which is found in most microorganisms, is an example of a pathway that is only active under certain conditions. As an anaplerotic reaction furnishing C_4-intermediates to the Krebs cycle, the glyoxylate shunt has a potentially very central position in the primary metabolism. It may be reasonable to assume that the glyoxylate shunt is inactive when the microorganism is growing in a batch cultivation with glucose as the carbon source since the enzymes in many organisms are known to be repressed by glucose. However, in a C-limited chemostat with glucose as the carbon source, the residual glucose concentration is very low, and it may be speculated that the glucose repression is less efficient in this case. Using the combined approach of metabolite and isotope balancing, Marx et al. [21] were able to show that the glyoxylate shunt in *Corynebacterium glutamicum* played only a minor role when *C. glutamicum* was grown in a C-limited chemostat. When *Saccharomyces cerevisiae* is grown on acetate the glyoxylate shunt is an important metabolic pathway which can be quantified, relative to the Krebs cycle flux, using labeling experiments [36, 37].

The two pathways responsible for lysine production by *Corynebacterium glutamicum* (Fig. 5) provide another example of pathways that change in relative activities depending on the growth conditions. Using labeled glucose, Sonntag et al. [38] were able to give precise estimates of the relative activities of the two

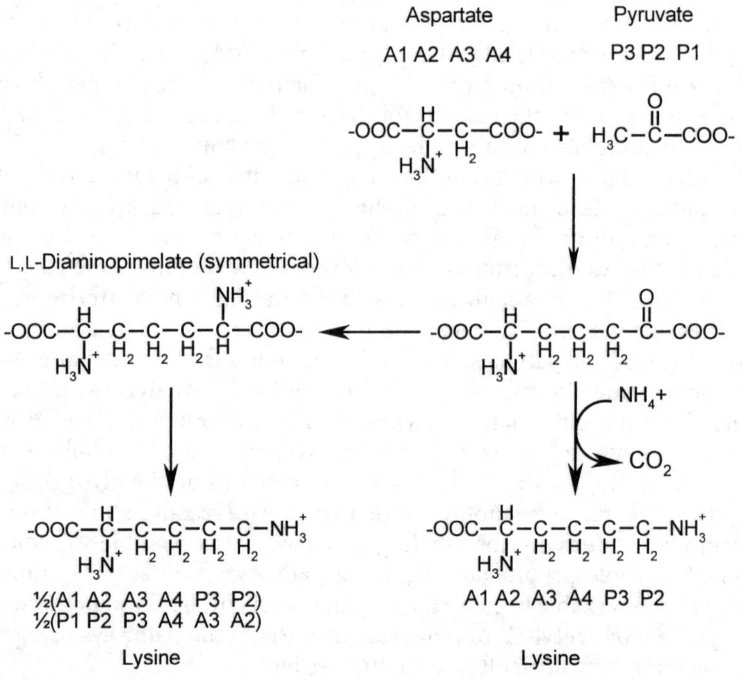

Fig. 5. Carbon transitions in the split pathway of lysine biosynthesis in *Corynebacterium glutamicum*. Not all the reaction steps or all the metabolites are shown. A_1, A_2, A_3, and A_4 refer to the carbon atoms C1–C4 of aspartate and P_1, P_2, and P_3 refer to C1–C3 of pyruvate

pathways leading from the common precursors, aspartate and pyruvate, to lysine. Comparing the relative activities of these pathways with the ammonium content of the extracellular medium showed a clear correlation indicating that the pathway containing diaminopimelate dehydrogenase, which has ammonium as a substrate, only takes place if the extracellular concentration of ammonium is above approximately 38 mM. The calculations of the flux ratio were based on measurements of the labeling pattern of both alanine, reflecting the labeling pattern of pyruvate, a substrate common to both pathways, and the labeling pattern of lysine, the product of both pathways. In this way no assumptions on the labeling patterns of the substrates, aspartate and pyruvate, had to be made, which is important because prediction of these labeling patterns is quite complicated.

3.2
Metabolic Channeling

The estimation of the relative activities of the lysine biosynthetic pathways relies on the assumption that the chemically symmetrical molecule, L,L-diaminopimelate, formed in one of the two pathways, is allowed to rotate freely after its formation, and that the next enzyme in the pathway therefore cannot distinguish between the two ends of the molecule. There is, however, some evidence supporting the hypothesis that symmetrical molecules may be channeled from one enzyme to the other giving rise to an orientation-conserved transfer of the symmetrical molecule. This is exemplified by the symmetric Krebs cycle intermediates, succinate and fumarate [39–43]. If metabolic channeling does take place, it will have a significant effect on the labeling patterns of the involved metabolites, which has to be taken into consideration when labeling analysis is used for pathway identification. Metabolic channeling is not only interesting because of its implications on the pathway identification, discussed above, but also because it is an example of enzyme–enzyme associations that, in the special cases of symmetrical molecules, can be investigated using labeling experiments.

Much of the evidence supporting the metabolic channeling model is based on analysis of the labeling pattern of pyruvate resulting from degradation of labeled propionate [40–42, 44]. The rationale, in the case of propionate metabolism, is that propionate is carboxylated to methylmalonyl-CoA which, in turn, is isomerized to succinyl-CoA (Fig. 6). Conversion of succinyl-CoA to succinate and subsequently to fumarate, malate and oxaloacetate, is expected to give symmetrically labeled oxaloacetate, since fumarate and succinate are symmetrical molecules. Pyruvate formed from decarboxylation of oxaloacetate is therefore expected to reflect the labeling pattern of a symmetrically labeled oxaloacetate molecule. The labeling pattern of alanine and, thereby, the labeling pattern of pyruvate, from a culture of *Saccharomyces cerevisiae* grown on 3-^{13}C propionate, showed a significantly higher ^{13}C-enrichment in C-2 than in C-3 which is not in accordance with scrambling in the Krebs cycle [42]. This result was interpreted as a consequence of metabolic channeling carried out by an enzyme complex, a so-called metabolon, consisting of succinate thiokinase, succinate

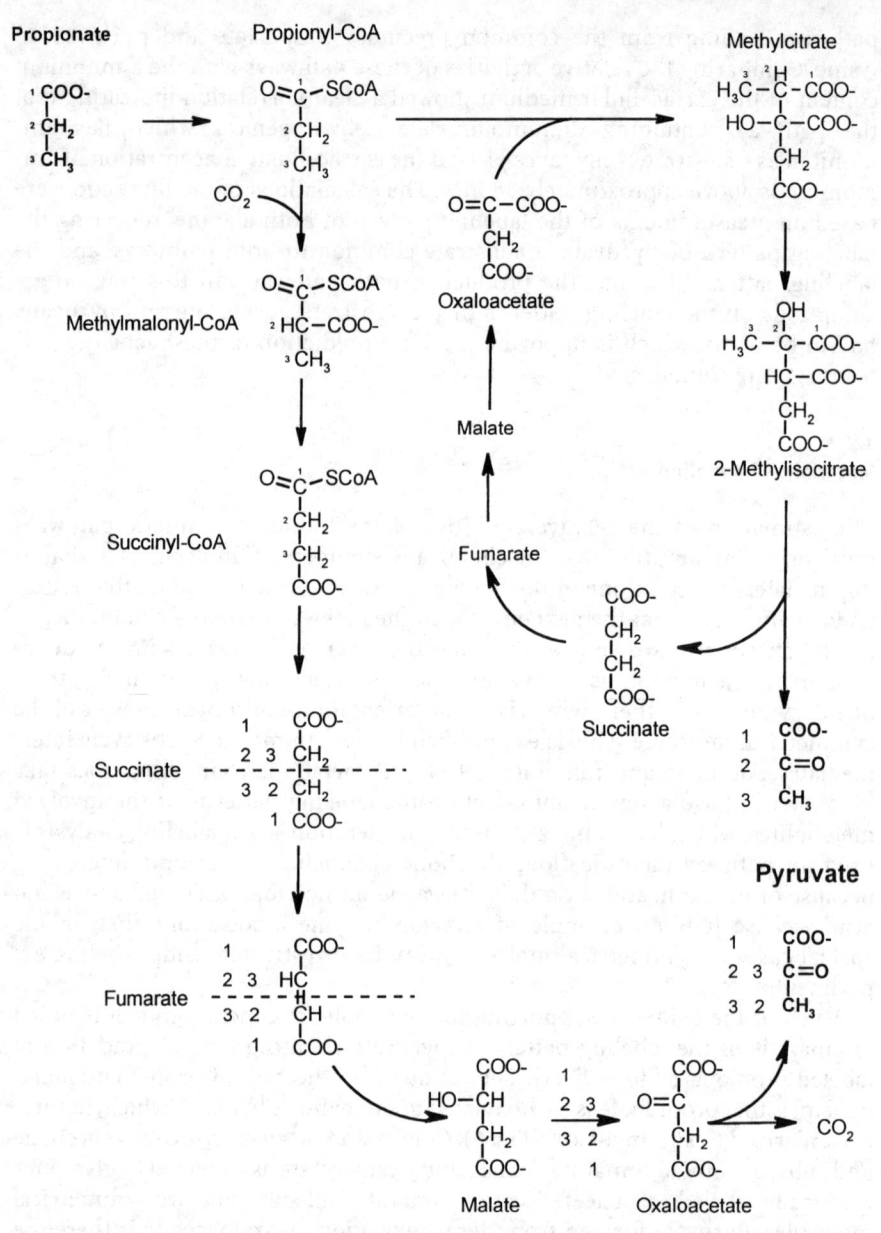

Fig. 6. Two pathways leading from propionate to pyruvate. The methyl citrate pathway gives retention of the carbon positions. The methylmalonyl-CoA pathway proceeds via the symmetrical intermediates succinate and fumarate, and the carbon atoms in pyruvate therefore reflect the scrambling that may take place as a consequence of the symmetry of these intermediates. If metabolic channeling takes place, only one of the two possible labeling patterns of malate is formed

dehydrogenase, and fumarase. The metabolic channeling gives rise to the orientation-conserved transfer of the carbon atoms in the asymmetrical succinyl-CoA via symmetrical intermediates, succinate and fumarate, to the asymmetrical molecule of malate.

However, an alternative explanation is that the propionate metabolism occurs via the 2-methyl citrate pathway, which gives rise to pyruvate with the observed retention of the carbon positions from propionate (Fig. 6) [45]. Enzyme assays on a chemostat cultivation of *Saccharomyces cerevisiae*, grown on 3-^{13}C propionate, showed high activities of the enzymes of the 2-methyl citrate pathway and negligible activity of propionyl-CoA carboxylase, the first enzyme in the methylmalonyl-CoA pathway [45]. Furthermore, the glutamate-labeling patterns arising from experiments where *S. cerevisiae* metabolized labeled pyruvate have been modeled without having to include metabolic channeling [36, 37]. These results support the assumption of scrambling at the stage of succinate and fumarate, i.e., no metabolic channeling takes place in the conversion of these metabolites. Observations of unexpected labeling patterns in other cell types include the studies of mammalian tissue [39–41]. In the study of C6 glioma cells [39], the orientation of the metabolic channeling was opposite to the one observed for *S. cerevisiae* [44] and other mammalian cell types [40, 41], indicating that the orientation of metabolic channeling is not necessarily a conserved trait in all cells.

It is, however, difficult to exclude the possibility of alternative reactions in these studies. In the case of the C6 glioma cells, it is suggested that oxaloacetate produced by pyruvate carboxylase may explain part of the observed labeling pattern [39]. In the case of metabolic channeling in connection with the propionate metabolism in mammalian cells [40, 41], there is a large number of possible pathways [35] that may support an alternative explanation of the observed labeling patterns. However, the existence of metabolic channeling cannot be excluded on this basis. Moreover, it is important to note that whatever reason there is for the observed labeling patterns, there has to be an explanation, and labeling analysis is a tool for suggesting possible explanations. It should also be noted that even though evidence of metabolic channeling supports the concept of complexes of functionally related enzymes, metabolons [46], the lack of evidence for metabolic channeling is not evidence against the concept of metabolons.

3.3
Compartmentation

The discussion of metabolic channeling in the examples above has focused on new pathways as alternative explanations for the results. However, the complexity of metabolic networks offers a variety of different features that influence the labeling patterns. Theories based on the existence of alternative pathways do not give a satisfactory explanation for the reports of time-dependent labeling asymmetry, attributed to metabolic channeling, which has been observed in both yeast and mammalian tissue [40, 43]. In these cases one has to consider other structures in the metabolic network, e.g., compartmentation of the pathways, that can cause a delay in the labeling of one metabolite relative to another metabolite, given that the metabolites have the same precursor. Both compartmentation and metabolic channeling aspects were considered in a study on various mutants of *Saccharomyces cerevisiae* [47], where the existence of two compartments in the mitochondria was suggested. One of the mutants in the study had a phenotype that made it unable to produce serine and one-carbon tetrahydrofolate in the cytosol. Thus, the only possible way to produce one-carbon units for biosynthesis was cleavage of glycine into carbon dioxide, ammonia, and one-carbon tetrahydrofolate, a reaction known to take place exclusively in the mitochondria.

The results showed that the glycine and one-carbon units used for biosynthesis of serine were taken from a different pool than the glycine and one-carbon units used for biosynthesis of choline. Moreover, as the fractional enrichment of the carbon atoms in C1 and C2 of choline, both derived from C2 of glycine, was 91%, and the fractional enrichment of C2 and C3 of serine, also both derived from C2 of glycine, was 66%, it appears that the one-carbon units formed from glycine are channeled directly to glycine molecules from the same pool (Fig. 7). Based on these findings, it was concluded that there may be two mitochondrial pools of glycine, serine, and one-carbon tetrahydrofolate in *S. cerevisiae* [47]. However, there are at present no biological or biochemical ex-

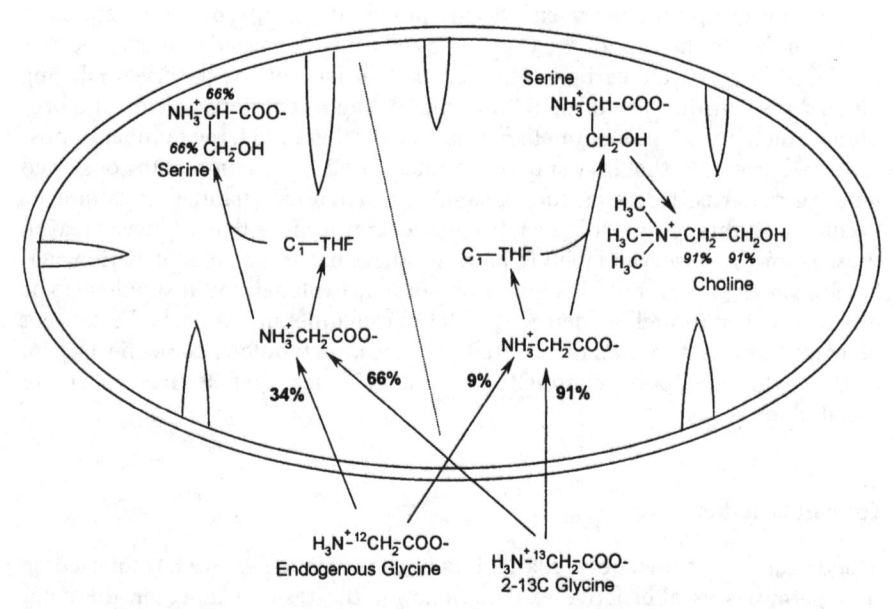

Fig. 7. Sub-mitochondrial compartmentation of the one-carbon unit metabolism in a mutant of *Saccharomyces cerevisiae* unable to produce C_1-tetrahydrofolate (C_1-THF) by other pathways than the mitochondrial cleavage of glycine to C_1-THF, carbon dioxide, and ammonia. ^{13}C-NMR analysis of choline and serine indicated the existence of two mitochondrial pools of glycine, serine, and C_1-THF. Values *in italics* refer to the fractional enrichment of the carbon atoms. Values **in bold** refer to the relative contributions of endogenous (unlabeled) and exogenous (labeled) glycine [47]

planation for the presence of these two compartments, but is seems reasonable that the explanation for the observation should be based on compartmentation phenomena.

It is, however, only occasionally that labeling experiments are able to give conclusive evidence of compartmentation of the metabolic pathways. This is mainly because it is only possible to distinguish between pathways taking place in different compartments if the precursors in these compartments are labeled differently. This is usually not the case because intracellular transport processes tend to equilibrate any labeling differences between precursor pools in the different compartments, and because most precursors are formed exclusively in one compartment, usually the cytosol, and then transported to the relevant compartments for further metabolic conversion. Thus, the precursors of the compartments are likely to be identically labeled, making it difficult to investigate the compartmentation using labeling techniques. However, since these transport processes require time to take place, information can be obtained from labeling experiments where the transients in the labeling patterns are followed. An example of this is the investigation of the penicillin biosynthesis in *Penicillium chrysogenum* [48].

In this study pulse-chase experiments using U-^{14}C-valine were carried out on high- and low-yielding strains of *P. chrysogenum* and the ^{14}C-labeling was measured in both proteinogenic valine and valine incorporated into the penicillin molecule. It was concluded that valine used for penicillin biosynthesis is derived from a compartment which is different from the compartment that contains the valine used for protein biosynthesis. In isotopic steady state situations the difference between the labeling patterns of valine used for penicillin and protein would have been identical because the effects of the transport process kinetics are eliminated in a steady state. However, discrepancies in steady state labeling patterns of the metabolites known to be derived from common precursors have been reported in the case of perfused mammalian hearts where metabolism of 1-^{13}C-glucose led to significantly higher ^{13}C-enrichment of lactate in the hearts than the enrichment of alanine. The labeling pattern of lactate in the perfusate was labeled to a higher extent than the lactate in the heart tissue, which was interpreted as the existence of a nonexchanging pool of lactate [49]. A metabolic steady state in perfused mammalian tissue is of course different from a steady state obtained in a continuous cultivation of a microorganism. However, if the observed metabolic system only contains metabolites with pool sizes that are small compared with the flux in and out of the pool, the metabolic network is in a pseudo-steady state, and the principles of mass and isotope balancing can be applied. Therefore, the methodology used in the study of tissue can also be used in the study of compartmentation in microorganisms.

3.4
Bidirectional Fluxes

An important difference between isotope balancing and metabolite balancing is the fact that the isotope label, e.g., due to reversible reactions, may be transported in the opposite direction to the net flux. Thus, reversible reactions, or more generally bidirectional pathways, have a significant influence on the labeling patterns. While bidirectional pathways have no implications on metabolite balancing, which only gives information on the net fluxes, information on bidirectionality of a pathway must be included in the analysis of labeling experiments. If bidirectionality is not included in the network it may seriously influence the interpretation of the network, especially when the pentose phosphate pathway is considered [50, 51]. The pentose phosphate pathway consists of an irreversible oxidative part, which is an important source of NADPH, and a nonoxidative part, where all the reactions are reversible. In the oxidative part, C1 of glucose 6-phosphate is lost as carbon dioxide. The usual method for estimating the flux through the oxidative pentose phosphate pathway is therefore to quantify the loss of label from glucose 6-phosphate, usually assumed to be labeled identically to 1-^{13}C-glucose in the medium. However, due to reversibilities of the nonoxidative pentose phosphate pathway and the reversible isomerization of glucose 6-phosphate into fructose 6-phosphate, not all glucose 6-phosphate molecules are labeled in the C1 position (Fig. 8).

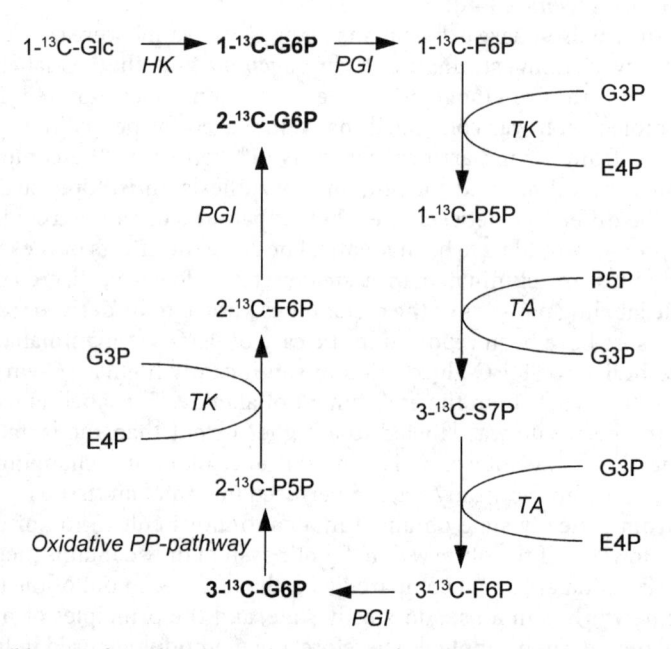

Fig. 8. Possible route for redistributing label from 1-^{13}C-glucose 6-phosphate to C2 and C3 of glucose 6-phosphate. Because C1–C3 of glucose can be labeled, glyceraldehyde 3-phosphate also becomes labeled in all three positions. Through a transaldolase-catalyzed reaction, glyceraldehyde 3-phosphate can be incorporated into C4–C6 of fructose 6-phosphate which may be converted to glucose 6-phosphate by phosphoglucose isomerase. In this way glucose 6-phosphate may, in principle, be labeled in all positions. *Glc* glucose; *G6P* glucose 6-phosphate; *F6P* fructose 6-phosphate; *G3P* glyceraldehyde 3-phosphate; *P5P* pentose 5-phosphate (i.e., xylulose 5-phosphate, ribose 5-phosphate, and ribulose 5-phosphate, which are readily interconvertible); *S7P* sedoheptulose 7-phosphate; *E4P* erythrose 4-phosphate; *HK* hexokinase; *TA* transaldolase; *TK* transketolase; *PGI* phosphoglucose isomerase; *PP* pentose phosphate

Consequently, not all glucose 6-phosphate molecules entering the oxidative pentose phosphate pathway contribute to the loss of label, leading to an underestimation of the flux through the pentose phosphate pathway. The complexity arising from reversibilities is illustrated by the fact that glucose 6-phosphate, in theory, can be labeled in all positions by the action of reversible reactions, without the need for any reactions catalyzed by phosphatases. Figure 8 shows a possible route for transferring the label from C1 of glucose 6-phosphate to C2 and C3 of glucose 6-phosphate. Glucose 6-phosphate labeled in the top three carbon atoms means that glyceraldehyde 3-phosphate can be labeled in all three position. Because glyceraldehyde 3-phosphate supplies the bottom three carbon atoms in a fructose 6-phosphate forming, transaldolase-catalyzed reaction, glucose 6-phosphate may, through the action of phosphoglucose isomerase, be labeled in the bottom three carbon atoms. Therefore, in principle, glucose 6-phosphate may be labeled in all positions. Although glucose 6-phosphate may contain labeled carbon atoms in all positions, some of the positions are only likely to carry a small fraction of the total labeling. Including reversible reactions complicates the computational analysis of the labeling patterns, but reversibilities have to be considered if optimal estimates of the fluxes are to be obtained [18].

NADPH balances are often essential for metabolite balancing based estimations of the net fluxes in a metabolic network. However, NADPH consumption and generation are often found in bidirectional reactions that cannot be quantified by metabolite balancing approaches. The mannitol cycle (Fig. 9) is an example of a pathway that can affect the NADPH balance, but has no net conversion of any metabolites, except for cofactors. In the mannitol cycle, NADH and $NADP^+$ are converted into NAD^+ and NADPH, respectively, at the expense of ATP [52]. Because mannitol happens to be symmetrical, the activity of the mannitol cycle will cause scrambling of the carbon atoms of fructose 6-phosphate, and the activity of the cycle may therefore be identified using labeling analysis. The mannitol cycle has been reported to be present in several fungi [52].

While the mannitol cycle is generating NADPH, another cycle, involving isocitrate and 2-ketoglutarate, may operate in the opposite direction [53] (Fig. 10). The net reaction is the conversion of NADPH and NAD^+ into $NADP^+$ and NADH, respectively. Using gas chromatography/mass spectrometry, evidence for the activity of the cycle was indicated by the appearance of citrate enriched in five positions, when U-$^{13}C_5$ glutamate was being metabolized in the perfused rat liver.

The cycle consisting of the reactions catalyzed by pyruvate carboxylase, malate dehydrogenase, and malic enzyme constitutes a third example of a pathway, with no net reaction, that may play a role in the redox metabolism. In this cycle, $NADP^+$ is reduced to NADPH by malic enzyme, and NADH is oxidized by malate dehydrogenase. The cycle was suggested to be active in *Corynebacterium glutamicum* during growth on 1-^{13}C fructose [55]. In this study, a combination of metabolite balances and labeling analyses showed that the NADPH-producing reactions in the catabolism were not able to supply sufficient amounts of NADPH for the biosynthetic reactions. Enzyme assays

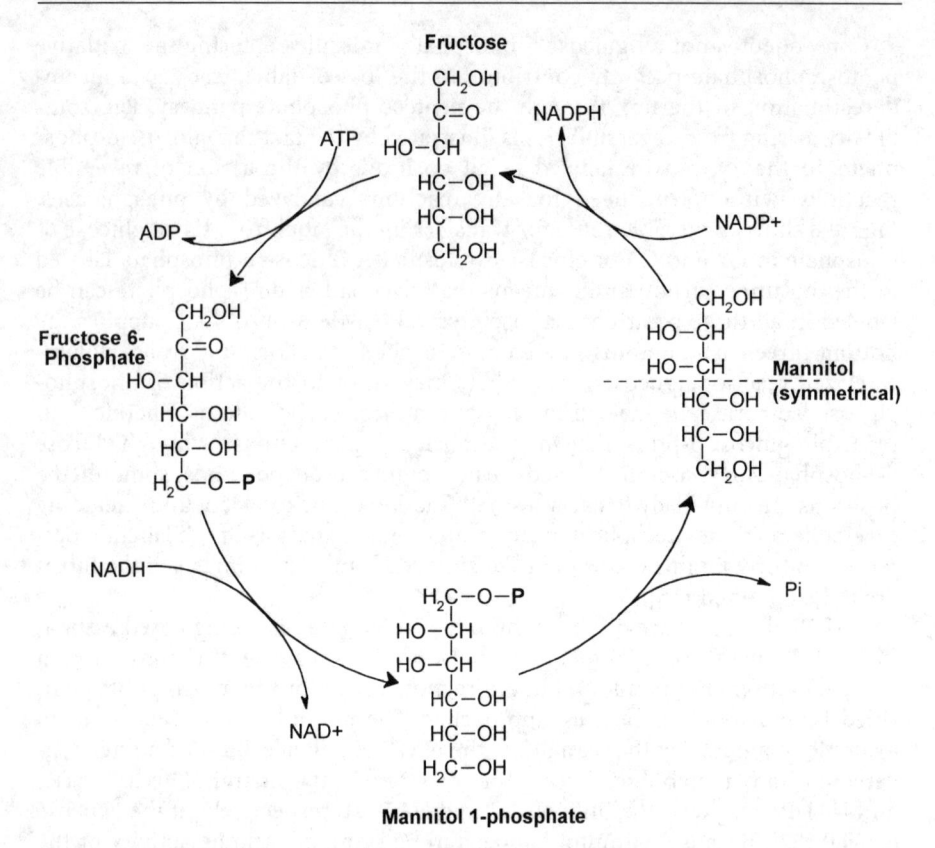

Fig. 9. Mannitol cycle. The net result of one turn of the cycle is the conversion of NADH and NADP$^+$ into NAD$^+$ and NADPH, respectively, at the expense of ATP. Since mannitol is a symmetrical molecule, the cycle causes scrambling of the carbon atoms of fructose 6-phosphate unless metabolic channeling takes place [52]

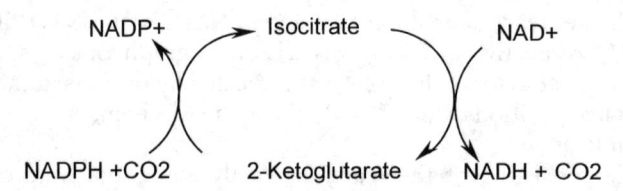

Fig. 10. Conversion of NAD$^+$ and NADPH into NADH and NADP$^+$, respectively. The cycle was suggested by Sazanov and Jackson [53] and evidence for its activity in the mitochondria of rat livers has been obtained using labeling experiments [54]

showed a higher level of malic enzymes compared with the level on growth on glucose, but the labeling analysis, that was based on the labeling pattern of glutamate, could not be used to verify a significantly higher enrichment in C1 and C2 of pyruvate, which would be the result of the transfer of highly labeled carbon atoms from malate to pyruvate. The lack of this observation may be because there is only a relatively small flux from malate to pyruvate compared to the flux to pyruvate from the Embden-Meyerhof-Parnas pathway. C5 of glutamate, originating from C2 of pyruvate, did show a slightly higher ^{13}C-labeling than the natural enrichment, which could be an indication of malic enzyme activity. However, as discussed above, the labeling in C2 of pyruvate may also be attributed to reversibilities in the pentose phosphate pathway.

4
Conclusions and Future Directions

Due to the complexity of the metabolism, investigation of metabolic networks necessitates powerful analytical methods, both experimentally and mathematically. Figure 11 illustrates some of the metabolic aspects, e.g., metabolic channeling and reversible reactions, that complicate the analysis of metabolic networks.

Using an integrated approach based on a combination of information from metabolite balancing and analysis of labeling patterns, many of these aspects

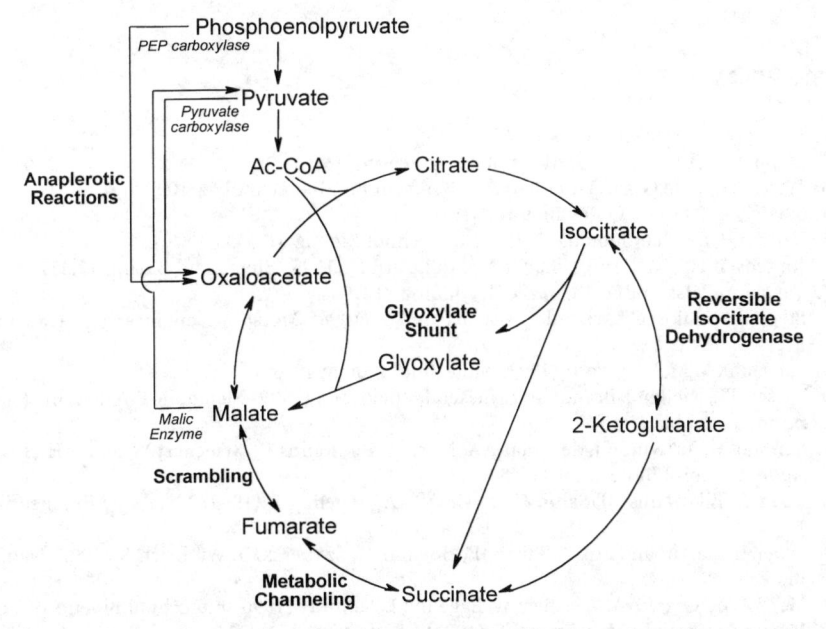

Fig. 11. Reactions in the tricarboxylic acid cycle and associated reactions. The unraveling of these reactions can be carried out using a combination of metabolite balancing and labeling experiments

can be investigated in a single analysis covering the entire primary metabolism. In particular, analysis based on isotopic steady state labeling patterns is mathematically tractable. Metabolic network analysis based on labeling patterns of proteinogenic amino acids from isotopic steady state cultures has been demonstrated to be a powerful approach to investigate the primary metabolism. This is partly because protein is found in significant amounts in all types of biomass, and partly because the amino acid precursors are derived from several parts of the central metabolism. The labeling patterns of the amino acids contain direct information on the labeling state of these precursors. Thus, information on the labeling patterns of, e.g., oxaloacetate and pyruvate need not be deduced from the labeling pattern of glutamate, but can be obtained directly from aspartate and alanine, respectively. Redundancy of information is crucial when questions regarding the network structure are to be addressed and redundant measurements are therefore of major importance in metabolic network analysis.

However, the labeling patterns obtained from an isotopic steady state represent the average metabolism, and not the metabolism at the level of a single cell. Since cells undergo fundamental changes during the cell cycle, it is reasonable to expect these events to be reflected in the central metabolism. Cell differentiation may be seen as a special case of compartmentation, and it is therefore likely that cell differentiation, like compartmentation, is amenable to investigation by metabolic network analysis. Thus, the future of metabolic network analysis may lie in extending the analysis of metabolism from a time-average description to a time-dependent description of the changes in the metabolism that take place as a consequence of, e.g., cell proliferation.

References

1. Bailey JE (1991) Science 252:1668
2. Stephanopoulos G (1999) Metabolic Engineering 1:1
3. Wiechert W, de Graaf AA (1996) Adv Biochem Eng Biotechnol 54:109
4. Szyperski T (1998) Q Rev Biophys 31:41
5. Vallino JJ, Stephanopoulos G (1993) Biotechnol Bioeng 41:633
6. Jørgensen HS, Nielsen J, Villadsen J, Møllgaard H (1995) Biotechnol Bioeng 46:117
7. Varma A, Palsson BO (1994) Bio/Technology 12:994
8. Stephanopoulos G, Aristodou AA, Nielsen J (1998) Metabolic engineering. Academic Press, San Diego
9. van Gulik WM, Heijnen JJ (1995) Biotechnol Bioeng 48:681
10. Nissen TL, Kielland-Brandt MC, Nielsen J, Villadsen J (1999) Metabolic Engineering (submitted for publication)
11. Sonntag K, Schwinde J, de Graaf AA, Marx A, Eikmanns BJ, Wiechert W, Sahm H (1995) Appl Microbiol Biotechnol 44:489
12. Marx A, Eikmanns BJ, Sahm H, de Graaf AA, Eggeling L (1999) Metabolic Engineering 1:35
13. Sauer U, Hatzimanikatis V, Bailey JE, Hochuli M, Szyperski T, Wüthrich K (1997) Nature Biotechnol 15:448
14. Marx A, de Graaf AA, Wiechert W, Eggeling L, Sahm H (1996) Biotechnol Bioeng 49:111
15. Martin G, Chauvin M-F, Baverel G (1997) J Biol Chem 272:4717
16. Chatham JC, Forder JR, Glickson JD, Chance EM (1995) J Biol Chem 270:7999
17. Zupke C, Stephanopoulos G (1994) Biotechnol Prog 10:489

18. Wiechert W, de Graaf AA (1997) Biotechnol Bioeng 55:101
19. Wiechert W, Siefke C, de Graaf AA, Marx A (1997) Biotechnol Bioeng 55:118
20. Schuster R, Schuster S, Holzhütter H-G (1992) J Chem Soc Faraday Trans 88:2837
21. Marx A, Striegel K, de Graaf AA, Sahm H, Eggeling L (1997) Biotechnol Bioeng 56:168
22. Fernandez CA, des Rosiers C (1995) J Biol Chem 270:10037
23. Schmidt K, Carlsen M, Nielsen J, Villadsen J (1997) Biotechnol Bioeng 55:831
24. Schmidt K, Nielsen J, Villadsen J (1999) J Biotechnol (in press)
25. Klapa MI, Park SM, Sinskey AJ, Stephanopoulos G (1999) Biotechnol Bioeng 62:375
26. Park SM, Klapa MI, Sinskey AJ, Stephanopoulos G (1999) Biotechnol Bioeng 62:392
27. Rollin C, Morgant V, Guyonvarch A, Guerquin-Kern J-L (1995) Eur J Biochem 227:488
28. Peters-Wendisch PG, Eikmanns BJ, Thierbach G, Bachmann B, Sahm H (1993) FEMS Microbiol Lett 112:269
29. Peters-Wendisch PG, Wendisch VF, de Graaf AA, Eikmanns BJ, Sahm H (1996) Arch Microbiol 165:387
30. Peters-Wendisch PG, Wendisch VF, Paul S, Eikmanns BJ, Sahm H (1997) Microbiology 143:1095
31. Koffas MAG, Ramamoorthi R, Pine WA, Sinskey AJ, Stephanopoulos G (1998) Appl Microbiol Biotechnol 50:346
32. Chalk PA, Roberts AD, Blows WM (1994) Microbiology 140:2085
33. Selig M, Xavier KB, Santos H, Schönheit P (1997) Arch Microbiol 167:217.
34. Siebers B, Wendisch VF, Hensel R (1997) Arch Microbiol 168:120
35. Textor S, Wendisch VF, de Graaf AA, Müller U, Linder MI, Linder D, Buckel W (1997) Arch Microbiol 168:428
36. Tran-Dinh S, Beganton F, Nguyen T-T, Bouet F, Herve M (1996) Eur J Biochem 242:220
37. Tran-Dinh S, Bouet F, Huynh Q-T, Herve M (1996) Eur J Biochem 242:770
38. Sonntag K, Eggeling L, de Graaf AA, Sahm H (1993) Eur J Biochem 213:1325
39. Portais J-C, Schuster R, Merle M, Canioni P (1993) Eur J Biochem 217:457
40. Sherry AD, Sumegi B, Miller B, Cottam GL, Gavva S, Jones JG, Malloy CR (1994) Biochemistry 33:6268
41. Malaisse WJ, Zhang TM, Verbuggen I, Willem R (1996) Biochem J 317:861
42. Sumegi B, Sherry AD, Malloy CR (1990) Biochemistry 29:9106
43. Sumegi B, Sherry AD, Malloy CR, Srere PA (1993) Biochemistry 32:12725
44. Sumegi B, Podanyi B, Forgo P, Kover KE (1995) Biochem J 312:75
45. Pronk JT, van der Linden-Beuman A, Verduyn C, Scheffers WA, van Dijken JP (1994) Microbiology 140:717
46. Robinson JB, Inman L, Sumegi B, Srere PA (1987) J Biol Chem 262:1786
47. Pasternack LB, Laude DA, Appling DR (1994) Biochemistry 33:74
48. Affenzeller K, Kubicek CP (1991) J Gen Microbiol 137:1653
49. Chatham JC, Forder JR (1996) Am J Physiol 270:H224
50. Flanigan I, Collins JG, Arora KK, MacLeod JK, Williams JF (1993) Eur J Biochem 213:477
51. Follstad BD, Stephanopoulos G (1998) Eur J Biochem 252:360
52. Hult K, Veide A, Gatenbeck S (1980) Arch Microbiol 128:253
53. Sazanov LA, Jackson JB (1994) FEBS Lett 344:109
54. Des Rosiers C, Fernandez CA, David F, Brunengraber H (1994) J Biol Chem 269:27179
55. Dominguez H, Rollin C, Guyonvarch A, Guerquin-Kern J-L, Cocaign-Bousquet M, Lindley ND (1998) Eur J Biochem 254:96

Author Index Volume 1–66

Author Index Volumes 1–50 see Volume 50

Adam, W., Lazarus, M., Saha-Möller, C. R., Weichhold, O., Hoch, U. Häring, D., Schreier, Ü.: Biotransformations with Peroxidases. Vol. 63, p. 73
Allan, J. V., Roberts, S. M., Williamson, N. M.: Polyamino Acids as Man-Made Catalysts. Vol. 63, p. 125
Al-Rubeai, M.: Apoptosis and Cell Culture Technology. Vol. 59, p. 225
Al-Rubeai, M. see Singh, R. P.: Vol. 62, p. 167
Alsberg, B. K. see Shaw, A. D.: Vol. 66, p. 83
Antranikian, G. see Ladenstein, R.: Vol. 61, p. 37
Antranikian, G. see Müller, R.: Vol. 61, p. 155
Archelas, A. see Orru, R. V. A.: Vol. 63, p. 145
Argyropoulos, D. S.: Lignin. Vol. 57, p. 127
Arnold, F. H., Moore, J. C.: Optimizing Industrial Enzymes by Directed Evolution. Vol. 58, p. 1
Akhtar, M., Blanchette, R. A., Kirk, T. K.: Fungal Delignification and Biochemical Pulping of Wood. Vol. 57, p. 159
Autuori, F., Farrace, M. G., Oliverio, S., Piredda, L., Piacentini, G.: "Tissie" Transglutaminase and Apoptosis. Vol. 62, p. 129
Azerad, R.: Microbial Models for Drug Metabolism. Vol. 63, p. 169

Bajpai, P., Bajpai, P. K.: Realities and Trends in Emzymatic Prebleaching of Kraft Pulp. Vol. 56, p. 1
Bajpai, P. Bajpai, P. K.: Reduction of Organochlorine Compounds in Bleach Plant Effluents. Vol. 57, p. 213
Bajpai, P. K. see Bajpai, P.: Vol. 56, p. 1
Bajpai, P. K. see Bajpai, P.: Vol. 57, p. 213
Bárzana, E.: Gas Phase Biosensors. Vol. 53, p. 1
Bazin, M. J. see Markov, S. A.: Vol. 52, p. 59
Bellgardt, K.-H.: Process Models for Production of β-Lactam Antibiotics. Vol. 60, p. 153
Beyer, M. see Seidel, G.: Vol. 66, p. 115
Bhatia, P. K., Mukhopadhyay, A.: Protein Glycosylation: Implications for in vivo Functions and Thereapeutic Applications. Vol. 64, p. 155
Blanchette R. A. see Akhtar, M.: Vol. 57, p. 159
de Bont, J.A.M. see van der Werf, M. J.: Vol. 55, p. 147
Brainard, A. P. see Ho, N. W. Y.: Vol. 65, p. 163–192
Broadhurst, D. see Shaw, A. D.: Vol. 66, p. 83
Bruckheimer, E. M., Cho, S. H., Sarkiss, M., Herrmann, J., McDonell, T. J.: The Bcl-2 Gene Family and Apoptosis. Vol 62, p. 75
Buchert, J. see Suurnäkki, A.: Vol. 57, p. 261
Bungay, H. R. see Mühlemann, H. M.: Vol. 65, p. 193–206

Cao, N. J. see Gong, C. S.: Vol. 65, p. 207–241
Cao, N. J. see Tsao, G. T.: Vol. 65, p. 243–280
Carnell, A. J.: Stereoinversions Using Microbial Redox-Reactions. Vol. 63, p. 57
Cen, P., Xia, L.: Production of Cellulase by Solid-State Fermentation. Vol. 65, p. 69–92

Chang, H. N. see Lee, S. Y.: Vol. 52, p. 27
Cheetham, P. S. J.: Combining the Technical Push and the Business Pull for Natural Flavours. Vol. 55, p. 1
Cho, S. H. see Bruckheimer, E. M.: Vol. 62, p. 75
Chen, Z. see Ho, N. W. Y.: Vol. 65, p. 163–192
Christensen, B., Nielsen, J.: Metabolic Network Analysis – A Powerful Tool in Metabolic Engineering. Vol. 66, p. 209
Ciaramella, M. see van der Oost, J.: Vol. 61, p. 87
Cornet, J.-F., Dussap, C. G., Gros, J.-B.: Kinetics and Energetics of Photosynthetic Micro-Organisms in Photobioreactors. Vol. 59, p. 153
da Costa, M. S., Santos, H., Galinski, E. A.: An Overview of the Role and Diversity of Compatible Solutes in Bacteria and Archaea. Vol. 61, p. 117
Cotter, T. G. see McKenna, S. L.: Vol. 62, p. 1
Croteau, R. see McCaskill, D.: Vol. 55, p. 107

Danielsson, B. see Xie, B.: Vol. 64, p. 1
Darzynkiewicz, Z., Traganos, F.: Measurement of Apoptosis. Vol 62, p. 33
Davey, H. M. see Shaw, A. D.: Vol. 66, p. 83
Dean, J. F. D., LaFayette, P. R., Eriksson, K.-E. L., Merkle, S. A.: Forest Tree Biotechnolgy: Vol. 57, p. 1
Dochain, D., Perrier, M.: Dynamical Modelling, Analysis, Monitoring and Control Design for Nonlinear Bioprocesses. Vol. 56, p. 147
Du, J. see Gong, C. S: Vol. 65, p. 207–241
Du, J. see Tsao, G. T.: Vol. 65, p. 243–280
Dussap, C. G. see Cornet J.-F.: Vol. 59, p. 153
Dutta, N. N. see Ghosh, A. C.: Vol. 56, p. 111

Eggeling, L., Sahm, H., de Graaf, A. A.: Quantifying and Directing Metabolite Flux: Application to Amino Acid Overproduction. Vol. 54, p. 1
Ehrlich, H. L. see Rusin, P.: Vol. 52, p. 1
Elias, C. B., Joshi, J. B.: Role of Hydrodynamic Shear on Activity and Structure of Proteins. Vol. 59, p. 47
Elling, L.: Glycobiotechnology: Enzymes for the Synthesis of Nucleotide Sugars. Vol. 58, p. 89
Eriksson, K.-E. L. see Kuhad, R. C.: Vol. 57, p. 45
Eriksson, K.-E. L. see Dean, J. F. D.: Vol. 57, p. 1

Faber, K. see Orru, R. V. A.: Vol. 63, p. 145
Farrell, R. L., Hata, K., Wall, M. B.: Solving Pitch Problems in Pulp and Paper Processes. Vol. 57, p. 197
Farrace, M. G. see Autuori, F.: Vol. 62, p. 129
Fiechter, A. see Ochsner, U. A.: Vol. 53, p. 89
Foody, B. see Tolan, J. S.: Vol. 65, p. 41–67
Freitag, R., Hórvath, C.: Chromatography in the Downstream Processing of Biotechnological Products. Vol. 53, p. 17
Furstoss, R. see Orru, R. V. A.: Vol. 63, p. 145

Galinski, E. A. see da Costa, M. S.: Vol. 61, p. 117
Gatfield, I. L.: Biotechnological Production of Flavour-Active Lactones. Vol. 55, p. 221
Gemeiner, P. see Stefuca, V.: Vol. 64, p. 69
Gerlach, S. R. see Schügerl, K.: Vol. 60, p. 195
Ghosh, A. C., Mathur, R. K., Dutta, N. N.: Extraction and Purification of Cephalosporin Antibiotics. Vol. 56, p. 111
Ghosh, P. see Singh, A.: Vol. 51, p. 47
Gilbert, R. J. see Shaw, A. D.: Vol. 66, p. 83
Gomes, J., Menawat, A. S.: Fed-Batch Bioproduction of Spectinomycin. Vol. 59, p. 1

Gong, C. S., Cao, N. J., Du, J., Tsao, G. T.: Ethanol Production from Renewable Resources. Vol. 65, p. 207–241
Gong, C. S. see *Tsao, G. T.*: Vol. 65, p. 243–280
Goodacre, R. see *Shaw, A. D.*: Vol. 66, p. 83
de Graaf, A. A. see *Eggeling, L.*: Vol. 54, p. 1
de Graaf, A. A. see *Weuster-Botz, D.*: Vol. 54, p. 75
de Graaf, A. A. see *Wiechert, W.*: Vol. 54, p. 109
Grabley, S., Thiericke, R.: Bioactive Agents from Natural Sources: Trends in Discovery and Application. Vol. 64, p. 101
Griengl, H. see *Johnson, D. V.*: Vol. 63, p. 31
Gros, J.-B. see *Larroche, C.*: Vol. 55, p. 179
Gros, J.-B. see *Cornet, J. F.*: Vol. 59, p. 153
Guenette M. see *Tolan, J. S.*: Vol. 57, p. 289
Gutman, A. L., Shapira, M.: Synthetic Applications of Enzymatic Reactions in Organic Solvents. Vol. 52, p. 87

Häring, D. see *Adam, E.*: Vol. 63, p. 73
Hall, D. O. see *Markov, S. A.*: Vol. 52, p. 59
Hall, P. see *Mosier, N. S.*: Vol. 65, p. 23–40
Harvey, N. L., Kumar, S.: The Role of Caspases in Apoptosis. Vol. 62, p. 107
Hasegawa, S., Shimizu, K.: Noninferior Periodic Operation of Bioreactor Systems. Vol. 51, p. 91
Hata, K. see *Farrell, R. L.*: Vol. 57, p. 197
Hembach, T. see *Ochsner, U. A.*: Vol. 53, p. 89
Herrmann, J. see *Bruckheimer, E. M.*: Vol. 62, p. 75
Hill, D. C., Wrigley, S. K., Nisbet, L. J.: Novel Screen Methodologies for Identification of New Microbial Metabolites with Pharmacological Activity. Vol. 59, p. 73
Hiroto, M. see *Inada, Y.*: Vol. 52, p. 129
Ho, N. W. Y., Chen, Z., Brainard, A. P. Sedlak, M.: Successful Design and Development of Genetically Engineering Saccharomyces Yeasts for Effective Cofermentation of Glucose and Xylose from Cellulosic Biomass to Fuel Ethanol. Vol. 65, p. 163–192
Hoch, U. see *Adam, W.*: Vol. 63, p. 73
Hórvath, C. see *Freitag, R.*: Vol. 53, p. 17
Hummel, W.: New Alcohol Dehydrogenases for the Synthesis of Chiral Compounds. Vol. 58, p.145

Inada, Y., Matsushima, A., Hiroto, M., Nishimura, H., Kodera, Y.: Chemical Modifications of Proteins with Polyethylen Glycols. Vol. 52, p. 129
Iyer, P. see *Lee, Y. Y.*: Vol. 65, p. 93–115
Irwin, D. C. see *Wilson, D. B.*: Vol. 65, p. 1–21

Jeffries, T. W., Shi, N.-Q.: Genetic Engineering for Improved Xylose Fementation by Yeasts. Vol. 65, p. 117–161
Johnson, E. A., Schroeder, W. A.: Microbial Carotenoids. Vol. 53, p. 119
Johnson, D. V., Griengl, H.: Biocatalytic Applications of Hydroxynitrile. Vol. 63, p. 31
Joshi, J. B. see *Elias, C. B.*: Vol. 59, p. 47
Johnsrud, S. C.: Biotechnolgy for Solving Slime Problems in the Pulp and Paper Industry. Vol. 57, p. 311

Kaderbhai, N. see *Shaw, A. D.*: Vol. 66, p. 83
Kataoka, M. see *Shimizu, S.*: Vol. 58, p. 45
Kataoka, M. see *Shimizu, S.*: Vol. 63, p. 109
Kawai, F.: Breakdown of Plastics and Polymers by Microorganisms. Vol. 52, p. 151
Kell, D. B. see *Shaw, A. D.*: Vol. 66, p. 83
King, R.: Mathematical Modelling of the Morphology of Streptomyces Species. Vol. 60, p. 95
Kirk, T. K. see *Akhtar, M.*: Vol. 57, p. 159

Kobayashi, M. see Shimizu, S.: Vol. 58, p. 45
Kodera, F. see Inada, Y.: Vol. 52, p. 129
Krabben, P. Nielsen, J.: Modeling the Mycelium Morphology of Penicilium Species in Submerged Cultures. Vol. 60, p. 125
Krämer, R.: Analysis and Modeling of Substrate Uptake and Product Release by Procaryotic and Eucaryotik Cells. Vol. 54, p. 31
Kuhad, R. C., Singh, A., Eriksson, K.-E. L.: Microorganisms and Enzymes Involved in the Degradation of Plant Cell Walls. Vol. 57, p. 45
Kuhad, R. Ch. see Singh, A.: Vol. 51, p. 47
Kumar, S. see Harvey, N. L.: Vol. 62, p. 107

Ladenstein, R., Antranikian, G.: Proteins from Hyperthermophiles: Stability and Enzamatic Catalysis Close to the Boiling Point of Water. Vol. 61, p. 37
Ladisch, C. M. see Mosier, N. S.: Vol. 65, p. 23–40
Ladisch, R. M. see Mosier, N. S.: Vol. 65, p. 23–40
Lammers, F., Scheper, T.: Thermal Biosensors in Biotechnology. Vol. 64, p. 35
Larroche, C., Gros, J.-B.: Special Transformation Processes Using Fungal Spares and Immobilized Cells. Vol. 55, p. 179
LaFayette, P. R. see Dean, J. F. D.: Vol. 57, p. 1
Lazarus, M. see Adam, W.: Vol. 63, p. 73
Leak, D. J. see van der Werf, M. J.: Vol. 55, p. 147
Lee, S. Y., Chang, H. N.: Production of Poly(hydroxyalkanoic Acid). Vol. 52, p. 27
Lee, Y. Y., Iyer, P., Torget, R. W.: Dilute-Acid Hydrolysis of Lignocellulosic Biomass. Vol. 65, p. 93–115
Lievense, L. C., van't Riet, K.: Convective Drying of Bacteria II. Factors Influencing Survival. Vol. 51, p. 71

Maloney, S. see Müller, R.: Vol. 61, p. 155
Mandenius, C.-F.: Electronic Noses for Bioreactor Monitoring. Vol. 66, p. 65
Markov, S. A., Bazin, M. J., Hall, D. O.: The Potential of Using Cyanobacteria in Photobioreactors for Hydrogen Production. Vol. 52, p. 59
Marteinsson, V. T. see Prieur, D.: Vol. 61, p. 23
Mathur, R. K. see Ghosh, A. C.: Vol. 56, p. 111
Matsushima, A. see Inada, Y.: Vol. 52, p. 129
McCaskill, D., Croteau, R.: Prospects for the Bioengineering of Isoprenoid Biosynthesis. Vol. 55, p. 107
McDonell, T. J. see Bruckheimer, E. M.: Vol. 62, p. 75
McGovern, A. see Shaw, A. D.: Vol. 66, p. 83
McGowan, A. J. see McKenna, S. L.: Vol. 62, p. 1
McKenna, S. L.: McGowan, A. J., Cotter, T. G.: Molecular Mechanisms of Programmed Cell Death. Vol. 62, p. 1
McLoughlin, A. J.: Controlled Release of Immobilized Cells as a Strategy to Regulate Ecological Competence of Inocula. Vol. 51, p. 1
Menachem, S. B. see Argyropoulos, D. S. : Vol. 57, p. 127
Menawat, A. S. see Gomes J.: Vol. 59, p. 1
Merkle, S. A. see Dean, J. F. D.: Vol. 57, p. 1
Moore, J. C. see Arnold, F. H.: Vol. 58, p. 1
Mosier, N. S., Hall, P., Ladisch C. M., Ladisch M. R.: Reaction Kinetics, Molecular Action, and Mechanisms of Cellulolytic Proteins. Vol. 65, p. 23–40
Moracci, M. see van der Oost, J.: Vol. 61, p. 87
Mühlemann, H. M. Bungay, H. R.: Research Perspectives for Bioconversion of Scrap Paper. Vol. 65, p. 193–206
Müller, R., Antranikian, G., Maloney, S., Sharp, R.: Thermophilic Degradation of Environmental Pollutants. Vol. 61, p. 155

Mukhopadhyay, A.: Inclusion Bodies and Purification of Proteins in Biologically Active Forms. Vol. 56, p. 61
Mukhopadhyay, A. see Bhatia, P. K.: Vol. 64, p. 155

Nielsen, J. see Christensen, B.: Vol. 66, p. 209
Nielsen, J. see Krabben, P.: Vol. 60, p. 125
Nisbet, L. J. see Hill, D. C.: Vol. 59, p. 73
Nishimura, H. see Inada, Y.: Vol. 52, p. 123

Ochsner, U. A., Hembach, T., Fiechter, A.: Produktion of Rhamnolipid Biosurfactants. Vol. 53, p. 89
O'Connor, R.: Survival Factors and Apoptosis: Vol. 62, p. 137
Ogawa, J. see Shimizu, S.: Vol. 58, p. 45
Ohta, H.: Biocatalytic Asymmetric Decarboxylation. Vol. 63, p. 1
van der Oost, J., Ciaramella, M., Moracci, M., Pisani, F. M., Rossi, M., de Vos, W. M.: Molecular Biology of Hyperthermophilic Archaea. Vol. 61, p. 87
Oliverio, S. see Autuori, F.: Vol. 62, p. 129
Orru, R. V. A., Archelas, A., Furstoss, R., Faber, K.: Epoxide Hydrolases and Their Synthetic Applications. Vol. 63, p. 145

Paul, G. C., Thomas, C. R.: Characterisation of Mycelial Morphology Using Image Analysis. Vol. 60, p. 1
Perrier, M. see Dochain, D.: Vol. 56, p. 147
Piacentini, G. see Autuori, F.: Vol. 62, p. 129
Piredda, L. see Autuori, F.: Vol. 62, p. 129
Pisani, F. M. see van der Oost, J.: Vol. 61, p. 87
Pohl, M.: Protein Design on Pyruvate Decarboxylase (PDC) by Site-Directed Mutagenesis. Vol. 58, p. 15
Pons, M.-N., Vivier, H.: Beyond Filamentous Species. Vol. 60, p. 61
Pons, M.-N., Vivier, H.: Biomass Quantification by Image Analysis. Vol. 66, p. 133
Prieur, D., Marteinsson, V. T.: Prokaryotes Living Under Elevated Hydrostatic Pressure. Vol. 61, p. 23
Pulz, O., Scheibenbogen, K.: Photobioreactors: Design and Performance with Respect to Light Energy Input. Vol. 59, p. 123

Ramanathan, K. see Xie, B.: Vol. 64, p. 1
van't Riet, K. see Lievense, L. C.: Vol. 51, p. 71
Roberts, S. M. see Allan, J. V.: Vol. 63, p. 125
Rogers, P. L., Shin, H. S., Wang, B.: Biotransformation for L-Ephedrine Production. Vol. 56, p. 33
Rossi, M. see van der Oost, J.: Vol. 61, p. 87
Roychoudhury, P. K., Srivastava, A., Sahai, V.: Extractive Bioconversion of Lactic Acid. Vol. 53, p. 61
Rowland, J. J. see Shaw, A. D.: Vol. 66, p. 83
Rusin, P., Ehrlich, H. L.: Developments in Microbial Leaching – Mechanisms of Manganese Solubilization. Vol. 52, p. 1
Russell, N. J.: Molecular Adaptations in Psychrophilic Bacteria: Potential for Biotechnological Applications. Vol. 61, p. 1

Sahai, V. see Singh, A.: Vol. 51, p. 47
Sahai, V. see Roychoudhury, P. K.: Vol. 53, p. 61
Saha-Möller, C. R. see Adam, W.: Vol. 63, p. 73
Sahm, H. see Eggeling, L.: Vol. 54, p. 1
Saleemuddin, M.: Bioaffinity Based Immobilization of Enzymes. Vol. 64, p. 203
Santos, H. see da Costa, M. S.: Vol. 61, p. 117
Sarkiss, M. see Bruckheimer, E. M.: Vol. 62, p. 75

Scheibenbogen, K. see Pulz, O.: Vol. 59, p. 123
Scheper, T. see Lammers, F.: Vol. 64, p. 35
Schreier, P.: Enzymes and Flavour Biotechnology. Vol. 55, p. 51
Schreier, P. see Adam, W.: Vol. 63, p. 73
Schroeder, W. A. see Johnson, E. A.: Vol. 53, p. 119
Schügerl, K., Gerlach, S. R., Siedenberg, D.: Influence of the Process Parameters on the Morphology and Enzyme Production of Aspergilli. Vol. 60, p. 195
Schügerl, K. see Seidel, G.: Vol. 66, p. 115
Schuster, K. C.: Monitoring the Physiological Status in Bioprocesses on the Cellular Level. Vol. 66, p. 185
Scouroumounis, G. K. see Winterhalter, P.: Vol. 55, p. 73
Scragg, A.H.: The Production of Aromas by Plant Cell Cultures. Vol. 55, p. 239
Sedlak, M. see Ho, N. W. Y.: Vol. 65, p. 163–192
Seidel, G., Tollnick, C., Beyer, M., Schügerl, K.: On-line and Off-line Monitoring of the Production of Cephalosporin C by Acremonium Chrysogenum. Vol. 66, p. 115
Shapira, M. see Gutman, A. L.: Vol 52, p. 87
Sharp, R. see Müller, R.: Vol. 61, p. 155
Shaw, A. D., Winson, M. K., Woodward, A. M., McGovern, A., Davey, H. M., Kaderbhai, N., Broadhurst, D., Gilbert, R. J., Taylor, J., Timmins, E. M., Alsberg, B. K., Rowland, J. J., Goodacre, R., Kell, D. B.: Rapid Analysis of High-Dimensional Bioprocesses Using Multivariate Spectroscopies and Advanced Chemometrics. Vol. 66, p. 83
Shi, N.-Q. see Jeffries, T. W.: Vol. 65, p. 117–161
Shimizu, S., Ogawa, J., Kataoka, M., Kobayashi, M.: Screening of Novel Microbial for the Enzymes Production of Biologically and Chemically Useful Compounds. Vol. 58, p. 45
Shimizu, K. see Hasegawa, S.: Vol. 51, p. 91
Shimizu, S., Kataoka, M.: Production of Chiral C3- and C4-Units by Microbial Enzymes. Vol. 63, p. 109
Shin, H. S. see Rogers, P. L.: Vol. 56, p. 33
Siedenberg, D. see Schügerl, K.: Vol. 60, p. 195
Singh, R. P., Al-Rubeai, M.: Apoptosis and Bioprocess Technology. Vol 62, p. 167
Singh, A., Kuhad, R. Ch., Sahai, V., Ghosh, P.: Evaluation of Biomass. Vol. 51, p. 47
Singh, A. see Kuhad, R. C.: Vol. 57, p. 45
Sonnleitner, B.: New Concepts for Quantitative Bioprocess Research and Development. Vol. 54, p. 155
Sonnleitner, B.: Instrumentation of Biotechnological Processes. Vol. 66, p. 1
Stefuca, V., Gemeiner, P.: Investigation of Catalytic Properties of Immobilized Enzymes and Cells by Flow Microcalorimetry. Vol. 64, p. 69
Srivastava, A. see Roychoudhury, P. K.: Vol. 53, p. 61
Suurnäkki, A., Tenkanen, M., Buchert, J., Viikari, L.: Hemicellulases in the Bleaching of Chemical Pulp. Vol. 57, p. 261

Taylor, J. see Shaw, A. D.: Vol. 66, p. 83
Tenkanen, M. see Suurnäkki, A.: Vol. 57, p. 261
Thiericke, R. see Grabely, S.: Vol. 64, p. 101
Thömmes, J.: Fluidized Bed Adsorption as a Primary Recovery Step in Protein Purification. Vol. 58, p. 185
Thomas, C. R. see Paul, G. C.: Vol. 60, p. 1
Timmens, E. M. see Shaw, A. D.: Vol. 66, p. 83
Tolan, J. S., Guenette, M.: Using Enzymes in Pulp Bleaching: Mill Applications. Vol. 57, p. 289
Tolan, J. S., Foody, B.: Cellulase from Submerged Fermentation. Vol. 65, p. 41–67
Tollnick, C. see Seidel, G.: Vol. 66, p. 115
Traganos, F. see Darzynkiewicz, Z.: Vol. 62, p. 33
Torget, R. W. see Lee, Y. Y.: Vol. 65, p. 93–115

Tsao, G. T., Cao, N. J. Du, J., Gong, C. S.: Production of Multifunctional Organic Acids from Renewable Resources. Vol. 65, p. 243–280
Tsao, G. T.: see Gong, C. S.: Vol. 65, p. 207–241

Viikari, L. see Suurnäkki, A.: Vol. 57, p. 261
Vivier, H. see Pons, M.-N.: Vol. 60, p. 61
Vivier, H. see Pons, M.-N.: Vol. 66, p. 133
de Vos, W. M. see van der Oost, J.: Vol. 61, p. 87

Wang, B. see Rogers, P. L.: Vol. 56, p. 33
Wall, M. B. see Farrell, R. L.: Vol. 57, p. 197
Weichold, O. see Adam, W.: Vol. 63, p. 73
van der Werf, M. J., de Bont, J. A. M. Leak, D. J.: Opportunities in Microbial Biotransformation of Monoterpenes. Vol. 55, p. 147
Weuster-Botz, D., de Graaf, A.A.: Reaction Engineering Methods to Study Intracellular Metabolite Concentrations. Vol. 54, p. 75
Wiechert, W., de Graaf, A.A.: In Vivo Stationary Flux Analysis by ^{13}C-Labeling Experiments. Vol. 54, p. 109
Wiesmann, U.: Biological Nitrogen Removal from Wastewater. Vol. 51, p. 113
Williamson, N. M. see Allan, J. V.: Vol. 63, p. 125
Wilson, D. B., Irwin, D. C.: Genetics and Properties of Cellulases. Vol. 65, p. 1–21
Winson, M. K. see Shaw, A. D.: Vol. 66, p. 83
Winterhalter, P., Skouroumounis, G. K.: Glycoconjugated Aroma Compounds: Occurence, Role and Biotechnological Transformation. Vol. 55, p. 73
Woodward, A. M. see Shaw, A. D.: Vol. 66, p. 83
Wrigley, S. K. see Hill, D. C.: Vol. 59, p. 73

Xia, L. see Cen, P.: Vol. 65, p. 69–92
Xie, B., Ramanathan, K., Danielsson, B.: Principles of Enzyme Thermistor Systems: Applications to Biomedical and Other Measurements. Vol. 64, p. 1

Subject Index

Acetate 74
Acremonium chrysogenum 115, 117, 121, 128
Acridine Orange 136, 169, 170, 171
ACV-synthase (ACVS) 117–120, 126, 129, 130
Adenosine diphosphate 192
Adenosine monophosphate 192
Adenosine triphosphate 192
Air-segmented continuous flow-wet chemical analyser 118
Amino acids analysis 192
(L-Aminoadiplyl)-L-cysteinyl-D-valine (ACV) 115, 120, 122
Amyloglucosidase 118, 127
Anaplerotic reactions 218
Arthrospores 117
Artificial neural network (ANN) 65, 66, 166
–, back-propagation 71
Aspergillus sp. 158
Atom mapping matrices 214
Atomic absorption spectroscopy 191
Aureobasidium pullulans 158

Back propagation network (BPN) 127
Bacterial contaminations 80
Bacterial infection 80
Baker's yeast, manufacture 77
BCECF-AM 171
Bicarbonate 8, 51
Bidirectional fluxes 225
Bingham model 128
Biomass 74, 94, 118, 128
Biomass concentration 4, 16, 41, 43, 49, 193
Biomass elemental composition 191
Biomass macromolecular composition 191, 202
Biomonitor 21
Bioprocess monitoring 87
– –, on-line 1ff

Bioreactor monitoring, electronic noses 65
Biosensors 27, 31
–, optical 33
Biosystems, dynamics 47
Bottleneck 115, 117, 126, 130
Bypass 6

Calcofluor White 170
Calorimetry 22
Capacitance 17, 21
Capillary 30
Carbohydrates, quantitiative analysis 192, 202
Carbol Gentian 168
Carbon balances 51
Carbon dioxide evolution rate 14
– –, partial pressure 12
– –, production rate (CPR) 118, 128
Carboxylic acids 115, 123
Casein hydrolysate 77
CCD 142
Cell disintegration 115, 124
Cephalosporin C (CPC) 115–120, 127, 128
Cephalosporinase 119
Chemometrics 196, 205
Chemotaxis 177
Circularity 152
Closed loop process control 1
Clump 162
Cluster (analysis) 202
Coagulation factor VIII, recombinant human (hFVIII) 79
Compartmentation 223
Concentration 5
Conductive polymer sensors 67
Confocal laser scanning microscopy 138
Confocal Raman microscopy 204
Convex bounding polygon 155
Corn steep liquor (CSL) 117, 139
Counting 150
Cultivation state, visualization 77

Cyanobacteria 160
Cyclo-adenosine monophosphate 192

DAPI 169
Data analysis 104
Data mining 166
Data partitioning 106
Data pre-processing 104
Deacetoxycephalosporin C (DAOC) 120, 126, 128
Deacetoxycephalosporin C-hydroxylase (HYD) 117, 119, 120, 126, 129, 130
Deacetoxycephalosporin C-synthase (expandase, EXP) 117–120, 126, 129, 130
Deacetylcephalosporin C (DAC) 117, 119, 120, 126, 128
Deacetylcephalosporin C-acyltransferase (ACT) 117, 119, 120, 127, 129
Deoxyribonucleic acid (DNA) 115, 118, 125
–, analysis 192
Diaphragm 7
Dielectric spectroscopy 193
Dielectrics 95
–, nonlinear 98
Diffuse reflectance absorbance (IR technique) 202
Discrete wavelet transformation (DWT) 127
Discriminant analysis 166
Dispersive infrared spectrometry 201
Dissolved inorganic carbon (DIC) 118
Dissolved organic carbon (DOC) 118
Dissolved oxygen concentration (DO) 117, 118
DNA, quantitiative analysis 192
Dot-blot 195

Electron microscopy 139
Electrophoresis, two-dimensional 195, 198
Elemental limitation 191
Elongation 152
Entanglement 162, 165
Entner-Doudoroff pathway 218
Entropy 145
Enzyme activities 194
Epifluorescence 137
Escherichia coli 75, 136, 167, 171
Ethanol 74
Euclidian distance map (EDM) 148, 150, 156, 159, 163, 165, 167
Exhaust gas 12

Field effect transistor (FET) 119
Field flow fractionation 41
Filament 159
Filter 25, 35, 42
Flow cytometry 28, 38, 103, 187, 188
Flow injection analysis (FIA) 25, 34, 119, 189, 193
Fluorescein diacetate (FDA) 169
Fluorescence 14, 15, 17, 38
–, culture 193
Fluorescence spectroscopy 115, 127
Fluxes, specific 47
Fourier descriptor 154
Fourier transform 106
Fourier transform infrared (spectroscopy) 201
Fractal dimension 155
French-press 124
Fuchsin 136, 168
Fura-2 172
Fuzzy logic 166

GC 28
Genetic programming 89, 102, 106
Glucose 74
Glucose oxidase (GOD) 50, 119
Glycerol 74
Gram-positive bacteria 80
Green fluorescent protein (GFP) 136
Growth hormone, recombinant human (hGH) 79
Growth rate, specific 75

Heat yield coefficient 21
Hematoxylin 168
Heterogeneity of a culture 189, 196, 205
Historic markers (for past physiological events) 195, 196
HPLC analysis 115, 119–123, 126
Hue Saturation Intensity (HSI) 143
Hydrolase 129
p-Hydroxybenzoic acid hydrazid (pHBAH) 118

Identification of organisms 196, 198, 200, 202
Image analysis 128, 187, 188
In-situ microscope 140
Infrared absorbance 13
Infrared spectroscopy 201
INT 170
Ion selective electrode 118
Isopenicillin N (IPN) 119, 120
Isopenicillin N-epimerase 119, 120

Subject Index

Isopenicillin N-synthease (IPNS, cyclase) 117, 119, 120, 126, 129
Isotope balancing 212, 220
Isotopomers 213

Kluyveromyces marxianus 158

Labeling 149
β-Lactamase 119
LC 28
Lignocellulose hydrolysates 77
Limitation 5
Lipid analysis, gas chromatography 196
– –, quantitative 202
Lipid pattern 196
Lysine biosynthesis 220

Magnetic resonance 39
Mannitol cycle 228
Mass spectrometry 29, 31
– –, pyrolysis 200
Media quality, prediction 77
Membrane 10, 13, 30, 37
Messenger RNA 194, 195
Metabolic channeling 221
Metabolic network analysis 209, 217
Metabolite balancing 210, 220
Metabolites, prediction 74
–, small key 192
Metal oxide semiconductor sensors 67
Methyl citrate pathway 222
Methylene Blue 136, 168
Methylmalonyl-CoA pathway 222
Microthrix parvicella 158
Mid infrared analysis 87, 90
Mixed population 196
Morinda citrifolia 158
Morphology 117, 127, 189
–, mathematical 144
MS 29, 31
Multisensor array, chemical 65
Multispectral imaging 171
Multivariate analysis 84, 89, 90, 98, 104
Mutarotation 50

NAD(P)H 15
Near infrared analysis 87–90
Neutral Red 136, 168
Nicotinamid-Adenin-Dinucleotid (phosphat) (NAD(P)H in reduced form 127
Nicotinic acid adenine dinucleotide (phosphate) 193
Noses, artificial 65

Nuclear magnetic resonance spectroscopy (NMR) 193

On-line monitoring 115, 117–119, 189
Optical density 17, 34
Oxygen 11
Oxygen transfer rate (OTR) 118, 128

Paramagnetic properties 11
PAS/MA 13
Pathway identification 217
Pattern recognition 37
PCA 73
Peanut flour (PF) 117, 128, 130
Pellet 162, 165
Penicillin N (PEN) 117, 120, 128, 130
Penicillium sp. 158, 160, 162, 168, 177
Pentose phosphate pathway 226
Peroxidase (POD) 119
pH 7
Photoacoustic spectroscopy 13
Physiological status 186
Population distribution 188, 189, 193, 198, 204
Pressure 8
Principal component analysis (PCA) 127, 166
Productivity, volumetric 5
Propidium iodide 136, 169, 170
Propionate metabolism 222
Protease 126, 129
Protease inhibitor 126
Protein pattern 198
Protein, quantitiative analysis 192, 202
Protozoa 167
Pt-100 7
Purity 5
Pyrolysis mass spectrometry 85, 200

QCM sensors 68
Quartz crystal microbalance sensors 67

Radius of gyration 154
Raman (micro-)spectroscopy 204
Raman spectroscopy 92
– –, surface enhanced 204
Rates, specific 4
Ratio imaging 171
Real time 46
Recombinant human carboanhydrase II 75
Redox metabolism 211
Redox potential 15
Reference electrode 7, 9
Respiratory quotient (RQ) 118, 128

Rheology 127, 128
Ribonucleic acid (RNA) 115, 125, 128
– –, levels 194, 195
– –, quantitiative analysis 192, 202
Rugosity 152

Saccharomyces cerevisiae 74, 136, 140, 144, 145, 149, 153, 157, 170, 171
Sampling 23
Segmentation 145
Self organized map (SOM) 127
Sensors 1ff, 67
–, acoustic 34
–, calorimetric 33
Separation 148
Sodium dodecyl sulphate polyacrylamide gel electrophoresis (SDS/PAGE) 198
Software sensors 35
Solubility 10
Somatic embryo 153
Spatial Gray-level Dependence Matrix (SGLDM) 174
Spores 158

Streptomyces sp. 145, 158, 160, 162, 165, 168, 177
Stress markers 195
Stress response protein 195
Sub-populations 188, 189, 193, 198, 204
Swollen hyphal fragments 127

Temperature 6
Texture 174
Thickness 163
Trichoderma sp. 158, 165
Trypan Blue 168

Validity 45
Variable ranking 106
Variance 145
Variant formation 198
Viability 168
Vibrational spectroscopy 87
Volume 156

Yeast cell cycle 191

Computer to Film: Saladruck, Berlin
Binding: Buchbinderei Lüderitz & Bauer, Berlin